Integrating
Project
Delivery

Integrating Project Delivery

Martin Fischer

Howard Ashcraft

Dean Reed

Atul Khanzode

WILEY

Contents

Foreword

By William McDonough
FAIA, Int. FRIBA

What we have come to call integrated project delivery is really about how we think about and frame the things we do best when we work together. It is, in effect, the art of marshaling collective intelligence, creativity, and imagination, and advancing that composition toward highly effective outcomes—buildings that support the cultures of human and ecological health and continuously improve over time.

I've found that a values-first approach to project design is the best way to harness collective intelligence and drive continuous improvement. Vision and values set a positive course for projects and effectively organize creative, collaborative work toward long-term goals. Often, however, businesses put a handful of metrics first and let all the other parts of their plan follow, with less-than-optimal results. Projects driven mainly by a few metrics tend to limit their promise on a relentless focus on meeting benchmarks and not enough on quality and innovation. The specific metric targets can become so consuming that larger guiding principles of positive behavior get lost. Plus, many current metrics tend to measure effectiveness by decrease—lower CO_2 emissions, fewer cubic tons of pollutants—rather than by the increase of positive effects, such as more energy sent back to the grid, more water purified on-site or more personal benefit for a building's inhabitants or the community within which it resides.

The distinction is especially important at this point in history when "high-performance" is frequently defined by doing less harm or "getting down to zero." To be sure, zero emissions, zero pollutants, and zero workplace health damage are laudable, widely shared goals. The orientation to zero grew out of important, real concerns on the part of business leaders for worker health and safety: a desire to create an environment that encourages employees to always be vigilant. In many companies with advanced safety protocols, workers are encouraged to act if they see unsafe conditions, with no fear of retribution. CEOs love zeros when they come before "accidents" and "spills." Increasingly, business leaders and organizations devoted to sustainability are striving for zero waste and zero carbon. And we applaud them. But when we talk to business leaders about new projects, we propose that zero become not a culminating point but a midpoint in a process of innovation and improvement, a pivot from doing less bad to doing a lot more good.

Vision and values make that pivot possible. In practice, we can start with values and principles drawn directly from the mission of clients and the specifics of place; they focus, frame, and organize the work we do together. If a CEO wants to build a safe workplace, we might encourage her to consider building one that not only does no harm but improves her employees' health, the health of the community in which the company operates, and the health of all the customers touched by the business. Such a workplace is not just optimal, it's exemplary.

As advocated in this book, positive engagement guided the integrated design and construction of the team we led for a new building at NASA's Ames Research Center in Silicon Valley, a 50,000 square-foot, beyond-state-of-the-art facility known as Sustainability Base—a name echoing and honoring Tranquility Base, site of the first moon landing. NASA is nothing if not positive and ambitious—one does not get to the moon on metrics alone—and right from the start the Sustainability Base project team declared its intention to design an environment that fully supports human and planetary well-being. That clearly stated value defined and organized the project's goals, strategies, and tactics as well as the ongoing collaborative research of some of the world's best scientists, engineers, and technicians. The result has been characterized as the highest-performing government building in America, delivered on schedule within a normal federal budget. Lawrence Berkeley National Laboratory's Stephen Selkowitz has called Sustainability Base a "genius building" due to its integration of collective intelligence and commitment to continuous improvement over time.

We live in an age of constant metrics. Recently, a corporate sustainability director of a Fortune 100 company pointed out to me that some $9 billion was spent in 2012 on "sustainability consulting." He noted the irony that we are watching the scorekeepers. Why are we benchmarking metrics rather than investing our time and energy in collective improvisation and innovation? That is the field of integrated project delivery. Think about it like a game. Statistics are exciting for some, but most of us are excited by the actual game. On the field of play, we act as a team, improvising at high speed toward a common goal. Detailed metrics of kicks and yards and hits are secondary to the quality of the game itself and the success in achieving the goal of winning the game—for the great satisfaction that it brings to people who are dedicated to the purposeful hard work. The purpose of the team's effort is to know how to play and improvise and to support each other and keep moving toward the goal. That's why people watch the game, and that is why the hard work of playing is fun.

The quest to create a building and a landscape that energetically support life led to a cascade of innovations. The Sustainability Base's self-monitoring system is so smart it knows precisely how much energy each inhabitant is using and predicts energy needs based on weather and work patterns; it can anticipate the building's thermal characteristics days in advance with great precision. Advanced daylighting fully illuminates the building up to 320 days a year, while a system of louvered windows provides continuous flows of fresh air. Water will be maintained in closed loop systems, purified to drinking quality for continual reuse or leaving the site after rainfall at the same rate, volume, and cleanliness as natural flows. In time, the building will use only renewable energy and will send its surplus back to the grid.

One might expect such a high-tech building to be extremely costly, but all the base building systems were produced within the budget set for a federal building. In keeping with NASA's guidelines, special systems that the design group wished to incorporate had to be offset by monetary benefits (for example, in energy consumption) so the extra investment could be recouped within a 7-to-10-year payback period. Those extras totaled only 6 percent of the budget, all with documented payback periods.

The outcome of our values-first, integrated approach was especially pleasing to NASA. Dr. Steve Zornetzer, the associate director of Ames Research Center and the project leader, noted that "the collaborative process yielded a highly sustainable and beautiful design—optimized for building performance and representative of our values . . . . I see this as a prototype of a twenty-first-century building. This is the way we're going to have to think about building in the future."

Sustainability Base was designed and structured for continuous quality improvement, and research into its operations is ongoing. The scientists and engineers who work there fully enjoy the fact that the work of progress is always a work in progress. They know the project will meet and surpass its goals, which, by definition, makes complete sense, when aspiration of continuous improvement is the goal. That is a good approach to integrated project delivery, too. Striving for positive goals and continuous improvement creates abundant opportunities for designers, engineers, contractors, and owners to fruitfully collaborate throughout the design and construction process. Let's seize those moments, strive together, and make every building we create an inspiring contribution to the health and well-being of our planet.

William McDonough is a globally recognized leader in sustainable development. McDonough is trained as an architect, yet his interests and influence range widely, and he works at scales from the global to the molecular. *Time* magazine recognized him as a "Hero for the Planet," noting: "His utopianism is grounded in a unified philosophy that—in demonstrable and practical ways—is changing the design of the world."

In 1996, McDonough received the Presidential Award for Sustainable Development, and in 2003 he earned the first U.S. EPA Presidential Green Chemistry Challenge Award for his work with Shaw Industries. In 2004, he received the National Design Award for exemplary achievement in the field of environmental design. McDonough is the architect of many recognized flagships of sustainable design, including the Ford Rouge truck plant in Michigan; the Adam Joseph Lewis Center for Environmental Studies at Oberlin College; and NASA's Sustainability Base, one of the most innovative facilities in the federal portfolio. He was the founding chair of the World Economic Forum's Meta-Council on the Circular Economy (2014–2016).

Foreword

By Phillip G. Bernstein

FAIA, RIBA, LEED® AP, VP Strategic Industry Relations, Autodesk, Inc.

Twelve years ago, an otherwise obscure industry association of institutional owners called the Construction Users Roundtable (CURT) gathered representatives of their building supply chain—architects, engineers, builders, facilities managers, and technology vendors—to ask what seemed to be a simple question: Why are construction documents so inadequate for purposes of construction? The well-understood litany of woes plagued many of the projects undertaken by this very experienced group of clients, who had focused their concerns on one of the more fraught exchanges in every project: the transfer of design intent to construction execution. After much discussion and further soul-searching, the CURT team concluded that the challenges of working drawings were just symptoms of a larger pathology characterized by "lack of cooperation and poor information integration."* And rather than suggest that the supply chain itself was responsible, they declared that it was in fact they themselves who could best catalyze the changes necessary to optimize construction. Writing in what is now their well-known white paper:

> The goal of everyone in the industry should be better, faster, more capable project delivery created by fully integrated, collaborative teams. Owners must be the ones to drive this change, by leading the creation of collaborative, cross-functional teams comprised of design, construction, and facility management professionals.

CURT's manifesto galvanized the U.S. construction industry by further declaring that integrated project structure, open information sharing, and building information modeling (BIM) were the key components of their vision for a radically revised way of delivering projects that performed.

These demands for radical change emerged as the industry was becoming aware of other process disruptions about which there was interest but little certainty. Concepts like Lean (a Japanese manufacturing strategy) and project alliance delivery (from Australia's infrastructure industry) promised new methods of optimization and decision making. Desktop computing was becoming powerful enough that the change from computer-aided design (CAD) software to BIM looked at least possible, if not likely. In fact, the putative collaborative transparency of BIM—where everyone had access to three-dimensional information—intimated some change in project delivery approaches and was often conflated with an emergent idea of "integrated project design." Was BIM a technology, a process or a delivery method? The supply chain really hadn't worked out what any of these things meant, or even what to call what is now well-known as integrated project delivery (IPD).

---

*Collaboration, Integrated Information and the Project Life cycle in Building Design, Construction and Operation, CURT WP-1202 (August 2004).

For an otherwise slow-moving supply chain, change came rapidly after CURT's white paper. By 2007 IPD had emerged as a distinct delivery typology joining the pantheon of hard bid, construction management, and design/build, moving in parallel with the rapid digitization of building. Various industry associations were floating provisional contract prototypes, and forward-thinking owners had begun to organize and execute early projects based on principles of IPD. In a market largely characterized by painfully incremental project-by-project improvements, IPD was a sea change in both approach and attitude. Analog information gives way to digital, transactional contracts to relational, and scopes of service and delivery objectives move from lowest bid to commitment to outcomes. For the first time, building process is focused exclusively on getting results rather than lowest first cost.

That premise is at the heart of this book, written by innovators operating at the intersection of theory and practice, technology, and delivery. Designed as a technically rigorous but accessible reference guide to both the motivations and protocols of IPD, the authors have inverted the traditional relationship of process and product and embrace the idea that integrated delivery means putting the clear objective of a high-quality, high-performing building as the highest aspiration of any project, and simultaneously obliterating the commoditized exchanges of consideration that denominate almost every transaction in the supply chain.

This change is much more significant than the evolution of IPD as "the next big thing" in project delivery typology. Normative practice draws a strong distinction between the performance standards for designers (who provide a "service" through professional judgment) and builders (who deliver a "product" based on information that is a result of designer services). The authors turn that construct largely on its head though a fundamental premise that underlies this entire book: the project team must, in its entirety, focus on delivering a *valuable* product to the owner, and not simply fulfill the minimum requirements so often seen in typical projects and delivery models. And there are systematic procedures and techniques by which that singular focus on value can be achieved, documented in great detail herein.

It is often argued that typical industry approaches achieve that same goal when project teams are "high functioning" and collaboration is both expected and achieved. It is almost impossible to predict, however, when a typical project might reach this lofty goal, and most agree that more often than not it isn't reached. Project performance and liability statistics support this conclusion, and stories about wildly successful construction projects are far outnumbered by tales of woe. As this text makes clear, delivering a valuable building means establishing clear goals and processes for achieving that outcome, aligning the interests of the team with that of the building itself, and using tools that support both prediction (necessary to make commitments of performance) and collaboration (necessary to provide absolute information clarity).

But how to get this done? Like BIM, the industry has latched upon the buzzword *integration* as a potential strategy. But having witnessed (and, in some cases, attempted to assist) the industry's embrace of BIM, it's clear to me that fundamental innovations in our industry need both exemplars and roadmaps. This text is designed to accomplish both through detailed procedural explanations and case studies demonstrating the implementation of those procedures. Both narrative and reference guide, it describes both the underlying theory and practice of integration, accessible to all the players of the supply chain. CURT would be astonished at what their original provocation has wrought.

While the economic collapse of 2008 likely slowed what would have otherwise been a much more rapid embrace, today IPD is a viable option for many projects where traditional delivery methods would yield suboptimal results, and project teams are properly prepared for its challenges, demands, and benefits. Like sustainability and BIM, the concept has become hip, and many designers and builders claim to have "been doing IPD for years." This is both charming and dangerous inasmuch as it speaks to both the potency of IPD and its related brand value. But without careful ground rules and proper procedural platforms, IPD may never escape the delicate stage of emergent innovations in their relative infancy. The text that follows is a definitive reference for those who see the processes that create buildings as deserving of modern, high-performance methodologies that best realize the tremendous value that buildings create for our society.

Phil Bernstein, FAIA, is a Vice President for Strategic Industry Relations at Autodesk, where he is responsible for the company's future vision and strategy for technology as well as cultivating the firm's relationships with strategic industry leaders and associations. Formerly a principal with Pelli Clarke Pelli Architects, he teaches Professional Practice at Yale, where he received both his BA and his MArch. He is co-editor of *Building (in) the Future: Recasting Labor in Architecture and BIM in Academia*, a senior fellow of the Design Futures Council, and former chair of the AIA National Contract Documents Committee.

Preface

"It is extension of application that discloses inadequacy of a theory, and need for revision, or even new theory. Again, without theory, there is nothing to revise. Without theory, experience has no meaning. Without theory, one has no questions to ask. Hence without theory, there is no learning."

—W. Edwards Deming

WHAT THIS BOOK IS ABOUT

The ideas and practices presented in this book are different than what most people think and do to design and build projects today. These different practices are needed because project teams need a strategy and a set of actions that allow them to overcome the challenges of the current "divide and conquer" approach to designing and building projects. Buildings have become more technically complex, the regulations that need to be considered more multifaceted, and the social and business expectations and pressures more intense. The increased technical complexity and the multifaceted regulatory constraints require the inclusion of experts that understand the specific technical systems to project teams. This increased specialization has led to fragmented project delivery, in part because the project management tools used on many projects are good at dividing the work up into chunks but less good at making sure that everyone's work fits together. The increased business and social pressures on building performance, however, demand a strategy to overcome this fragmentation, a strategy to integrate project teams and their work. That is why this book about integrating the delivery of building projects is needed now.

Because the practices for integrating project delivery are new, they can seem difficult. They certainly require different attitudes, skills, and behaviors. The new game is a plus-sum—not a zero-sum—game. The project team becomes a virtual enterprise. Companies become business partners rather than entities that cooperate when it serves their interests. People stop working in silos and exchange information frequently instead of periodically. This accelerates as experts learn that they can trust each other, which makes it possible for them to truly collaborate. Trust requires companies to operate open-book and people to admit mistakes, lack of confidence, and uncertainty. It also requires a culture that doesn't punish those who are honest. Project team members must work very hard to stay on the same page to be aligned in their work. Everyone on an integrated team commits to working in this new way, to give it their best. It's appropriate and necessary that this commitment extend to sharing risk of failure and rewards for great performance.

This is why we always recommend using a contract that supports integration. We believe that this new method of project delivery is better precisely because it requires agreement to business terms that do not allow project participants to "succeed" on their own. All partners, whether they are the owner, a designer, or a constructor, know that they can only succeed or fall short together. In this way, an integrated project delivery (IPD) agreement goes a long way to solving the "motivation" problem that plagues well-intentioned efforts to improve performance. Everyone on an integrated team, from the owner representatives to the workers who put the work in place, needs to know what it means to integrate their efforts.

This book explains a system we believe will enable people with different expertise and experience to consistently create valuable high-performance buildings. Our starting point is a recognition that the architectural and building systems and components of modern facilities are interdependent and must be integrated in design and construction to perform well. We explain theory and describe it in practice for each element of integration.

Our goal has been to write something of value for both experienced practitioners and students of the industry. It is our intention and hope to help readers understand four things:

1. What the elements of integration are.
2. How they interconnect.
3. Why they are all necessary.
4. How they have been and can be put into practice.

The book focuses mostly on design and construction. This is not to say that the use and operations phase of buildings is not as important. After all, buildings are designed and constructed to be used. However, without design and construction, there is no building to use and operate. We believe that the concepts we present are also useful to orchestrate the operations and repurposing phases of buildings, although the specifics and the examples would be different. While the examples throughout the book are from building projects, we believe that the concepts apply to other types of facilities, such as infrastructure and industrial projects.

HARDER WITHOUT A MAP

Everyone needs a map when venturing into unfamiliar territory. We wish we had had one when we started out. Now we do and offer that here, convinced that this will help advocates for integration, coaches, and leaders describe a complete system to project team members. Why do we think this? Because we've seen people on almost every project team struggle to understand why and how they can improve outcomes through integration. While most people we've met, especially the owner's team, are not satisfied with the way things are done now, they know what these are and how to do them. People with industry experience new to integrated delivery have only done things the old, fragmented way. Our sense is that most teams have tried to implement various techniques, methods, and software tools, especially building information modeling (BIM), without knowing how to plan, collaborate, and share their knowledge as partners. The industry culture of starting to work before looking at work processes, making do (Howell & Ballard, 1997; Koskela, 2004; Macomber & Howell, 2004), and attributing blame for failures undermines team learning and continuous improvement. Worst of all, the focus on reducing cost has short-circuited honest efforts to understand customer value well enough so it can be translated into tangible project objectives. It's no wonder the industry continually falls short in delivering what the customer really wanted.

Many, if not all, of the real-life examples in this book came about because a few people under-stood collaboration and were willing to show others. They encountered resistance and succeeded in overcoming it through education and persuasion. In many cases, the owner's project manager had to force the issue and make it clear that the only good option for doubters was to genuinely try to work differently or go to work on a conventional project.

We have chosen to describe what teams have done and achieved without describing the drama behind the scenes so that we could connect their experiences and accomplishments to the entire Simple Framework and its elements—the map we made—in just enough detail so that our map could be useful. Although we expect adoption of integrated practices to be challenging for some time, we are confident that the map the Simple Framework provides will make this easier and lead to much better outcomes.

HOW THIS BOOK IS ORGANIZED

We have tried to answer one or two big questions in each chapter, as shown in Table P.1.

TABLE P.1   Chapter Questions

Chapter	Title	Question
1	What Would Make Us Proud?	What do we want to do and what can we do?
2	Transitioning to Integrated Project Delivery: The Owner's Experience	What do owners who have used IPD think about what they can do to improve outcomes?
3	A Simple Framework	What is the roadmap, the strategy to successfully produce a high-performance building?
4	Defining High-Performing Buildings	What is a high-performing building?
5	Achieving Highly Valuable Buildings	What makes a high-value building?
6	Integrating the Building's Systems	How can systems be integrated to achieve a high-performing building?
7	Integrating Process Knowledge	How can process knowledge be integrated?
8	Integrating the Project Organization	What is an integrated project organization, and how is it created?
9	Leading Integrated Project Teams	What is an integrated project delivery team, and how do you create, lead, and manage one?
10	Integrating Project Information	What does it mean to integrate project information, why is this so important, and how can we do this?
11	Managing with Metrics	How do we define and uphold the client's value goals for their unique high-performing building over the course of a project?
12	Visualizing and Simulating Building Performance	How do we enable stakeholders to visualize and understand how their building will perform through every step of design, long before it is built?
13	Collaborating in an Integrated Project	What does it mean to collaborate in an integrated project?
14	Co-locating to Improve Performance	How can we leverage co-location to improve behaviors and outcomes?

TABLE P.1   (*continued*)

Chapter	Title	Question
15	Managing Production as an Integrated Team	How do we manage production as an integrated project team?
16	Avoiding the Pitfalls of Traditional Contracts	Why is it so difficult to use traditional contracts to support project integration?
17	Contracting for Project Integration	How does an integrated form of agreement support integrated organization and behaviors?
18	Delivering the High-Performing Building as a Product	How can high-performing, valuable buildings be developed and delivered as a product?

GETTING THE MOST OUT OF THIS BOOK

There are at least two ways to read the book. The first is to read about integration in the order it is required, a "why and what" order. We recommend this for people who are not yet practitioners, such as college and university students or outsiders looking into the construction industry. Simply read the chapters in the sequence in which they are presented.

The second way is to follow the order in which projects become integrated, a "how and what" approach. We suspect that experienced professionals may find this easier and prefer to start with the problems they are trying to solve. Because the Simple Framework elements are connected and reinforce each other, we believe readers with experience can begin anywhere and go in whatever direction they want.

Our Journey Writing This Book

We began our journey to write this book in a conversation with the Wiley editor for the *BIM Handbook* at the 2008 Georgia Tech BIM Symposium. Writing a good book about IPD in a year shouldn't be that difficult, we reasoned. After all, we had extensive experience using BIM, Lean construction, and IPD on projects, and considered ourselves thinkers and leaders in each of these domains.

Almost eight years later, we are now writing this preface. What happened? Why has it taken so long? There's an old saying that goes, "Be careful what you wish for because you just might get it." And in this case, we wished for IPD to expand and it did. It became our "day job" that squeezed out time for writing, forcing us to work in "fits and starts." But lack of time isn't the only reason this project has taken so long.

As we delved deeper into the subject, we realized that existing frameworks were incomplete. We initially emphasized organization, process, and behaviors, which is what very good thinkers had already done. Leaders of the American Institute of Architects California Council (AIACC) had focused on these issues in their publication *Integrated Project Delivery: Working Definition* (Eckblad et al., 2007). The authors of the paper "Managing Integrated Project Delivery" (Thomson, Darrington, Dunne, & Lichtig, 2009), published by the Construction Management Association of America (CMAA), looked at those same aspects through the lens of Lean and relational contracting.

We thought that these were insightful and very useful documents that did not need to be reinterpreted by us. In fact, we recommend the entire series of papers on IPD produced by AIACC and the CMAA paper and use them in our own practices.

We wanted to expand and go deeper using the book format. And we were very clear that we should not produce a recipe book, even for what seemed like a delicate soufflé. We decided to look at IPD from every angle. We "peeled the onion" and ended up creating an illustration we called the "IPD Universe" that contained the "Magic Formula," which didn't vary much after version 9 was drafted in late January 2010. All we did after that was add more horizontal swim lanes for additional pieces of the IPD puzzle; the last version (March 2012) had 14. We were pleased with arriving at the "Magic Formula," shown in Figure P.1, even though we realized that it was a description rather than an explanation of integrated delivery.

The Magic Formula did and still does make perfect sense to us. IPD has several elements:

- *Value definition*. The first is value definition, where owner organization and user needs and constraints are understood. Understanding stakeholder values is important and reflected in setting performance goals. Owner representatives and delivery team members must turn these goals into performance objectives that can be measured qualitatively or quantitatively.

- *Framework*. Second is the framework consisting of a relational contract, an integrated organization, designing for performance within cost constraints (target value design), and creating an infrastructure of sharing information that any team member created.

- *Environment*. Third is the environment, made up of the right people—meaning people who are willing to think and work differently than on most project teams today. Leaders must be willing to ask people to leave if they can't change, including themselves. They must be willing to use BIM to visualize and simulate performance as a team, rather than delegate that to a group of specialists. Everyone must be "in." Team members must co-locate, to be in proximity to respond to questions and work through problems as quickly as possible. Everyone must allow others to see their work plan, progress, and issues. This transparency extends to overall team performance on scope and budget, schedule, safety, quality, and especially the current state of design, overall and in detail.

- *Interactions*. This environment inevitably produces many high-quality interactions among team members in which they exchange ideas while getting to know each other as individuals.

- *Network of knowledge*. People soon realize who knows what and whether they are willing to share their knowledge. A strong network of knowledge forms along with a network of commitments as people learn how to act as good "suppliers" and "customers" by being clear about what they need and when and what they're capable of providing.

VALUE DEFINITION	FRAMEWORK	ENVIRONMENT	INTERACTIONS	NETWORK OF KNOWLEDGE
• Enterprise Needs & Constraints • Stakeholder Values • Performance Goals • Objectives & Metrics	• Relational Contract • Delivery to Target Cost • Integrated Organization • Information Infrastructure	• Right People • Virtual World • Proximity • Transparency	• Quantity • Quality	• Connections Across Boundaries • Clarity of Customer Supplier Relationships

FIGURE P.1 The Magic Formula for integrating project delivery.

The "Magic Formula" is a coherent and concise picture of how the elements of IPD interact. However, something big was missing: the thing teams were creating for their customer. The Formula focused on everything but the product, that is, the building. If the end point was the facility, perhaps we should "pull back" from the outcome to determine how a team can achieve it. But to do that, we needed to define the type of building that people wanted and needed.

We called it the "high-performing building." It seemed obvious that it had to be useful for the people who would work inside and visit for services. Our definition certainly would include form and aesthetics. The building should be economical to operate. If our planet is to remain habitable, the building should be constructed and run without depleting or harming the environment. It would have to meet planning and building regulations. The building would certainly have to contribute to the success of whoever paid for it, however those people or their organization defined success. That was on the outcome side. On the input/construction side, this high-quality, high-performing building would have to be built for the money and time available. Otherwise, the building would exist only in people's minds as a need and aspiration.

We knew that all of the elements of the Magic Formula were necessary to create high-performing buildings within the constraints that all but a very few owners face. Our challenge was to find a way to explain this. We had all participated in "pull-planning" sessions in which we helped project teams make value flow by identifying and sequencing only the work needed to create something essential for the success of the project. So it was simply a question of pulling. But what should the order be? The answer came to us in a matter of minutes on September 27, 2012, when Martin Fischer grabbed a marker, walked up to a large whiteboard, and drew the Simple Framework almost exactly in the way we have described it in this book. Martin talked as he drew. The scene is artistically recreated in Figure P.2.

A high-performing building can only be achieved through a building with integrated building systems, which can only be produced through an integrated process, which depends on an integrated team with the right people, which needs integrated information, i.e., BIM+ to function effectively and efficiently. Simulation and visualization are the primary ways in which BIM+ informs the integrated team. Collaboration and co-location are the primary ways that allow the integrated team to integrate processes. Production management methods enable the productive design, fabrication, and construction of the integrated building system. Outcome metrics define the performance of the building and validate the integrated building system. All of this is supported by the appropriate agreement or framework.

At each step of the way, Martin reflected on what the best strategy, action, or tool would be to achieve the next step in the diagram. For example, integrating a building's technical systems so that they work in concert and not against each other is the best strategy we could think of to achieve a high-performing building. This then required metrics to stipulate the desired performance of a building to measure its performance—as a whole and for the individual systems.

Once Martin drew the diagram, the steps to the high-performance building and their relationship to the Magic Formula seemed obvious. We now had our definition of success for an owner and the project team, the reason why a strong network of knowledge (an integrated organization carrying out integrated processes supported by integrated information) is needed and why the practices we observed on projects—co-locating the project team, formally defining workflows and weekly production plans,

FIGURE P.2 The Simple Framework explained. Courtesy of CDReed.

creating a building information model that combines the work of many disciplines, and so on—made sense.

This Simple Framework could carry all of our thinking in a comprehensive yet comprehensible package. In this new approach, each element would be a chapter. Each chapter would follow a "standard chapter structure." We would explain the Framework element, describe success, explain how teams can implement the element, provide real-life examples, describe how the element fit with others in the Framework, and end with a reflection on what this means for delivering high-performing buildings, looking forward. We quickly drew up a plan to finish the book within six months because we could now see what had to be done. We were incredibly optimistic and naive about writing! Explaining each piece of the Framework and finding, researching, and describing as many examples as we could turned out to be a lot more work than we ever thought. We failed to account for two critical factors: the first was that estimating time for highly iterative work is very difficult; the second was the fact that time to write became very scarce. We began to make progress, but it was agonizingly slow.

We kept at it, however. Insights kept coming, seemingly on their own when we were ready to embrace them, as with the Magic Formula and Simple Framework. We spent countless hours talking to skilled and experienced practitioners, who explained what they did and how they did it. Visionary and courageous owners shared their stories, as did others blazing trails to the future on their projects and in their companies.

We've come to see the Simple Framework as a model that could be made into a system, where each element depends on all the others and can only produce a breakthrough together but not apart. Models are like the act of planning. Their value is in preparing and orienting people so they can see, think, and act differently than they did before. All models are abstractions of reality, however. To connect the Simple Framework to actual practice, we have interspersed examples of how the principles have played out on real projects. Because individuals and teams can control and change only so much on a project, we felt it was critical that we also describe each of the Simple Framework elements and explain how they interact and contribute to integrating project delivery. That way, readers can start integrating parts of their projects wherever they find opportunity.

REFERENCES

Eckblad, S., Ashcraft, H., Audsley, P., Bleiman, D., Bedrick, J., Brewis, C., ... Stephens, N. D. (2007). *Integrated project delivery: A working definition*. Sacramento, CA: AIA California Council.

Howell, G., & Ballard, G. (1997). Lean production theory: Moving beyond "can-do." In L. Alarcón (Ed.), *Lean construction* (pp. 17–23). Rotterdam, Netherlands: A. A. Balkema.

Koskela, L. (2004, August 3–5). Making-do—the eighth category of waste. Paper presented at the 12th Annual Conference of the International Group for Lean Construction, Helsingør, Denmark.

Macomber, H., & Howell, G. (2004, August 3–5). The two great wastes in organizations. Paper presented at the 12th Annual Conference of the International Group for Lean Construction. Helsingør, Denmark.

Thomsen, C., Darrington, J., Dunne, D., & Lichtig, W. (2009). *Managing integrated project delivery*. McLean, VA: Construction Management Association of America (CMAA).

Acknowledgments

My contributions to this book would not have happened without the support, insight, and challenges from many people. First and foremost, my wife, Mary, and our son, Brandon, tolerated way too many hours on the laptop, in meetings, and away from home for visits to companies and projects to learn about innovative practices. My co-authors' critical thinking, trust, and tolerance was the biggest factor in the book getting published. I'm thankful to Angelo Pozzi, who taught me how to be a scientist as an engineer and showed me that one must always try to find a way to do practically relevant and theoretically sound work. He also introduced me to Stanford University. Frank Lobdell and Terry Winograd taught me how to see and to take a human perspective. Ray Levitt introduced me to model-based thinking in a world where anecdotes and exceptions abound. He also showed me how to include the organization and process perspectives along with the product perspective. John Kunz reinforced this model-based approach in countless ways. The members of the Center for Integrated Facility Engineering (CIFE) at Stanford challenged the status quo as well as ideas we had while also providing opportunities to learn from the initiatives they pursued to create better buildings in a better way. Finally, the students in my classes and in my research group continuously challenged the established approaches and ways of thinking and demanded better explanations. Catherine Engberg, a student in the first undergraduate class I taught at Stanford, summed up the challenge of managing building projects perfectly when she said at the end of the class that she "was surprised how easy it was to comprehend each individual concept covered in the class, but how complex it was to combine the application of the concepts to address real-world challenges." Her statement motivated me to make the combined application of project management concepts easier. I hope that this book is a step in this direction.

Martin Fischer

I would like to thank the many people who have assisted me in developing the material for the book. First and foremost, my wife, Marilyn, whose comments and advice clarified my thoughts and improved the text. I would also like to thank my partners, who have given me the freedom to explore and develop new areas in project delivery. And, finally, I am deeply indebted to the many clients and IPD teams I have been privileged to serve, each of which has taught me important lessons and, in that way, has contributed to this book.

Howard Ashcraft

Many people have helped each of us along our long journey to write about integrating project delivery. There's no way I could have stayed with this project without my wife, Carol, listening and encouraging me not to give up on so many walks together. On top of that, she produced some delightful artwork for the book.

Then there's Eric Lamb, who waited patiently for eight years to see the final product, asking only, "How are you doing?" Without Eric's support, Atul and I couldn't have taken time to work on the book. Everyone in DPR, from Doug Woods and George Pfeffer to carpenters and laborers, kept telling us how much they were looking forward to the book. Talk about pressure!

Thanks to Yumi Clevenger for her sincere efforts to help us, even though it was obvious that we weren't ready for that at the beginning. Lyzz Schwegler deserves a gold star for the heroic work she did to organize and push us into action. She showed us what we should do to write a book rather than talk about one. Pamela Maydanis made time she didn't have to provide images and permissions for the cover and two chapters.

We are honored and indebted to William McDonough, Phil Bernstein, Stuart Eckblad, Eric Lamb, Ted van der Linden, Mike Humphrey, Mike Messick, John Andary, Doug Kot, Bruce Cousins, Adam Rendek, and Matt Grinberg for writing the forewords, afterwords, and features that appear throughout the chapters.

Many people took valuable time to explain what happened on their projects and gave us photos and illustrations to help us tell those stories. Stuart Eckblad and Damon Chandler are at the top of that list, followed by Dylan Connelly, Grant Walker, Eric Miller, Digby Christian, David Chambers, George Hurley, Ralph Eslick, Ken Lindsey, Mark Napier, Kelly Griffin, Jason Herrera, Ray Trebino, Osman Chao, Jack Poindexter, Andy Hill, Cassie Robertson, Ryan Ferguson, Jamie Hammond, John Haymaker, Ari Pennanen, Rick Drake, Ben Schwegler, Ellen Belknap, Tom McCready, Brent Nikolin, Blake Dillsworth, and Scott Eastman. And there are so many others.

Thanks to Markku Allison, Michael Bade, Stewart Carroll, Andrew Arnold, Forest Flager, and Zig Rubel for describing their work to move the industry forward in our final chapter.

Significant change requires leaders who are willing to step forward to stake their position and reputation on bringing it about. In addition to the other visionary leaders who shared what they've learned in Chapter 2, I thank Owen Matthews, Peter Beck, Chuck Greco, David Pixley, David Long, Will Lichtig, Eric Miller, Eric Ahlstrom, Dave Umstot, Wylie Bearup, Gary Aller, Kip Edwards, Blake Dilsworth, Ron Migliori, Zach Sargent, Samir Emdanat, John Pemberton, Craig Russell and David Van Wyk for being those people for Lean and integrated project delivery.

Finally, there are the committed visionaries who see possibilities for doing things differently and better long before others, and remain committed to creating that different future. I've been lucky enough to have been taught and influenced by some of those people, namely Greg Howell, Glenn Ballard, Hal Macomber, Chauncey Bell, Ray Levitt, John Kunz, and Martin Fischer.

Last, but not least, I and my co-authors are indebted to our editors at John Wiley & Sons for helping us move forward even when that was slow and difficult.

Dean Reed

I am very grateful to my co-authors, Martin, Howard, and Dean, for providing me the opportunity to work with them on this book. Martin has been a mentor for me for many years and we could not complete this book without Martin's ability to synthesize complex ideas into simple and easy-to-understand theories and explain it in an easy-to-understand text. Dean's insights into integrated projects and Lean construction methods and his ability to ask probing questions on how things work the way they do on projects has led to many remarkable ideas in this book. Howard has been the driver and editor of our manuscript and we would never be over the finish line without the tireless efforts from Howard over the last six months.

I would like to thank all the people with whom I have had the pleasure of working at DPR Construction over the past 19 years. DPR's project teams have contributed immensely to our understanding of projects, teams, and intricacies of how a building comes together. I would like to thank Eric Lamb and George Pfeffer, both members of DPR's Management Team. I owe many ideas in this book to the discussions I have had working on various project teams with Eric and George. There are many other people at DPR I would like to thank: other members of DPR's Management committee: Doug Woods, Jimmy Dolen, Peter Salvati, Mike Ford, Mike Humphrey, Greg Haldeman, Jody Quinton, and Michele Leiva have supported my efforts through and through. DPR's Director of Marketing, Yumi Clevenger, and marketing coordinator, Lyzz Schwegler, advised us throughout the past few years on the book and helped with the early versions of our manuscript, and I am very grateful for the help they have provided. Other individuals that I would like to thank are Ralph Eslick, George Hurley, Osman Chao, Ray Trebino, Jack Poindexter, Joseph Yau, Steve Spence, Rob Westover, Zach Murphy, Patrick Cusson, Steve Helland, Whitney Dorn, Zach Pannier, Mark Whitson, Brain Gracz, Hock Yap, David Ibarra, Mark Thompson, and Rodney Spencley. I am indebted to my team members in the Technology, Consulting, and Innovation Group at DPR. They have worked shoulder to shoulder in accomplishing some great things at DPR and discussions with them on how to really implement BIM/VDC in our industry have contributed to this book. I would especially like to thank Jim Washburn, Radhika Menon, Everardo Villasenor, Margarethe Pfeffer, Saurabh Tiwari, Justin Schmidt, Chris Rippingham, Durga Saripally, Sangwoo Cho, Kaushal Diwan, Alice Leung, Tony Dong, Frank Wang, Mojtaba Taiebat, Hannu Lindberg, Jean Goyat, Akanksha Sinha, Adi Subramaniam, Alaina Kmitta, Moawia Abdelkarim, Andrew Arnold, Adam Rendek, Ryan Meacham, Scott Widmann, Jiun Chiang, Andrew Fisher, and Alan Watt. I am sure I have missed many people at DPR, and it is impossible to list all their names, but I am forever grateful for all the insights and help you have provided.

Many people in the industry and academia have helped shape the ideas in this book. These include Dr. Ray Levitt of Stanford University. Ray's recommendations on readings in organizational behavior were invaluable in shaping our ideas on teams and simulation. Dr. Glenn Ballard of UC Berkeley, a pioneer in the field of Lean construction helped shape our ideas on target value design and interaction of BIM and Lean. Dr. John Kunz, past executive director of CIFE at Stanford University, helped us with the conceptual understanding of controllable factors and distilling the project delivery approach through the VDC framework of Product, Organization, and Process (POP). Dr. Paul Teicholz, former director of CIFE, has provided valuable advice about what it takes to write a book as well as his insights into the integrated approach using VDC methods. Dr. Bob Tatum, Emeritus Professor, Stanford University, encouraged understanding of coordination processes and systems. We have benefited tremendously from the interactions we have had with Digby Christian of Sutter Health and Stuart Eckblad of UCSF Mission Bay Hospitals. Both provided tremendous leadership on complex integrated projects and greatly contributed to our ideas about the role of an owner on integrated projects. I would also like to thank industry mentors who have helped in many discussions we have had with them. Dan Gonzales of Design Village; Kathleen Liston of Moderna Homes; Samir Emdanat and Bob Mauck of Ghafari Associates; Dr. Greg Howell of Lean Construction Institute; Hal Macomber of Lean Consulting; Will Lichtig of Boldt; Andy Fuhrman; Greg Luth of GPLA Structural Engineers; Rob Leicht and John Messner of Penn State University; Dr. Chuck Eastman of Georgia Tech; Rafael Sacks of Technion; Dr. Marcus Schreyer of Max Bogl in Germany; Dr. Reijo Hanninen of Granlund in Finland; Sheryl-Staub-French of the University of British Columbia; Timo Hartmann of Twente University, Netherlands; Andreas Ask of NCC Sweden; Niles Wingeso Falk of MT Hojgaard, Denmark;

Burcu Akinci of Carnegie Mellon University; Phil Bernstein, Carl Bass, and Amar Hanspal at Autodesk; and Zig Rubel of Aditazz.

I could never have finished this book without the support of my friends and family. My dad, Raghunath, and mother, Vibhavari, Uncle Ramesh, and Aunt Vijaya always encouraged me to finish what I started and inculcated the love of education in me. My inspiration to pursue more also stems from the advice of my grandfather, the late Ramchandra Buldeo, who would have been thrilled to be here today. My brother, Vivek, and cousins Sanjay, Ravi, and Kiran have been supporting me for a long time and continue to do so. Finally, I could never have finished this effort without the love, encouragement, and support from my wife, Leena, and my lovely daughter, Sania. They have tolerated my many early mornings and being away on weekends to me to finish the book and have provided me the extra energy I have needed to get through this endeavor.

Atul Khanzode

Implementing new and innovative project delivery methods for large, world-class academic medical facilities can be a challenge. Integrated Project Delivery for public projects is such a challenge. We have included a brief case study and numerous examples from just such a project, one that deserves to be described in much greater detail.

Approval of new methods requires the trust of many. Trust that the change will lead to better outcomes and value. The University of California and the University of California San Francisco Medical Center accepted that challenge for the Mission Bay Hospitals project. The results are extraordinary for quaternary care, the patients and staff.

Acknowledgment to trust new ideas must go to the leadership of both the University of California and the UCSF Medical Center. Their commitment to innovation and alternative delivery methods has accelerated the design and construction industry's adoption of collaborative and integrated methods for public projects. Projects are completing with lower costs, shorter schedules, and higher quality. Extraordinary!

Martin Fischer, Howard Ashcraft, Dean Reed, Atul Khanzode

CHAPTER 1

What Would Make Us Proud?

"You're here because you know something. What you know you can't explain, but you feel it. You've felt it your entire life, that there's something wrong with the world. You don't know what it is, but it's there, like a splinter in your mind, driving you mad."

—Morpheus, *The Matrix* (1999)

1.1 CURRENT STATE OF FACILITY PERFORMANCE

Buildings should perform to or even exceed our expectations. We should be able to set high standards during design that are met during operations and use. We have some success defining and predicting first cost, design-construction duration, and structural and water-tightness performance. Project teams often achieve desired performance for these objectives today. Facilities don't tend to fall down—they tend to keep the water out, and they tend to be delivered on time and on budget, although achieving all of these goals simultaneously is somewhat less common. But when we include other project goals, it is far less common that project teams meet them.

Consider, for example, building energy performance. This operational parameter is particularly significant to life cycle cost, sustainability, and carbon emissions and affects user comfort and functionality. A variety of modeling tools exist to predict performance during the design phase; yet it is rare that a building performs during the use phase as simulated during design.

A recently completed building at Stanford University exemplifies this issue. It has better energy performance than buildings of similar size and function, but, in the first years of operation, used significantly more energy than the design team projected during the design phase (Kunz, Maile, & Bazjanac, 2009). The difference between projected and actual performance is caused by operational inefficiencies (which have been and are being corrected), by unrealistic or unconfirmed assumptions in building use, changes in use of some of the spaces to more energy-intensive activities that were not incorporated into the building energy performance simulations, and shortcomings of the energy simulation and prediction tools. Even though the building performs relatively well, the project team did not meet the aggressive energy performance goals set during design or, at least, failed to notify the users and the client that actual energy use would be above expectations. Hence, the building owners and users were surprised by the lower than expected energy performance.

Unfortunately, the literature reports similar experiences for other buildings for which well-organized clients with lots of design, construction, and operations experience hired very strong design and construction teams, yet failed to achieve the energy performance goals set during the project's design (Scofield, 2002, 2009).

Nilsson and Elmroth's (2005) analysis of 23 retrofitted buildings in the Malmö project revealed three main reasons for the poorer than expected energy performance: (1) vendors indicated overly optimistic window performance, that is, the window calibration in the vendor's laboratory characterized the windows too optimistically with respect to their insulation performance in the context of the actual buildings; (2) the energy analysis program did not consider thermal bridges properly; and (3) the buildings leaked too much air, largely due to tolerances in the structural and façade systems that did not create a building envelope that supported the aggressive energy performance targets.

It could, of course, be that these examples of the performance of individual projects are isolated aberrations of otherwise strong performance. The accounts of the performance of buildings and infrastructure on larger scales suggest otherwise.

A study by the American Physical Society (APS) concludes that buildings that follow the LEED (Leadership in Energy and Environmental Design) guidelines for green buildings (U.S. Green Building Council [USGBC], 2002) don't end up, on average, having better energy performance than buildings that don't follow the LEED guidelines (APS, 2008). As a final example in a potentially much longer list of far from sustainable performance of the built environment, in the United States, energy that went into making the materials that end up as construction and demolition waste each year—which is the largest contributor to landfills—is, according to our rough calculations, about equivalent to the electricity California uses each year. In other words, each year the U.S. construction practices throw away materials with embodied energy content equivalent to the electricity used in California. And it is not just energy performance that is suffering. The American Society of Civil Engineers (ASCE) has been giving poor grades for the state of the infrastructure in the United States for many years (ASCE, 2009). Walter Podolny and others report about several bridges that had to be retrofitted or rebuilt after just 20 or 30 years of service in Europe (Podolny et al., 2001), which is a life cycle performance and durability that is lower than society typically expects from infrastructure projects.

In summary, the experience of well-intentioned facility owners and highly capable design and construction teams and the accounts of the performance of built facilities from various sectors of the industry and various parts of the world suggest that today's approaches to delivering and operating constructed facilities do not give us what we want. Just consider the well-documented high impact of the built environment on energy consumption, emissions of carbon dioxide (CO_2) and other greenhouse gases (GHGs;), (National Science and Technology Council [NSTC], 2008), contributions to landfills, and so on.

Although better software and measurement tools would be welcome, we believe that the fundamental problem lies in how design, construction, operation, and use are integrated. In most cases, we do not fully use our human assets and fail to organize information and work into an optimized flow. Unless we do so, we will continue to be surprised and disappointed by building performance and will forfeit the opportunities created by better models and visualization tools. We need to change. Building owners should take a hard look at the performance of their facilities, and service providers (architects, engineers, builders, etc.) should thoroughly consider the impact of their current practices and then develop an inspired and inspiring strategy for dramatically higher-performing facilities. Only then can we achieve the performance we predict and the buildings we deserve.

1.2 WHAT IF?

What if every building and every piece of infrastructure truly worked? What it they were all designed not simply to fill a need, but to enhance our way of life? What if they were finished on time and on budget, without doing harm to people or the environment? What if every building performed as highly as possible, with all systems working in concert to support its purpose? But they don't. And those of us in the architecture, engineering, and construction (AEC) industry know that something isn't right.

The AEC industry is responsible for building the world's physical wealth,[1] from truly magnificent structures to modest dwellings. The industry is arguably one of the oldest in the world, and many people come to it for deeply personal reasons, a long family tradition, or an inspiring experience. This field attracts motivated, incredibly hardworking professionals—early risers who work in rain, snow, and heat, and who believe it is one of the best industries in the world. The problem isn't the people; it is how they and their work are organized.

In just the past 20 years, buildings and infrastructure have become vastly more complex than they were for most of human existence. Advances in mechanical, electrical, plumbing, conveying, information, and other systems have led to rapidly increasing specialization, dramatically increasing the coordination required to engage the many specialists in a timely, efficient, and effective manner. Construction projects also suffer from variability, unpredictability, and uncertainty, such as which specific system will eventually be selected, who is involved in the building process, how facilities and their systems and parts are produced and assembled, and a host of external factors such as weather, market conditions, and so on. Each project brings together a different set of players who might or might not have worked together before; every project is unique in some way.

Despite our very best efforts, we consistently end up with a product that satisfies few, including ourselves. In almost every building, a well-meant shortcut is taken somewhere during design, construction, or operation that results in a product that is less than the original vision, and less than what the users actually require. Too many projects squander time, money, energy, labor, materials, knowledge, and other precious resources largely because of how they are organized and carried out; too often, the AEC industry is characterized by unmet expectations (KPMG, 2015). Owners and project participants are, of course, aware of this track record. Sadly, we see many project organizations set up to avoid failure, which is seen as a success, instead of striving to create a great building that sustains its users in their endeavors.

When we look critically about the process used to deliver a building, we see a huge amount of fragmentation. In an attempt to tackle highly complex problems, the industry has responded by breaking projects into small, isolated pieces, and focusing on producing each of those at the lowest possible cost. But we rarely, if ever, consider how to put all these pieces together over time to create the best building possible by thinking about how a team is organized, how information flows within a project, or how to define the vision and goals of the project and keep them alive for its duration. We keep our sights on the bits and pieces rather than raising our eyes to the project as an integrated whole. For those of us who have spent our careers in the industry, it seems hard to believe that so many projects are delivered in such a shockingly suboptimal way. And yet, many of us have had a sneaking feeling all along that there must be a better approach. We see workers idle for lack of materials, materials pile up in parking lots with no one to install them, hundreds of thousands of dollars in rework due to poor planning or late or incomplete communication. In the end, we see a building that costs more to operate than it should, and doesn't meet the owner's and users' original vision.

The knowledge and experience of each professional and each company is not integrated in a consistent and timely manner, and consequently innovative ideas and opportunities—which could lead to creating a better building—are missed, overlooked, or ignored. Current contractual agreements, rather than reinforcing the need to bring the team members together to create innovative solutions, drive them apart to work in independent silos. It should come as no surprise that the current organization and process often lead to an adversarial relationship between the parties, rather than a relationship of cooperation and coordination. Too often, the focus is on short-term opportunities, which rarely yield the best choices for the project as a whole.

For example, on one bridge project, the owner and the design team decided to concentrate on a single design option to shorten the design period and allow construction to start at the earliest date. This approach caused the team to overlook an alternative design that could have dramatically reduced the construction time while also improving safety for the workers and the ship traffic that passes below the bridge. Shortcutting design to accelerate schedule meant the design ultimately had to be drastically modified, which cost the owner redesign and review time—squandering some of the benefits of the improved design.

Figure 1.1 shows the bridge being built with the modified design. The first design called for the bridge deck between the support piers to be cast in situ. That method would have been much more

FIGURE 1.1   Prefabricated box girder deck being set in place. Courtesy of Michael Veegh.

expensive to build, compared to precast box girders that can be assembled on land and then floated out to the bridge and lifted in place. The second option (using the precast box girders) is what the team eventually adopted, but only after time and money were lost pursuing the initial design.

What if we could create buildings that actually exceed expectations, instead of fall short? What if we could finish projects on time and under budget, with no harm, and no rework? What if we could produce a building that is beautiful, efficient, useful, and cost effective, all at the same time? We have come to accept less than this, but we don't have to. What we physically construct may well be our most permanent and lasting legacy, and it should be something we are proud of.

We believe that an integrated approach is the most promising strategy to produce a truly high-performance building. Through harnessing the talents of many individuals from different disciplines, creatively shaped incentives, working collaboratively, and sharing knowledge, we can produce long-lasting value for owners and occupants. Figure 1.2 shows the result of the integrated UCSF Mission Bay Hospitals project team's effort to produce a product—in this case a patient room headwall—that integrates architectural, structural, mechanical, electrical, plumbing, and medical features and that is buildable and operable and effectively serves its final customer, the patient.

On the UCSF Mission Bay Hospitals project, the team decided early on (in Design Development) to prefabricate headwalls in patient rooms off site. They did this because headwalls are complex, composed of many different systems, including high and low voltage, mechanical, data, medical gas, and so forth. Project foremen explained the many things that could go wrong in installing and testing the many utility systems that make up each headwall. Offsite fabrication meant they could build and test utilities in a controlled environment and simply pop the complete wall into place on site.

Although we can avoid an integrated process, we can't avoid interactions. Once built, every building functions as a whole with technical systems that are either supporting or fighting each other. But we can also choose to produce buildings with highly integrated systems by using teams and integrated work processes. The aim of this book is to describe and explain how to set up and manage projects in such an integrated way.

1.3   A WAY FORWARD

Essentially, an integrated team operates as a virtual organization committed to the project. It is not organized to optimize the outcome of individuals or their companies. The integrated team allocates resources and makes decisions guided by the core project values. Ideally, it is a synergistic system that is greater than the sum of its parts, built on the assumption that major building projects are not inherently a zero-sum game; it is possible for everyone involved to win together.

Integrated project delivery (IPD) has several major components. At the outset, owners will define their values and goals so that the entire team can understand, quantify, and track what they are working toward. Next, an organizational structure is created and planned intentionally to enable good work. The organization emphasizes cross-functional teams at a working level; transparency and information

FIGURE 1.2   Headwall installation. Licensed by The Regents of the University of California on behalf of its UCSF Medical Center; courtesy of DPR Construction.

flow; swift identification of problems, response, and adaptation to them; and decision making as close to the frontline as possible.

Coupled with the structure is a physical place to work, which encourages desired team behaviors and makes them possible. The physical space allows face-to-face interaction, builds enduring relationships, enables questions to be answered rapidly, promotes personal accountability, and vastly increases information sharing so that every team member stays on the same page. Finally, work practices that include a high degree of collaboration, simulation, and visualization enable the right work to be done the right way, and built just once. Meaningful metrics, displayed through a project dashboard, gives each team member the real-time information necessary to know how the project is performing and to take steps to improve.

This book delves into the fundamental concepts of integrating project delivery. We discuss how to set up an organization and processes that best leverage each discipline's knowledge and experience to deliver a building that meets or exceeds the owner's goals; in other words, set up an organization focused on achieving a successful project, not an organization set up to avoid failure. We also outline how to set up the various roles and responsibilities of the project participants and create a true team culture, along with the skills that designers, engineers, and contractors need to develop to effectively and efficiently integrate their projects.

While the AEC industry has been abuzz for some time about the potential rewards of an integrated delivery approach, many firms now show signs that they have the capacity and capability to make IPD a reality (Bell, 2012). We truly believe that our industry could be on the brink of a historic shift, but it will take the audacity, fearlessness, and conviction of thousands of boots on the ground to make it happen.

Integration Results

The Lean Construction Institute recently sponsored research conducted by Dodge Data & Analytics to benchmark owner satisfaction and project performance and the effect of Lean principles on project outcomes. Along the way, the research team also looked at how project delivery methods compared (Mace, et al., 2016).

The researchers sent questionnaires to 81 owners, covering 162 projects. An additional 10 IPD projects were studied in detail in a team led by Renée Cheng (Cheng, et al., 2016). The owners were asked to separate their projects into typical versus best project. Figure 1.3 displays their findings regarding project delivery type.

Neither the "typical" nor the "best" columns sum to 100% because the researchers excluded project delivery types that did not achieve 20% of the sample size. The importance of the data lies in the difference in columns for a given project delivery type. In design-bid-build, it was almost four times more likely that a project would be "typical" rather than "best." Construction management at risk fared better, although still did not have an even likelihood of success. Design-build is significantly more likely to result in a "best" project. The star of the study, however, was Integrated Project Delivery, which was over 20 times more likely to result in a "best" project.

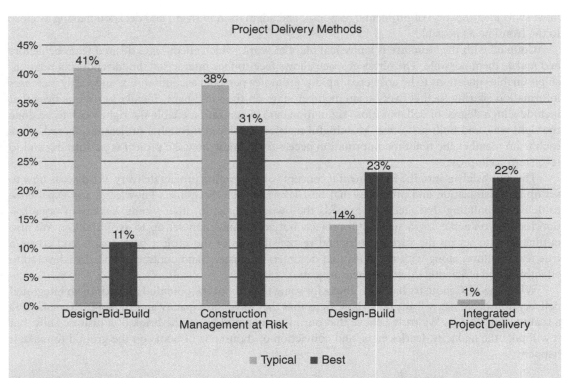

FIGURE 1.3   Comparison of project delivery methods. Courtesy of Lean Construction Institute.

This study is consistent with prior work by the research team led by Renée Cheng that published data based on IPD project participants' perceptions of the value of IPD as compared to their experience with other project delivery types (Cheng, et al., 2015).

The Dodge researchers reported several interesting relationships between Lean practices and "best" project outcomes. First, they found that in 76 percent of the "best" projects, the key project team members, including trades, were engaged during conceptual design or earlier. In contrast, only 34 percent of the "typical" projects had engaged the key project members by that time. They also looked at Lean methods and found that all of the Lean methods studied were used at least twice as much on "best" than on "typical" projects. Their results are particularly striking with regard to target value design (40 percent "best," 6 percent "typical") and use of co-location (44 percent "best," 6 percent "typical"). The Dodge results strongly complement the recommendations in this book as to the use of integrated project delivery, target value design, co-location, and production management tools.

This is not to say that integrating project delivery is easy—change rarely is. As one senior IPD project manager said, at first, you feel like you are continually failing. The team seems to be constantly finding problems, and it may seem easy to get discouraged or get the impression that a traditional process runs more smoothly. The fact is that the issues that surface early in IPD will eventually come to light on any project. But when they do, it may be too late to implement elegant and efficient solutions. IPD forces problems into the light, and enables the team to address them before they get worse.

Yes, IPD can be difficult, but the reward is that the resulting product should be superior, with little room for suboptimization. IPD thrives on exchange of ideas and information, which are often in conflict with each other; it is a challenging environment for many people to work in. However, the current process is extremely unsatisfying and frustrating and lets us be, essentially, lazy. An integrated approach may be more demanding, but ultimately, it is more fulfilling.

At the end of the day, would we be proud to tell our children that we helped create millions of tons of waste[2] or that we helped design and build a building that's merely adequate, or that someone died while working on our project,[3] or that we helped create a building that's sufficient now, but will be obsolete in 10 years? We believe the AEC industry can and must do much better—and an integrated approach offers the best blueprint to do that.

NOTES

1. We are indebted to our colleague John Kunz, a computer scientist, for pointing out this important fact to us.
2. The EPA estimated that the United States generated 170 million tons of construction and demolition waste from construction, demolition, and renovation for nonresidential and residential buildings in 2003 alone, equal to 3.2 pounds of building-related materials waste per capita per day (USEPA, 2009).
3. According to the Bureau of Labor Statistics, 798 workers were killed on the job in the construction industry in 2011—over 14 people per week (USDOL, 2011).

REFERENCES

American Physical Society (APS). (2008). *Energy future: Think efficiency*. Retrieved October 4, 2010, from http://www.aps.org/energyefficiencyreport/report/aps-energyreport.pdf.

American Society of Civil Engineers (ASCE). (2009). *Report card for America's infrastructure*. Retrieved October 4, 2010, from http://apps.asce.org/reportcard/2009/grades.cfm.

Bell, J. (2012, August 10). Designers and contractors try out integrated project delivery, Portland Business Journal. Retrieved on October 8, 2016, from http://www.bizjournals.com/portland/print-edition/2012/08/10/designers-and-contractors-try-out.html.

Cheng, R., Allison, Markku, Sturts-Dossick, C. Monson, C, (2015) *IPD: Performance, Expectations, and Future Use: A Report On Outcomes of a University of Minnesota Survey*. University of Minnesota. Retrieved October 16, 2016, from http://ipda.ca/site/assets/files/1144/20150925-ipda-ipd-survey-report.pdf.

Cheng, R., Allison, M., Sturts-Dossick, C., Monson, C., Staub-French, S., Poirier, E., (2016). *Motivation and Means: How and Why IPD and Lean Lead to Success*, University of Minnesota. Retrieved from http://arch.design.umn.edu/directory/chengr/.

KPMG, *Climbing the curve: 2015 global construction project owner's survey*. Retrieved February 28, 2016, from https://www.kpmg.com/BR/PT/Estudos_Analises/artigosepublicacoes/Documents/Build-Construction/global-construction-survey-2015.pdf.

Kunz, J. (2010, September 16). *VDC for sustainable facility design, construction, and operation*. Presentation at CIFE-SPS VDC Certificate Program Introductory Course, Center for Integrated Facility Engineering (CIFE), Stanford, CA.

Kunz, J., Maile, T., & Bazjanac, V. (2009). *Summary of the energy analysis of the first year of the Stanford Jerry Yang & Akiko Yamazaki Environment & Energy (Y2E2) Building*. Technical Report #183, Center for Integrated Facility Engineering, Stanford, CA.

Mace, B., Laquidara-Carr, Donna, Jones, S. (2016). *Benchmarking Owner Satisfaction and Project Performance*. Retrieved from http://www.leanconstruction.org/learning/.

National Science and Technology Council (NSTC). (2008, October). *Final report: Federal research and development agenda for net-zero energy, high-performance green buildings*. National Science and Technology Council, Committee on Technology, Subcommittee on Buildings Technology Research and Development. Retrieved October 7, 2010, from http://www.bfrl.nist.gov/buildingtechnology/documents/FederalRDAgendaforNetZeroEnergyHighPerformanceGreenBuildings.pdf.

Nilsson, A., & Elmroth, A. (2005). The buildings consume more energy than expected. In *Sustainable city of tomorrow: B01—Experiences of a Swedish housing exposition* (pp. 107–118). Stockholm, Sweden: Formas.

Podolny, W., Cox, W. R., Hooks, J. M., Miller, M. D., Moreton, A. J., Shahawy, M. A. … Tang, M.-C. (2001). *Performance of concrete segmental and cable-stayed bridges in Europe*. Report FHWA-PL-01-019, Office of International Programs, Office of Policy, Federal Highway Administration, U.S. Department of Transportation. Retrieved October 4, 2010, from http://international.fhwa.dot.gov/Pdfs/conc_seg_cabstay_euro.pdf.

Scofield, J. H. (2002). Early energy performance for a green academic building. *ASHRAE Transactions, 108* (Part 2), 1214–1230.

Scofield, J. H. (2009). Do LEED-certified buildings save energy? Not really … *Energy and Buildings, 41,* 1386–1390.

U.S. Department of Labor, Bureau of Labor Statistics (USDOL). (2011). *Revisions to the 2011 census of fatal occupational injuries (CFOI) counts*. Retrieved from http://www.bls.gov/iif/oshwc/cfoi/cfoi_revised11.pdf.

U.S. Environmental Protection Agency, Office of Resource Conservation and Recovery (USEPA). (2009). *Estimating 2003 building-related construction and demolition materials amounts*. National Service for Environmental Publications, EPA530-R-09-002.

U.S. Green Building Council (USGBC). (2002). *Green building rating system for new construction and major renovations* (LEED-NC), Version 2.1. Retrieved October 4, 2010, from http://www.usgbc.org/Docs/LEEDdocs/LEED_RS_v2-1.pdf.

CHAPTER 2

Transitioning to Integrated Project Delivery: The Owner's Experience

"The way to get started is to quit talking and start doing."

—Walt Disney

Throughout this book we take a project focus because buildings are designed and built as projects. We also frequently note the owner's central role in building projects in general and with integrated project delivery (IPD) specifically. The owner is often the party calling for change; the owner assembles the IPD team, defines the project goals, and, as you will read below, has a leadership role that is critical to project success. But how do owners view their role? What do owners have to do within their own organizations so that projects can be delivered successfully with IPD? How do they respond to skepticism within their organizations from above and below? What skills must their project leaders have? How do they manage architecture, engineering, and construction (AEC) partners that have limited experience with IPD? What do owners believe they are gaining through IPD? What has surprised them, and what advice would they give to owners contemplating IPD?

We decided to ask them.

This chapter summarizes conversations with owners who have embraced IPD and who are leading their teams and the industry. In groups of two to three, they met telephonically to discuss the owner's role in IPD. Their conversations were recorded and transcribed and are summarized below.

Because we chose to draw from projects that were well under way or completed, the sampling is biased toward health care projects. But the IPD market is rapidly expanding into institutional, academic, commercial, high-tech, and entertainment projects. To broaden the perspective, we provided a draft of this chapter to owners in these sectors, and they have provided critical review and comment.

The interviewees did not downplay the difficulties and frustrations in adopting IPD. But their comments were overwhelmingly positive. The tenor of their comments is very similar to a recent study of IPD projects undertaken by the University of Minnesota in association with the Integrated Project Delivery Alliance (Integrated Project Delivery Alliance/University of Minnesota, 2015). As stated in the report's Key Findings, the responses strongly supported IPD as a superior delivery method—a result that was consistent across all demographics. What this report confirms is that the opinions and observations of our interviewees are not outliers. They are representative of the owner's experience with IPD.

In alphabetical order, the interviewees were:

Michael Bade, Vice Chancellor and Campus Architect, University of California, San Francisco. UCSF is a leading academic health care university with world-class research, teaching, and clinical groups. UCSF has two primary campuses: Parnassus and the new Mission Bay campus.

Brenda Bullied, Director of Facilities Innovation and Planning, Lawrence & Memorial Hospital. Lawrence & Memorial (L&M) is a regional hospital system in New London, Connecticut, that operates a 300-bed acute care hospital as well as related outpatient facilities. L&M used IPD for a cancer center affiliated with the Dana-Farber Cancer Institute, a medical office building, and an inpatient renovation. Brenda was the project manager responsible for these projects.

Digby Christian, Director of Integrated Lean Project Delivery, Sutter Health. At the time of the interview, he was the Senior Project Manager for the Sutter Health Eden Medical Center (SHEMC), a $320 million project, 130 bed, 230,000 square foot, that opened on December 1, 2012. This project was delivered under an 11-party integrated project delivery contract and won ENR California's Best Projects Award for 2012, Structural Engineers of Northern California Excellence Award (2014), AIA Technology in Architectural Practice IPD Process Innovation with BIM Award (2010), Tekla Global BIM Awards Competition (2010), and the Fiatch CETI Award (2008), as well as other awards. Digby Christian is currently overseeing the California Pacific Medical Center (CPMC) Van Ness and Geary and St. Luke's Campus projects, a Lean IPD project exceeding $1 billion in value.

Wendy Cohen, Regional Executive, Kitchell (formerly Director of Facilities & Development, Palomar Health). Wendy was the primary owner representative for Palomar Medical Center (PMC) West, a very successful, $956 million, 360-bed, cutting-edge acute care hospital in Escondido, California.

Crista Durand, At the time of the interview, Ms. Durand was the Vice President Strategic Planning, Lawrence & Memorial Hospital. Crista provided senior leadership on the L&M projects. Crista was responsible for all L&M facilities and is the Senior Management Team representative on L&M's IPD projects. Crista is currently President of Newport Hospital.

Stuart Eckblad, FAIA, Director of Design and Construction, University of California Medical Center, Mission Bay, San Francisco. The Mission Bay Medical Center complex is composed of a new 289-bed children's, women's, and cancer hospital; the Gateway Medical Building, which will contain ambulatory services for pediatric, women, and cancer patients; and an Energy Center to run the 878,000-gross-square-foot hospital complex. The project construction cost exceeds $800 million, and it was completed in late summer 2014.

Sean Graystone, Superintendent, House of the Temple, Supreme Council 33° Ancient and Accepted Scottish Rite of Freemasonry, Southern Jurisdiction. Sean is spearheading the functional renovation of a turn-of-the-century architectural and historical masterpiece. Completed in 1915, the House of the Temple was designed by John Russell Pope, constructed by Norcross Brothers, and has an innovative Rafael Gustavino dome. Modeled after the tomb of Mausolus at Halicarnassus, the House of the Temple is considered one of the most important examples of American Neo-Classical architecture.

Chuck Hays, President and CEO, MaineGeneral Health System. MaineGeneral is the third-largest health care system in Maine, operating two hospitals as well as outpatient, long-term care, home health, and retirement facilities. In November 2013, it completed a new, LEED Gold, 650,000-square-foot regional hospital that provides state-of-the-art treatment and operational efficiency.

Roger Johnson, Senior Vice President, Enterprise Real Estate, TD Bank Financial Group. TD Bank is one of the largest financial institutions in North America. It has approximately 2,500 branches in Canada and the eastern United States. It is currently upgrading its offices in Toronto, an approximately $150 million effort, and beginning the rollout of new branch banks across Canada.

Mark Linenberger, Vice President, Linbeck Group; Senior Management Team leader on Cook Children's Medical Center (CCMC) IPD projects. Linbeck Group is a construction manager with a long history of innovation, and is currently the construction manager for both of Cook Children's Medical Center's IPD projects. Linbeck introduced CCMC to IPD.

Robert Mitsch, Vice President Facility Planning and Development and Real Estate Services, Sutter Health. Sutter Health is one of the largest health systems in California, serving over 100 communities and operating 24 hospitals as well as other health care facilities. At the time of the interviews, it was in the latter portion of an $8 billion construction program to upgrade the seismic capacity of existing facilities, replace facilities, and build new facilities to accommodate growth. It had four major IPD projects in construction and was in the late design stages for several others, the most expensive of which had a construction cost exceeding US$1 billion.

George Montague, Vice President Real Estate, Cook Children's Medical Center (CCMC). CCMC completed a 120,000-square-foot medical office and administrative building with an adjacent parking garage. Based on its favorable experience with the current IPD project, CCMC is now undertaking a $300 million, 300,000-square-foot South Tower hospital expansion and a new central utility plant.

William Seed, Currently a Lean/IPD consultant. At the time of the interview, he was a Staff Vice President of Design and Construction, Universal Health Services. Universal Health Services has 240 campuses spread across 37 states. These include 26 surgical/medical hospitals and 200 behavioral health hospitals. Universal Health Services executes all projects over $5 million using Lean IPD.

David Tam, MD, FACHE, Chief Administrative Officer, Palomar Health. In 2012, Palomar Health completed a very successful, $956 million, 360-bed, cutting-edge acute care hospital in Escondido, California. The project was awarded a National Best of the Best Projects in 2012.

We expect that the situations and examples shared by these owners and the terminology used will sound familiar to the readers with several years of experience. Readers without industry experience should note how these owners attempted to integrate the information, the team, the process, and the design and execution plan for the project and how they lead the definition of the performance targets for their projects.

2.1 THE ROAD TO IPD

There is no one road to IPD; however, certain patterns emerged from the interviews, and it became clear that regardless of the path taken, IPD is an owner-driven process. Sean Graystone explained: "It seems to me that owners—smart owners—are driving the transition to Lean and IPD."

Frustration with existing project delivery systems was the most common reason for turning to IPD. Several owners mentioned prior projects that were disappointing, and they knew their organizations had to develop better-performing systems. They simply could not repeat their prior experiences. Sutter Health, for example, was embarking on an $8 billion seismic retrofit and expansion program and needed a delivery system that was predictable and reliable—which they were not gaining from Construction Manager at Risk, even with design assist services. William Seed was faced with turning around a program at UHS that was not efficiently delivering their facilities. "I was very frustrated with the entire delivery process of how projects get delivered. It was just a continuous battle." In some instances, such as Lawrence & Memorial Health, dissatisfaction with existing project delivery was coupled with a belief that IPD fit their operational focus on process transformation. They needed a change, and IPD made sense.

Others arrived at IPD through research. MaineGeneral's CEO, Chuck Hays, recognized that they were undertaking a project much larger than anything they had previously attempted (and that it would be the largest construction project in Maine) and decided to research the best method for delivering a complex hospital project. MaineGeneral engaged a consultant to research project delivery options and created a team to evaluate available options, which recommended IPD. Sutter Health sent teams across the country and even to Europe to evaluate project delivery options. TD Bank's Roger Johnson was also frustrated by the existing project delivery process. "I have been in real estate and construction my entire life, and I was frustrated with the traditional process of design and build, or any variation on that, and the conflicts that are just built into the process." He learned about IPD by reading *The Commercial Real Estate Revolution* (Miller, Strombom, Iammarino, & Black, 2009) and immediately recognized that the dysfunctions chronicled by the authors were plaguing TD Bank's delivery process and that IPD provided a path for change. William Seed first discovered IPD in a seminar about Lean IPD on Sutter Health Projects and its potential for removing dysfunction and improving results.

Values were also important. Lawrence & Memorial was drawn to the values of IPD, which matched L&M's focus on process transformation. Sutter Health also believed it fit their organization. According to Robert Mitsch, "We think it matched our value system as an organization in terms of transparency and holding each other accountable."

The upshot is that, whether driven by frustration, research, process transformation, or a combination of these factors, these owners followed a deliberate path to IPD.

2.2   THE OWNER'S ROLE

The group agreed that owners have a critical role in IPD. When asked what it took to be a good IPD owner, they identified five key characteristics: clarity, commitment, engagement, leadership, and integrity.

2.2.1   Clarity

The owner must be able to define what it wants and what the IPD team must achieve. At a minimum, this requires clearly expressing programmatic needs at project inception and continuously throughout the project. Christa Durand was emphatic about owner clarity. "The owner must be very clear about their expectations for the project and what they want."

But clarity should also exist at a strategic level. As Digby Christian stated, "You have to understand why you want to do the project." Michael Bade explained, "An IPD project demands an owner who can state the objective of a project strategically. When an owner isn't engaging strategically, the project is looked at in simplistic, programmatic terms. One of the most powerful aspects of IPD, in my mind, is that the IPD team helps the owner test its own assumptions." He went on to describe the Mission Hall project where the team recognized the owner's need for a transformative medical academic space. Without increasing the budget, they created linked collaborative spaces within the building. These "town hall" spaces were not in the owner's initial program but added value by enhancing the building's research and academic program. By understanding the owner's strategic needs, they were able to create a facility that effectively responded to those needs, not just to the numbers in the programming document.

2.2.2   Commitment

All of these owners expressed commitment to IPD and a willingness to support the process with training and resources. Robert Mitsch observed, "The biggest thing that people who are just starting IPD don't understand is the commitment an owner must make to this delivery method." Digby Christian believes the owner's commitment is critical for behavioral change. "The owner has to demonstrate full commitment to the project so there's no opportunity to do an end run around the process. Not everyone who shows up on the project is going to be comfortable doing IPD, and not everyone believes it will be better, so the owner needs to be strongly committed to 'doing things this way' and have a vision for what 'doing things this way' means." George Montague observed, "I think it has been really helpful to have a higher level of involvement to do several things; one is just to demonstrate the owner's commitment to IPD and to change behavior." Several owners commented that if the owner's commitment is lacking, team behavior will not change.

Ideally, commitment runs from top to bottom. Robert Mitsch noted that Sutter Health's CEO, Pat Fry, was deeply committed to their Lean IPD process. Digby Christian believed that support was critical to his project's success, stating, "It really helps knowing that the guy at the top who runs the whole system wanted this to happen. So we were rock solid from my level all the way up to seven levels above in management." But commitment needs to be refreshed. Several owners built support, but then as managers and executives changed, they had to reeducate the new leaders in order to maintain that support. Within large organizations, support needs to be continuously refreshed.

2.2.3   Engagement

None of these owners were passive. Wendy Cohen stated: "The owner needs to be fully engaged and an equal partner at the table—which is different from traditional delivery methods." Although the level of staffing varied among the owners, all maintained daily presence on the project. Their managers were empowered to lead teams and seek solutions, did not shirk their leadership role, and took responsibility for their projects. William Seed remarked, "What drives value into the program is owner engagement." Michael Bade also saw the need for deep engagement and recognized that the owner's engagement needed to be balanced and challenged by the team. "IPD is so much in the owner's self-interest, but a lot of owners manage by sitting back, and you can't do that with a Lean IPD process. You have to be an engaged and knowledgeable participant and actively work to ensure that your teams are tracking and trying to achieve your objectives. But you are also placing yourself in the position of having them challenge you. It is a reciprocal relationship."

All of the owners reported that they were intensely involved in the projects. Brenda Bullied noted, "I had no idea how involved I would be as an owner—these projects are part of my life and I would say this is true for every person on the team—we live and breathe these projects differently than any other kind of construction project." Roger Johnson agreed: "I think you actually take more of a role—at least we took more of a role in the project than we would in a typical design-bid-build." And Crista Durand found that these projects required more of her time than traditional projects—but it was worth it.

Staffing varied between owners. Robert Mitsch and Digby Christian planned an active leadership and administrative role on Sutter Health projects, and they staffed accordingly. In contrast, MaineGeneral thought they would have to hire six more people to manage their project, but they only hired one. Chuck Hays noted, "To me it just felt like it flowed much easier than having to beat on contractors who were behind or weren't putting enough manpower on the job." In their case, the leadership effort was distributed among the team, which may have resulted in less owner effort. William Seed agreed with both. "I'm going to agree with Chuck [Hays] that it is easier because you have empowered the team and distributed the leadership, but I also agree with Bob [Mitsch] that you spend more energy at the proper time to empower people to make those decisions and that is why you are seeing the benefits you are seeing."

Stuart Eckblad did not think that his team put in more effort than they did on conventional projects, but he agreed that the type of work differed. "Going back to the whole issue of owner involvement, it is a different kind of involvement. … I use my 10 hours of the day differently." Digby Christian and Mark Linenberger agreed with this observation. Crista Durand stated that IPD required a much higher level of responsiveness and involvement at her level. William Seed echoed the deeper level of engagement but also noted that it was easier, too. "My department and my staff appreciate that we don't go home with stomachaches and headaches anymore. We appreciate that our intellectual capital is now used to drive improvement and increase value as opposed to fighting battles over what is in and out of a contract or the value of a change order."

The more engaged an owner is in the IPD process, the more it will achieve.

2.2.4   Leadership

All of the owners led their projects, but as described below, their leadership styles and interactions varied significantly. Although leadership is crucial, it manifests itself in many ways. Moreover, as Sean Graystone stated, "An owner has to know when to lead and when not to lead."

William Seed stressed the need for leadership. "The more leadership the owner takes, the better value and the more success is driven into the project. I would say the owner's leadership is absolutely critical at the beginning of the project to set what we call conditions of satisfaction that the team can actually measure and use to make decisions." Robert Mitsch explained that the need for leadership became apparent as they engaged more deeply. "What we really did not understand early on was how important our role was. … It is not something you can set forth on a piece of paper and expect everybody to follow the process. It was something we had to participate in fully and we actually had to lead in order for IPD to be successful."

Leadership styles varied widely. Several of the leaders had project managers that administered and managed the overall project. They took explicit control of running meetings and the project, overall.

For others, leadership was distributed among team members. George Montague used a distributed approach. "I have been present, but in our meetings we have passed around the responsibility of leading and facilitating the meetings. I think one reason is that it has forced people to engage and be involved." At MaineGeneral, the Project Management Team, consisting of the owner, contractor, and architect, jointly managed the project. Palomar Health's PMC West project was similar.

Wendy Cohen summarized the approach for the PMC West project in the following way. "The leadership was a joint effort on our project with a core group, which included one person from the general contractor, one person from the architect, and one from the owner's side. These were people that had the day-to-day responsibility for executing the project. This team created a close relationship that built significant trust over time. Most decisions were made jointly or by one of the three on the team. The trust was a key factor here, so that each person of this team could come together when needed or make decisions independently when needed. It was also important that we were seen as one unit and that it wasn't possible to divide and conquer us by different aspects of the project team."

But regardless of structure, the owner was actively engaged, clear in its vision, and willing to step up if team leadership faltered.

There was general agreement that IPD projects require champions who measure project processes and behaviors against IPD ideals. In part, these people are those within the team who are willing to call out when the process is not occurring properly. Deep change comes when a team can declare a breakdown in process, reflect on the cause, and then implement solutions. For many teams, analysis, reflection, and change are new skills, and many of the owners used outside coaching to build these behaviors. Others have used a combination of outside training to assist staff or to "train the trainer." William Seed has used outside facilitators and coaches, but also provides coaching by his own staff. One of his project managers is assigned to coach on other project managers' projects. He believes it is important to have someone independent of the project manager. "It is a lot easier for somebody from outside of the job to come in and feel the culture or the pulse of the job and help guide it back into the place that it ought to be functioning."

The owners focused on mentoring and teaching, as well as managing. At Sutter Health, successful project managers check in on other projects to see how they are working. They provide experience, expertise, and assistance to the less experienced project managers. William Seed expects Universal Health's project managers to have mentoring and teaching roles. "Frankly, I think my project managers are more Lean coaches than they are project managers. They don't manage the project's budgets—the team manages the budgets. They don't wrestle with quality—the team wrestles with quality."

IPD leadership is not imposed from above, nor is it forced consensus. Active questioning and debate are critical to the process. Michael Bade stated, "It is those points of disagreement that are the genesis of creative solutions, and people need to learn the skills of disagreeing with each other in a friendly way and inviting each other to define the problem first and to solve it second. In IPD leadership you stimulate people to think more deeply and you reward them with recognition and kudos and stuff like that when they actually push farther than they have done before."

Traditional project delivery uses a "propose/dispose" model with the team proposing solutions to the owner's program and the owner accepting, rejecting, or critiquing those proposals. IPD puts the owner in an active role, engaging with the team at a strategic level and influencing or jointly developing the solutions to project challenges. IPD leaders empower the design and construction team to take responsibility for the project, challenge the owner's assumptions, and deliver the project to the agreed goals. As the comments indicate, these owners understood the power provided to them by IPD and accepted the responsibility to use it effectively.

2.2.5   Integrity

The owner has a key role in setting the project tone. IPD is based on optimizing the entire project, and that includes the interests of the participants as well as the owner. An owner that acts only in its immediate self-interest or that doesn't act in accordance with its expressed principles will find that the participants do not fully engage in the IPD process. Wendy Cohen explained, "[T]he owner is key in setting the tone, and the owner needs to create the space and environment to develop trust-based relationships." Mark Linenberger agreed. "The owner can quickly create trust and thus creditability, by doing what we say we are going to do, every time—creating reliability." As Mark also noted, "The owner models the way." The challenge, as Wendy Cohen noted, is to "identify the right person from the owner that will be the point person and to ensure that the people that they enlist on their team have the ability to work in an IPD environment."

2.3   ORGANIZING THE OWNER

Wendy Cohen reflected that "IPD is simple in concept, but it becomes complex when applied to specific organizations." The PMC West project a nearly billion-dollar hospital for a public hospital district, was challenging because there were many stakeholders represented by a 40- to 50-person executive steering committee, the hospital management, and the publicly elected board of directors, each with its own interest in the project. It is an example of the difficulties in determining who is the "owner" in a large IPD project. Dr. David Tam was thrust into the project through his role as the district's chief administrative officer and had to carefully orchestrate the interactions between the broader owner groups and the project team. This involved the delicate task of determining when to involve stakeholders, when not to involve them, and how deeply to involve the organization into the IPD process. As Dr. David Tam observed, "In an IPD project, it is important that the 'owner' be carefully defined. The IPD owner is the person or persons that engender trust with the members of the integrated project." He and Wendy Cohen served this intermediary role, providing a consistent face to the IPD team. They did not try to "sell" the organization on IPD and only started explaining the process after the project became successful. In Wendy Cohen's words, she became a translator between the IPD project and a more traditional organization. By keeping IPD within the construction organization but providing results that the traditional organization valued, they kept IPD moving forward without requiring buy-in by the entire organization. It was a stealth IPD project.

Dr. David Tam observed that the success of PMC West changed the organization itself. "It wasn't just stealth that we did IPD and the organization didn't know it, but it was also stealth in the sense that the organization absorbed the principles of IPD and started using it in a variety of other ways. The project was an actual pivotal transformational chapter in the organization. It actually caused the organization to change. The project was so significant from all aspects that when it became clear that IPD was working for the construction process, it infiltrated the organization." When the project moved from construction to activation, IPD principles and Lean tools, such as pull planning, continued to be used.

The success of PMC West demonstrates how organizing a major project can affect the organization itself. It echoes comments by Robert Mitsch and Crista Durand about the interaction between the construction processes and the operational processes used by a health organization. Process transformation in one can inform the other.

2.4 RESISTANCE FROM WITHIN

William Seed's approach to adopting IPD was the simplest. "I never asked for permission. I just sort of did it. Then I showed I was doing better than they claimed to have been doing and overcame challenges to IPD." George Montague felt that IPD was a natural extension of their working relationship with key partners, such as Linbeck. Moreover, one of the hospital trustees was an architect who favored IPD, which made the rest of the board comfortable with changing project delivery methods. George Montague also stated that the success of their medical office building project made it easier to undertake the larger hospital renovation.

But most of the owners faced resistance from senior management or from operational staff. Moreover, the effort to overcome management resistance had to be periodically repeated because executives and board members would forget the lessons learned and the progress made. In a declining economy, these senior executives would argue for returning to low bid procurement. Moreover, several of the owners work within highly structured organizations that have strong compliance, internal audit, or legal protocols. For example, although Sutter Health had a track record of successful IPD projects, it still had to satisfy internal and external auditors and show that the projects were being properly managed and controlled, which required significant time and effort from project staff.

A respondent from a large organization emphasized the need to confront the issues raised by people who don't understand the IPD model. They can be concerned about the greater up-front costs and may believe that competition is the only way to assure the right price. Although persons who understand IPD may be frustrated by these doubts, it is important—particularly in a large organization—to recognize that these concerns are heartfelt, and to address them intelligently.

Roger Johnson recognized that building understanding was a first step in developing support in a large institution. He asked the firm auditors to read the first 100 pages of *The Commercial Real Estate Revolution* (which chronicles problems in design and construction coming from traditional project delivery). He then engaged the support of the bank's sourcing director, who had prior experience with Project Alliances in petrochemical exploration (similar to IPD), and a knowledgeable supporter of IPD processes. Sean Graystone faced a similar problem with the internal counsel, who began as a skeptic but, as she participated in training, began to understand the theory and saw the results—now she supports IPD.

As Michael Bade observed, resistance has also come from middle management. "There are a lot of people who are invested in the old way of doing things. They know how to fight with each other really well, and a process that demands that they step forward and become more of a positive participant is something that is outside their area of professional expertise." Roger Johnson had a similar experience. "There still some folks on the team who are not in leadership roles but probably one down from leadership roles who are much more comfortable in the traditional ways and are dragging their heels on really adopting where we want to go."

Christa Durand found that not everyone likes the level of personal accountability IPD imposes. "I think that where there has been resistance, it has been, quite candidly, with those departments that are not high performers. They have resisted it because there is more work and a high level of accountability." Lawrence & Memorial had to pass over several project managers because they did not have the skill set to perform at the level required. Roger Johnson used innovative thinkers with flexible management styles who could see the benefit of the process and volunteered to undertake the IPD projects. And when the projects progressed and were successful, there was a change within the bank. "There was resistance throughout, but when people started seeing the results, started seeing the activity of the team coming together and everybody looking to develop the best end product and not just line their own pockets, most of those people who were resisting in the beginning have become supporters. At this point in time and inside the bank today, there are actually very few naysayers or resisters to this process, whether it be in audit, sourcing, or on the construction team."

The upshot, from several owner leaders, is that you need to expect challenges from within, you need to be resilient, and you need to persevere.

2.5   RESISTANCE FROM THE AEC COMMUNITY

These owners were more critical of their own organizations than they were of resistance within the AEC community. In general, they were very pleased with the willingness of their partners to step up and take responsibility for the entire project. But it wasn't always easy to find good IPD partners. Robert Mitsch noted that as a pioneer, Sutter Health had to look hard to find partners who were committed to change. Even now, there are pretenders. "There are many in our industry who raise their hand and want to do IPD, but who have not spent any time or effort to understand what IPD really is." When interviewing for IPD partners, Chuck Hays had a similar experience. "It is interesting how many people say they do IPD, but when you dig into their experience, it is not even close." Vendors that claimed to have "always been doing IPD" were seen as lacking understanding or being disingenuous.

Several owners reported that they had to dismiss one or more participants because those participants weren't engaging collaboratively. In two instances, the dismissed participants were entities that the team "assumed" would be good partners and, therefore, skipped the vetting process. As William Seed described the error, "Let's not bother to go through the CBA [Choosing by Advantages] on this particular selection because we all think this is the key guy and we need to make a decision fast, so let's just pick this guy. Those are the ones that fail because you didn't do your diligence and you didn't vet them for the new processes or the new culture." Chuck Hays noted that the one contractor they had a problem with was the one they didn't vet because they assumed it would be a good fit. Firm culture was viewed as critical to team success.

Despite difficulties in finding partners, resistance, and backsliding, all of the owners felt that their contracting partners became project partners through the process. Brenda Bullied was surprised at how vendors became partners. "I think that what has surprised me is our subs have been fully engaged in this process and have actually turned around to becoming partners in the process. I didn't expect that to happen at all." Chuck Hays recounted MaineGeneral's experience of building a big IPD team in a small construction community. "After viewing two of the larger projects in Maine, everybody we talked to said you absolutely are going to have to pull labor from Boston and New York. You don't have a qualified workforce [in Maine]. We ended up selecting a Maine architect and a Maine construction manager that we trusted, but who couldn't possibly do this size project. They partnered with national firms. We trusted them to pick good partners, and they did pick good partners, and we told them that the winning team would be the one that could tell us how to use Maine labor effectively." Giving back to their community is an important value for MaineGeneral. On the new regional hospital, 96 percent of the trade contracts have been awarded to Maine companies. The designers, construction managers, and trades have all worked together to help local firms rise to the level required, and the community has responded.

The upshot is that getting a good team can be challenging. An owner must carefully select its team and must place a high value on culture and collaboration. William Seed advised other owners to "look for partners who are willing to innovate, bring the best ideas, and continually look for ways to solve problems as opposed to their traditional manners. This requires a different interviewing process and technique than the traditional background screening and financial review." Chuck Hays recommended focusing on fit. "I would advise a new IPD owner to pick cultural fit over everything. Pick partners that can be team players and who will work well with a group. Pick partners that will take responsibility for the whole project, not just their piece."

2.6 EDUCATION AND TRAINING

Brenda Bullied succinctly summarized the owner's experience. "Educate, educate, educate. Educate every staff member, even VPs. Educate nonstop about the process and why we are doing IPD."

Every owner invested heavily in educating their staff and their teams. William Seed built education into the projects he managed. "We have found training to be critical to pretty much everything we are doing. …There is going to be a learning curve and a training curve, and if you embrace that up front and prepare for it, I think you can do a better job of getting a team up to speed fast." Education was also used to reinforce IPD behaviors and processes. Sean Graystone used education to reinforce behavioral change. "How did we address the risk of reverting to traditional behavior? We educated."

Although every owner valued education, they used varied methods of educating their staff, and their teams ranging from structured project kickoff training to independent coaching, to internal training based on training from others. The owners that gained the most from education and training made it part of the fabric of the project. They did not expect instant transformation. As Michael Bade noted, "Change takes time, but it is possible."

Many of the projects, such as Stuart Eckblad's UCSF Medical Center, used formal kickoff meetings to begin building teamwork and start the process of organizing into teams and developing goals and processes. "We did do some pretty extensive training about how to get firms to be integrated for the collective good. We had a program at that time that was about a week long where the principals of the firms and the lead detailers worked together trying to figure out the best ways to solve issues." They went on to use quarterly surveys and refreshers to reinforce learning.

Virtually everyone used consultants for training, at least initially. Wendy Cohen reported that the Palomar project used a facilitator to get the project going. "The facilitator was initially involved in the creation of the team, namely, the project board, development of mission and core values, and the rollout to the larger team. The facilitator was very active in the beginning, maybe a few days a week. Once the team was established, their role was reduced to one or two times a month." TD Bank engaged a project facilitator to provide IPD and Lean training. Sean Graystone required his team to participate in Lean and IPD training and strongly encouraged them to attend courses and conferences given by the Lean Construction Institute. They also engaged an attorney experienced in IPD projects to meet with the project team to explain how IPD operated. UCSF engaged a civil engineering firm that has gone through a Lean transition to assist UCSF in developing its processes. Universal Health engaged Lean coaches on their projects, but now expects its IPD partners to not only participate in training but contribute to the cost. Some owners, such as UHS under William Seed, are looking to the team to carry part of the training cost. "I was very frustrated that it was always us that had to do the training for the contractors and the architects that were inexperienced and didn't really understand the process. So we had to bring coaches in. I now rely and insist on partners participating in the exercise, including paying for coaching. And I encourage them to bring coaches into their own organization outside of my projects, so they have their own learning track."

Many of the owners are developing internal training capability. William Seed always uses coaches, but they may be from his staff. One of his project managers has been tasked to coach other project managers, and he is developing another internal Lean coach. Sutter Health uses their experienced IPD project managers to evaluate projects and train less experienced personnel. MaineGeneral used some external training, but also set up a team to focus on Lean processes. Chuck Hays described their process. "We had an internal Lean group of three people that we cleared of other responsibilities and had them focus on our design processes. We also had a lot of BIM [Building Information Modeling] training. It was actually exciting because we were able to increase the knowledge of the subcontractors and consultants and make them more competitive by training them on 3-D modeling."

Owners also used consultants to jump-start their training process, but then continued on their own, with only intermittent assistance. Christa Durand stated: "We did a host of educational series, and when we hit key milestones with design or whatnot, we had special meetings to get outside assistance. ... Howard [one of the authors of this book], you were our facilitator. I think we took expertise and advice you gave to us and parlayed that into several education and training seminars for staff, clinical folks, contractors, architects, etc. We didn't hire a facilitator, we leveraged your skill set to 'train the trainer' and then took the show on the road. We think it worked effectively."

George Montague raised an issue regarding on-boarding of new personnel that resonated with other owners. "One thing we have found that was frustrating—but probably unavoidable—is that through the course of the project people come and go and you have to bring the newcomers up to speed." He felt that there were several important roles for a consultant: "(1) What I call shepherding [running the project processes and meetings]; (2) training, teaching and installing tools; and (3) helping with the continuous process of on-boarding."

Crista Durand summarized, "From a governance perspective and from a leadership perspective, I think it all comes down to education."

These teams emphasized training and were successful. Where training has been overlooked or short-changed, the teams have struggled to overcome traditional behavior.

2.7 THE IPD CONTRACT

The interviewees were not asked questions about the IPD contract, but there were some interesting comments. All of the participants, with the exception of Stuart Eckblad and Michael Bade, did projects using a multiparty (owner/architect/contractor) or polyparty (owner and the entire risk/reward team) contracts. Because UCSF is a governmental institution, it could only use the best value process permitted by statute, which did not allow a multiparty or polyparty contract. However, they used a system of design-build or interlocking contracts to attempt to create a risk/reward approach similar to a single IPD contract. The following comments reflect a few of the interviewees' observations about the effect of the contract and differences in contract structure.

IPD is actually different from other collaborative approaches. For example, until Sutter Health solidified their approach to contractually bind the parties to a common goal with shared risk and reward, they did not get the behavior they sought. Robert Mitsch reflected on Sutter Health's experience. "We didn't create a relationship [until Sutter Health adopted full IPD] between the contractor and architect that had any substance; so, although we legislated that the contractor was going to assist in the design process and this was going to hopefully eliminate gaps in the drawings [where they were not constructible], we found that all we did was legislate something and the old practices continued in design and the contractor would then just price it. Sutter Health then evolved to a full single IPD contract under the integrated form of agreement (IFOA), which they found made a difference. "When we are all in this together and if the whole thing blows up we are all at risk, then you get different behavior and different results." Recent research led by Renée Cheng and the University of Minnesota (2016) similarly concluded that the IPD agreement strengthened the team and improved resilience.

Stuart Eckblad had the interesting experience of starting a project somewhat traditionally with only the designers under contract (because of procurement issues specific to the project), but then adding IPD elements when the construction manager and trades were procured, by using interlocking risk/reward provisions in the contracts. "And what was interesting is as a result of the traditional process for the first 18 months, we were way over budget and having to consider dropping scope. About three months after we were able to get everyone on board [with interlocking contracts], we actually had a huge reduction in costs and got all of our scope."

In discussing whether to use a polyparty (more than just owner, architect, and contractor) or multiparty contract (owner, designer, and contractor), Digby Christian observed, "I'm finding it harder and finding it a little more complicated going into projects where the owner only contracts with the architect and construction manager. I am not one of those people who believes that increasing the number of signatories increases the complexity. It just creates a little more paperwork. In terms of running the project, I find that it makes it easier, and I am a big fan of making sure that you contract directly with all of the major risk drivers on the project and I will do it again on the next big complex project I get. I just find it much easier. You get direct relationships. You get to talk directly to the principals of each risk/reward participant, and you get to ensure that communication goes directly to everyone who has a stake in the project."

The owners that discussed contracts did feel that contractual provisions, particularly shared risk/reward among the team based on project outcome, did affect behavior. And if the owner has the capacity and capability to manage a polyparty contract, it may provide the owner with more information and control.

2.8   THE RIGHT LEVEL OF CHALLENGE

Several of the owners noted that setting aggressive targets was a factor in project success. Mark Linenberger, who as a construction manager brought IPD to Cook Children's Medical Center, noted, "We have to set the bar high and when everyone feels they can grab the bar, the old ways become apparent very quickly." Challenge encourages behavioral change. William Seed echoed the need to set tough targets. "We take a slightly different approach than others do in establishing our project target cost estimates, so we set the target aggressively and once the contracts are signed, the team literally has to find new ways in order to earn their profit."

Teams are motivated not to miss targets. Digby Christian found that teams didn't want to lose. "The team at Castro Valley was extremely determined not to lose money and not as driven to make more money. Because they were always in danger of losing money through the entire length of the project, we got a lot of behavioral change." This fits with current thinking about loss aversion. The Castro Valley project was significantly over budget at inception, but was brought into budget through hard work by the team. As Digby explained, "We got the team to understand that cost is a design constraint. Design is a constraint, but the budget is also a constraint. They had to get 10 percent out of their number."

Dr. David Tam used the metaphor of a "burning platform" that would force people to change. "From my perspective, IPD probably works best if there is a burning platform—a sense of urgency that requires people to get rid of their preconceived notions, adopt new processes, and jump into trusting each other. The challenge for the owner is how to create this sense of urgency, this sense that you are standing on a burning platform. Turning around the PMC West Project, facing immense challenges, required abandoning traditional behaviors and jumping in (PMC West was commenced fairly traditionally, and IPD was brought into the project by the new construction manager in order to put the project back on track).

Crista Durand also set high goals. On a project that already had a challenging budget, she observed, "I challenged our cancer team to bring in another million dollars from the budget. They all kind of looked at me, but we are three quarters of the way to achieving that goal."

Although there was strong support for setting tough goals, none of the owners felt that the goals they set were impossible. Thus, they kept the projects in the zone where teams were challenged, not discouraged. This is the sweet spot for project performance.

2.9   FRUSTRATIONS

Although the projects discussed were all very successful, there were still difficulties and frustrations. Robert Mitsch noted the difficulty in finding experienced IPD partners. Even within firms that have IPD experience, the level of experience can vary significantly from team to team. Michael Bade was frustrated by the inconsistency even within more experienced organizations. "I think the most frustrating thing is that these ideas are new to the industry and even in companies that have demonstrated some facility and familiarity it's not uniform across all their people so there's a constant kind of need to focus, to train, to clarify, to support, to define, to envision how these processes should work, to explain them, and to relentlessly advocate for them." Digby Christian had a similar observation. "The actual personnel count more than the companies. Because this way of working is so new, unless you can move the same people from project to project, you will probably have to train the new team." As companies expand their internal training and develop more experience throughout their personnel, this should become less necessary.

Change is difficult, and virtually everyone cited instances where team members fell back on old behaviors. Mark Linenberger noted that developing trust is a real and continuous challenge. It is easy to become protective of your interests and forget that in IPD the team wins or loses together. As Wendy Cohen noted, not everyone can make the change. "The most frustrating part of IPD is when other people on your team do not understand the concept. They tend to be fairly disruptive, and it becomes very obvious to the rest of the team members who have embraced the process. I think spending a lot of up-front time selecting the right people and building the team is the key here. I don't think you will ever have 100 percent buy-in, but I think you can increase your chance of success with being very specific about the type of people that are on the team."

It was not uncommon for owners to have to remove or replace members of their own team that couldn't adapt to a collaborative process. Similarly, participating companies and sometimes specific individuals had to be removed from the project. These decisions were often team decisions, and one owner reported that the IPD team was strengthened by the process of realizing they had a team member that didn't fit and then jointly taking action to reconfigure the team.

Even when you have the right team, there will be continual personnel change that must be addressed through on-boarding. Wendy Cohen found that it is important that the entire team, from project executives to field crew, understand the IPD process and how their behavior needs to change. "IPD is most effective when it drives down to the people actually doing the work." In the PMC West project, the team held regular dinners with the project foremen to create a connection between the field and project management.

Crista Durand and Brenda Bullied of Lawrence & Memorial were concerned about the loss of information between projects and have used almost the identical personnel to design and build two consecutive projects. They feel that any potential loss in competitiveness was far outweighed by having an experienced, effective, and integrated team, and the initial results are quite good. Other owners are considering "batching" smaller IPD projects to spread the training cost over several projects and to keep effective teams together.

2.10   TARGET VALUE DESIGN

Three of the participants discussed target value design on their projects. Digby Christian noted that the process went extremely well and that they treated cost as a design constraint, just like any other constraint that had to be considered as the design developed. They then set up budgets for each team member and challenged each of them to reduce their cost by 10 percent. Stuart Eckblad briefly explained how they approached Target Value Design. "Once we got everyone on board, we then went through each of those systems and developed subtargets. We then identified cost drivers within each subtarget. Now the teams could really focus on those drivers to find, for example, how design changes affected the length of duct work—a major cost driver. We were able to get the subs to come up with a better sound insulating system that both shortened the ducts and was much quicker to install. In other cases, we found different and less expensive ways to solve a problem. We couldn't get rid of the fire alarm system, but we did find a way to get rid of all the paging systems and use a more effective, quieter system that was also better for our patients."

UCSF used sub-targets around major interrelated systems, such as mechanical, electrical, and plumbing. Mark Linenberger explained that Cook Children's is using a comparable approach built around Construction Specifications Institute (CSI) or Uniformat divisions. "We are taking more of the approach that Stuart described as far as breaking it down to divisions and challenging each division initially with a certain reduction."

Mark Linenberger also had an intriguing observation about a "hidden contingency." "One of the things from our first experience relative to reduction in costs is this hidden contingency. It is not the contingency we typically carry but is the product of overestimating costs because of the way we have performed work in the past. Our historical units are based on the wrong behaviors." Mark's observation is echoed in a comment by Digby Christian. "My feeling—though I never did precise metrics on it—was that we got about a third of the money [cost reduction] from actual innovation and two-thirds from people that started to reduce the buffer that they had in their numbers because when they saw how much certainty there was and how much coordination, they got more and more comfortable." The interviewees also cited many examples where team members in IPD exceeded their traditional production rates, had less rework than they traditionally expect, or didn't have to do duplicate work (e.g., design drawings followed by submittals and fabrication drawings where design drawings and submittals could be replaced by the engineer directly working with the fabricator and using the reviewed fabrication drawings in lieu of the design drawings and submittals).

Mark Linenberger's and Digby Christian's observations raise interesting questions on whether historical production data needs to be adjusted when estimating potential IPD costs, and if it does, what effect this should have on cost targets and project profit. If, for example, the cost of an IPD project is less than a comparable project delivered through Design-Bid-Build or Construction Manager at Risk, should the participants' profits in the IPD project be lowered to meet the new cost basis, or maintained at the same level (with possibility of increase or decrease based on project performance). The general agreement was that project participants should not be "penalized" for being more efficient and that profits should be based on the value delivered, not the cost incurred.

2.11   RELIABILITY

For Sutter Health's Digby Christian, reliability was a critical advantage. "What has the organization gotten out of IPD? Reliability and predictability."

Although the owners were focused on improving the value of their projects, they were also interested in being able to accurately predict outcomes. For many owners, being able to deliver the project without any significant problems, delays, or budget upsets was critical to business planning. This was especially true for the owners that built repetitively. As Robert Mitsch explained, in addition to controlling spiraling construction costs, Sutter Health was looking for a methodology that delivered as promised. "For our high-complexity projects, there is no more reliable way [than Lean IPD] to deliver." William Seed echoed Sutter Health's experience, noting that UHS' experience (40 out of 45 projects delivered for less than budget) speaks volumes about reliability.

2.12   VALUE

We asked the owners to discuss metrics, and they responded by providing specific examples of traditional value measures—cost and schedule—as described below.

The owners also discussed qualitative improvements. MaineGeneral, for example, is achieving LEED Gold, although it was programmed to be LEED Silver. Sutter Health Castro Valley had a very low percentage of rework (0.5 percent for trades that normally have 7 to 10 percent), which speaks eloquently about the quality of installation enabled by the IPD process. Moreover, others stated that the finished systems were more rationally laid out and maintainable. Quality of design and creativity were discussed. Michael Bade cited the creativity of the team in developing the town hall concept to achieve a more useful building. Sean Graystone discussed the quality of the information developed by the team for the renovation of the historic and architecturally significant House of the Temple in Washington, D.C. "I don't have metrics, but I can testify to the quality of the information that we derive. The quality of the information we are getting is amazing." And William Seed noted how the process affected the owner's decisions. "I think this methodology [IPD] offers us a far better ability to make value decisions—how to spend the money, where to spend the money. By understanding real-time costs and real-time impact of one system to another, we can make better decisions." Better decisions result in more project value.

But everyone also had examples of how IPD had a direct impact on cost and schedule. A few examples follow:

Chuck Hays summarized MaineGeneral's experience on its $320 million regional hospital. "We are about nine months ahead of schedule, $3 million under budget, and still have a lot of our contingencies left. There have only been two change orders on the project, which is unbelievable to me." The project was completed after the interview, and it was completed 10 months early, had five change orders (all requested by the owner), and stayed under budget. In addition, this hospital is achieving a higher LEED (Leadership in Energy and Environmental Design) rating than planned and has several million dollars in scope additions within the target cost.

TD Bank's experience, as reported by Roger Johnson, is similarly impressive. "We have seen a statistically significant reduction in our construction costs and an absolute reduction in construction time. Comparing the baseline information we have on similar projects we have done without IPD and the ones we have done with IPD, there has been a greater than 15 percent reduction in cost and 20 percent reduction in time." Since their first project, TD Bank has gone on to do several more IPD projects.

William Seed described the success of a program of construction. "Our first project we delivered was $10 million under budget on a $130 million project. We are clearly delivering projects 15 percent below all of our market competitors. Of the last 45 projects, 40 of them were delivered for less than 100 percent of the budget."

Discussing the benefits of collaboration, Mark Linenberger pointed out, "we had a very high level of collaboration between the steel erector and fabricator. We didn't apply any new structural design elements, but because it was highly collaborative and very connected, the iron workers spent 1,000 fewer man-hours—it was almost a tenth of their planned man-hours in the erection." Robert Mitsch reported, "The project in Castro Valley was completed about four months ahead of schedule and under budget through the IPD Lean delivery model. It was about a $320 million dollar project."

Even where there were significant construction problems, IPD did well. Brenda Bullied described an experience during Lawrence & Memorial's second IPD project. "During the course of this renovation, we discovered an unknown condition that had we been in a standard design-bid process would have added six months and close to $1 million in construction costs. I think it was $600,000 to $700,000, and on a $16 to $17 million project, it was a significant savings by having that team in place to lend their expertise to the issues at hand and wanting to work through them. Traditionally, we would have come to a complete stall on the project, with time and money being wasted."

IPD has also compared favorably against similar projects. Dana-Farber Cancer Institute partners with local hospitals to create cancer treatment centers. Christa Durand discussed Lawrence & Memorial's experience compared to other Dana-Farber projects. "The cancer center project is moving so well. We are under budget, we are ahead of the construction schedule, and there are very few change orders. We have saved over six months of traditional design time and hundreds of thousands of dollars. ... Dana-Farber has done five of these community hospital partnerships, and from start to finish they have typically taken two to three years. Our project from start to finish will be eighteen months or less."

Dr. David Tam led the conversion of the PMC West project from Construction Manager at Risk to IPD. "First of all, I firmly believe that it saved our bacon and that without the IPD process we would not have been able to achieve our schedule or our budget. Second, we accomplished a high-quality product. I don't know whether it is a by-product of IPD or a planned outcome of IPD, but it really brings together pride of ownership by the entire team. ... For example, the majority of the trades did not see a certificate of occupancy being the end of their process. It became relatively easy to say that success of the project under IPD is when there was a patient in a bed and the hospital is licensed and operating. This has huge value to a hospital system." Wendy Cohen echoed his observation. "IPD created an environment where pride of ownership could happen."

Wendy Cohen felt that the lack of legal disputes was also significant. She noted that projects the size of PMC West often have lingering legal issues, but that none of the issues that arose were ever allowed to grow to where attorneys or company executives had to get involved. Dr. David Tam credited the IPD environment. "The environment that the IPD process cultivated prevented us from going there [legal disputes]." Similarly, none of the IPD projects undertaken by any of the other owners resulted in any claims or litigation.

The owners are also focused on value to their team members, and several, such as William Seed, reported that many projects had enhanced profitability for the team. "All of my project managers are deeply engaged in the process of bringing out improvements to the team because we want those teams to succeed for their own purposes, not just ours. About 45 of our projects have offered some level of enhanced profit for the team members."

2.13   WOULD YOU DO IT AGAIN?

All of the owners would do it again, and most already are. Lawrence & Memorial, under Crista Durand, has undertaken three projects. "We have three IPD projects and have embraced IPD so much, I want to do everything IPD. I know this sounds ridiculous, but even with the small [$3 to $5 million] projects, I don't want to go back to design-build. Once you taste great wine, you don't want to go back to the cheap stuff."

Universal Health has committed to using IPD on every project above $5 million. At the time of the interviews, Sutter Health had four major IPD projects in construction and two in design. Cook Children's embarked on a second IPD project that is six times larger than their first. UCSF is also planning on using IPD for other projects. Stuart Eckblad affirmed: "Without a doubt we would do it again." In fact, due to the success of the UCSF Mission Bay projects, the University of California system has had legislation[1] passed to allow the UCSF Mission Bay best-value process to be used on any project larger than $1 million.

Roger Johnson, Crista Durand, and Sean Graystone all stated that the process has met and exceeded their expectations. Sean Graystone summarized his experience. "So has it met my expectations? Absolutely, the product has exceeded what we have put into it by far. ... IPD is a system of subtle changes that leads to really large-scale results."

2.14   ADVICE TO OTHER OWNERS

Several of the owners had specific advice for those trying IPD for the first time.

> I think number one you really have to understand the value proposition. I think IPD is really interesting and it's not for all projects but if you really don't understand the value proposition you really can't manage it.
>
> **—Stuart Eckblad**

Get your hands dirty. You shouldn't use our contracts without translating them into your own world, and by the way, how do you want your world to be? You have to answer that question before you can effectively translate.

—**Michael Bade**

Definitely consider using IPD, but make sure that you are fully prepared for actively engaging as a leader on the project and leading your team to support IPD principles.

—**Wendy Cohen**

For someone that is just learning about IPD get good advisors and people who have had some experience—it is really helpful. It has been a great experience for us, and I think we will certainly share our enthusiasm for it if anyone were to ask about it.

—**George Montague**

(1) There are not going to be a lot of vendors who understand it and change is difficult. They are going to be resistant and if left unattended will resort to old means and methods. (2) You need to focus on the owner's role and making sure you are participating in the right way to support IPD and don't just rely on the contract. (3) Invest your time in learning about IPD and getting experienced people who have done it and who can lead your projects into an efficient mode more quickly than if you tried do it on your own.

—**Robert Mitsch**

Avoid preconceived notions. Ask questions at every step.

—**Dr. David Tam**

Look for partners who are willing to innovate and bring the best ideas and continually look for ways to solve problems.

—**William Seed**

I wish I had known about IPD a lot sooner in my career. I mean, quite frankly, I wish I had learned about IPD and taken the risk of plunging in sooner than now.

—**Crista Durand**

2.15 HUMANITY AND MORALE

Although it wasn't a specific focus of the interviews, many of the participants commented on the softer or behavioral side of IPD.

Chuck Hays noted: "The other thing that surprised me and that I didn't expect was that almost uniformly the people working on the job have said this is the best job they have ever worked on. It's really overwhelming with the number of people that have come up to me and told me, my husband or my son is working on this job and it's the best job they have ever worked on."

George Montague saw a fundamental change. "There was clearly a huge enthusiasm about finding better ways to—I was going to say a better way to solve the same problem but, in fact, a better way to do things because some of the old problems just weren't there anymore. It was just a new way of looking at things."

Sean Graystone talked about the value of returning control and craftsmanship to the craftsmen. "The subcontractors have to feel satisfied in their work; so do the craft people and the engineering people; so do all the museum consultants and everyone else." Michael Bade saw a similar motivation. "If people have fun and do work they can be proud of and feel that the situation they are put in and the structure is conducive to fun and results they can be proud of—that's what they want to have." As Brenda Bullied concluded: "What motivates people is doing a good job."

Michael Bade summarized the effect of the human equation on project outcome. "If the project team is working really well together, you are going to have a good outcome. First and foremost, are people respectful, are they engaged with each other, do they listen well, do they seek out problems to solve together, do they automatically recognize all of the dimensions of a problem and marshal the people who are responsible for those dimensions from the outset to solve the problem? It is really the soft stuff that is important, and projects where the people interact with each other that way will surprise you with the quality and the creativity of the results. It is projects where people are not engaged with each other in that way where the results remain pedestrian."

2.16   SUMMARY

None of these experienced owners suggest that IPD is the right process for every owner or every project. "The more complex the project," Robert Mitsch explained, "the more appropriate this delivery model is. On a simple project, you may not need a full-blown IPD model, but we have found that for our high-risk/high-complexity projects, there is no more reliable way to deliver."

As these experiences with IPD show, IPD needs a different kind of owner, an owner who is engaged and committed, can draw the team into its vision, and can provide leadership throughout the project. If an owner undertakes its projects as these leaders have, the results will be exceptional.

NOTE

1. California Pub. Cont. Code section 10506.4, et seq.

REFERENCES

Cheng, R., Allison, M., Sturts-Dossick, C., Monson, C., Staub-French, S., Poirier E. (2016). *Motivation and Means: How and Why IPD and Lean Lead to Success*, University of Minnesota. Retrieved from http://arch.design.umn.edu/directory/chengr/.

Integrated Project Delivery Alliance/University of Minnesota. (2015). Retrieved March 8, 2016, from http://www.ipda.ca/research-performance/industry-research/industry-research/.

Miller, R., Strombom, D., Iammarino, M., & Black, B. (2009). *The commercial real estate revolution: Nine transforming keys to lowering costs, cutting waste and driving change in a broken industry*. Hoboken, NJ: Wiley.

CHAPTER 3

A Simple Framework

"Would you tell me, please, which way I ought to go from here?"

"That depends a good deal on where you want to get to," said the Cat.

"I don't much care where—" said Alice.

"Then it doesn't matter which way you go," said the Cat.

"—so long as I get SOMEWHERE," Alice added as an explanation.

"Oh, you're sure to do that," said the Cat, "if you only walk long enough."

—Alice's Adventures in Wonderland

3.1 A ROADMAP FOR INTEGRATING PROJECT DELIVERY

By now, you might be wondering why you bought this book. "Working together, coordination, integration, value," you're thinking, "I already do all of that!" At face value, it might seem that way. But truly integrating project delivery is much more comprehensive than that.

To begin talking about integrated project delivery (IPD), we need a roadmap that will give us a bird's-eye view of the system, and help us talk about the overall motivation for designing this delivery system. Figure 3.1 represents, essentially, the components to successfully produce a high-performance building.

Let's start by working backward, from the product, that we have agreed to deliver. A "high-performance building" is one that enables its end users to be more effective; it is the right building for their needs. In a nutshell, a building must be useful to or **usable** by its occupants, it must be **buildable** safely within the time and money budgets available, it must be **operable** so that the building managers can create the right environment for the occupants with a commensurate expense, and, finally, a building must be **sustainable** in its economic, environmental, and social context. A high-performance building is able to demonstrate that it meets the values and objectives stated by the owner at the beginning of the project, using specific metrics developed to evaluate its achievement.

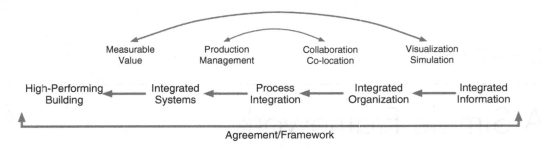

FIGURE 3.1   A simple framework for integrating project delivery.

A high-performance building is composed of highly integrated systems, where systems are designed to work together and complement each other. The nature of these integrated systems means that many people from disparate disciplines and trades must work together, collaborate, and exchange information. This integrated organization brings owners, architects, engineers, and builders together on a daily basis in the same physical space on site, where they are able to jointly solve complex problems. The owner, design, and build team needs an integrated process to produce an integrated building, and must take time to design ways to work together from the very beginning—these are the practices, processes, and methods they will use to work. Collaboration and co-location allow an integrated team to perform those integrated processes.

To work together effectively, the team must have a way of communicating reliably and efficiently. "Integrated information," which supports simulation and visualization, and the easy access to that information, are used heavily to create a transparent process, in which all members of the team understand the work at all times. Simulations and visualization enable team members to share their knowledge effectively, try out design ideas, separate fact from fiction, contrast good solutions with poor solutions, and communicate with other team members and stakeholders. No one person or discipline is the gatekeeper of information or interpretation. Integrated information also allows the team to simulate the performance of the building, allowing the design to be evaluated in terms of performance metrics before it has been built.

In other words, by building a comprehensive and appropriately detailed virtual prototype, the team members are not only able to understand the incredibly complex creature they are dealing with, but they are also able to evaluate whether the building will perform as they intend and make adjustments accordingly. To deliver the building that owners want and users need, project teams must be able to see the design they are creating and its performance vividly, early, and repeatedly. Integrated information, simulation, and visualization make that possible.

Fundamentally, IPD is motivated by the belief that no part of the design, construction, and operation process should be left to chance; it rejects the notion that a large-scale project is simply too big or too complex to control or that a small project is too simple to be improved. Most owners spend time at the beginning of a project defining their goals for the building, or where their "value" is. However, once the actual work begins, team members quickly lose their connection to the overall objectives, and those goals cease to be relevant. There must be a "reminder mechanism"—a way to ensure the team stays on track and doesn't lose sight of what they are working toward.

But because we do have control over much of the process, we can also track our efficacy. Meaningful metrics, not simply data collected for the sake of having numbers in a chart, must be used to track both how well a team is performing and how closely the building conforms to the goals and values of the owner. Metrics are essential to understanding and, if need be, correcting team performance during the process. For example, if an IPD team is aiming for a less than 24-hour turnaround on requests for information (RFIs), but a month after moving into a co-location space the average turnaround is three days, some sort of adjustment is clearly needed. By identifying and addressing the problem during the process, teams can learn, adapt, and become more efficient.

Upholding the entire IPD system is a contractual agreement and framework, which sets the "ground rules" for the project, and reinforces the idea that decisions can and must be made for the good of the project, not just for individual benefit. The contract will encourage and enable an integrated delivery system and allow organizations and individuals to share information, collaborate, innovate, and challenge each other without fear of retribution. Note, though, that an IPD contract alone is not sufficient for delivering a high-performing building. It also takes the practices described in this book.

Now, we can look at this "simple framework" for IPD in a little more detail (Fischer, Khanzode, Reed, & Ashcraft, 2012).

3.2   HIGH-PERFORMANCE BUILDINGS

One way to think of high-performance is whether the building meets the criteria of "buildable, operable, usable, and sustainable." Essentially, this means the building can be constructed in a safe, effective way; it is easy and efficient to maintain; it is well suited for whatever it is used for; and it does not harm people or the environment. Of course, the right aesthetics are part of this performance.

A truly high-performance building supports its end users in performing their activities as optimally as possible; it is the "right" building for what the users need. For example, a bridge should allow cars to cross it safely and quickly, even in inclement weather. A school should allow teachers to educate, inspire, and engage with students; a hospital should enable doctors and nurses to heal sick people; and so on. This may seem like obvious performance criteria; however, what sets a high-performance building apart is its level of success in terms of measurable value. A high-performing hospital not only allows doctors to heal some people; it is a place that promotes the maximum possible healing in every way. Nurses don't have to walk five miles a day, so they see more patients; operating rooms have enough space for necessary equipment; recovery rooms can be expanded or contracted to accommodate varying patient numbers; building materials used do not introduce contamination and other health risks; and so forth.

Values, goals, and end users are obviously unique to every project, so it is critical that the design and construction team understands the users' goals and vision for the building. Delivering a high-performance building begins with an intense effort to understand and define the purpose of the building, how to measure that purpose, and how to best achieve it. Crucially, stakeholders from every stage of the process must be involved in the design phase, since each stage shapes the building and its performance.

A high-performance building also efficiently uses energy, materials, and labor during both the delivery and operating phases, which lowers first and life cycle costs and other impacts. Most owners want to optimize life cycle operating costs (building maintenance, building operations, and business operations), building longevity, and first cost to construct, yet traditional practice focuses primarily on design and construction costs only. It may seem like a tall order to optimize the design of a building to fit all of these criteria, but every building will have some level of performance in these areas whether it is designed explicitly or not. In other words, even if we design a building solely for lowest construction cost and fastest delivery, the building obviously still has a life cycle cost, but we did not plan for it intentionally, and therefore have left it entirely to chance.

DPR San Francisco Net Zero Energy Building

DPR's San Francisco office is one example of a high-performance building, which meets the four criteria of buildable, operable, usable, and sustainable. The office was the first in the City of San Francisco to be certified as a net zero energy building (NZEB) by the International Living Future Institute's Living Building Challenge.[a] This certification is awarded only after proving energy consumption after one year of occupancy along with other criteria. The building has also earned LEED (Leadership in Energy and Environmental Design) v4 BD+C (Building Design and Construction) Platinum certification.

The owner, designers, and building team partnered early in the process and simulated various scenarios using computer models for airflow and energy use before selecting sustainability strategies for the project. By integrating the knowledge of designers and builders from the project's inception, and considering many alternatives before settling on the most optimal design, the DPR San Francisco office team was able to create a building that responds intelligently to the needs of its occupants, is net zero energy, and has very low embodied carbon content.[b]

The building has outperformed the goals of the project and is considered a net positive energy producer of renewable energy. The photovoltaic (PV) system will pay for itself in energy savings over the life of the lease, reducing the net cost of construction to $152/sq. ft.

The design and construction was achieved with a budget of $185/sq. ft., which is marginally above the market cost of standard construction, and consistent with standard Class-A tenant improvement (TI) budgets for projects in the San Francisco Bay Area.

Over nearly eight months, the team researched, designed, permitted, and constructed the energy efficient, high-performance 24,010-square-foot modern workplace with a number of sustainability features, including:

- A 118-kW roof-mounted PV system that utilized 343 *SunPower 345-watt* modules to generate renewable energy and provide power throughout the offices.
- Complete structural roof renovation to support the PV array and roofing replacement.
- Two large atrium skylights that were retrofitted using View Dynamic Glazing (electrochromic glass) to reduce both solar heat gain coefficient (SHGC) and control glare at the workstations.
- Roof-mounted solar thermal water heating system for showers and sinks.

- Nineteen *Solatube 750 DS* vertical skylights, which provide natural light to the workstations.
- Eight *Velux* solar-powered operable skylights over the atrium.
- Nine 8-foot Essence and four (4) Haiku *Big Ass Fans*® that efficiently and quietly provide air flow within the office.
- Three living walls designed and installed by *Habitat Horticulture* including a living wine bar (live plants growing beneath the glass bar top) that connect occupants to nature.
- Reclaimed redwood from the deconstructed Moffett Field Hangar One in Mountain View, California, and reclaimed Douglas fir from pilings salvaged from the San Francisco Transbay Transit Center Project.
- AER-DEC sink basins and low-energy, ultra-low-flow/flush plumbing fixtures; hand dryers by *Sloan Valve Company,* which promotes water savings and a touch-free environment.
- Building Management System (BMS) with a *Honeywell* Command Wall that resides on a 60-inch multi-touch screen for real-time performance adjustments.
- Building performance dashboard provided by *Lucid* showing building performance and is a great visual tracking tool (http://buildingdashboard.com/clients/dpr/sanfrancisco/).
- Shared learning lab, fitness center, and common restrooms, which are shared with a sub-tenant space of 4,000 square feet.

Figures 3.2, 3.3, and 3.4 show the systems, their interaction with people, and how the space appears to someone walking from front to back through the central corridor. Figure 3.2 shows what was done to turn a previously occupied 60+-year-old industrial/office space in a dense urban area in downtown San Francisco into a daylit, open, and functional NZEB.

The section view in Figure 3.3 shows the space relative to the large existing skylight, *Solatube* vertical skylights to bring daylight in, and the large Essence ceiling fans.

Figure 3.4 shows the wine bar (a common gathering place in every DPR office), conference rooms up- and downstairs, the skylight above, grand communicating staircase, Living Wall, and mezzanine railings.

Ted van der Linden, LEED AP BD+C, USGBC & ILFI Board Member, Sustainability Advisor, Strategic Consultant to the AEC Community at Theodore van der Linden, LL[1]
Mike Humphrey, Management Committee, DPR Construction[2]
Mike Messick, Senior Project Manager, DPR Construction[3]

[a]http://living-future.org/case-study/dprsanfrancisco
[b]http://www.ecobuildnetwork.org/images/pdf_files/The_Total_Carbon_Study_FINAL_White_Paper_published_20151113.pdf

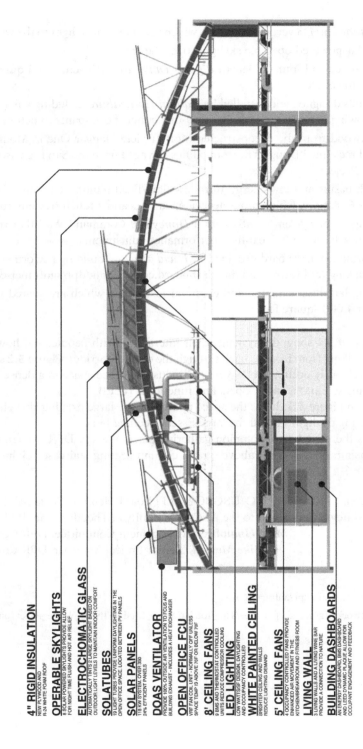

4" RIGID INSULATION
NEW PLYWOOD AND
R-24 WHITE FOAM ROOF

OPERABLE SKYLIGHTS
8 SOLAR POWERED SKYLIGHTS PROVIDE ALLOW
FOR NIGHT TIME FLUSH AND HOT AIR RELIEF

ELECTROCHOMATIC GLASS
AUTOMATICALLY TINTS 2 LARGE SKYLIGHT BASED ON
OUTDOOR LIGHT LEVELS TO MAINTAIN INDOOR COMFORT

SOLATUBES
19 LIGHT TUBES PROVIDE UNIFORM DAYLIGHTING IN THE
OPEN OFFICE SPACE. LOCATED BETWEEN PV PANELS

SOLAR PANELS
118 KW SYSTEM. OVER 950
24% EFFICIENT PANELS

DOAS VENTILATOR
PROVIDE 100% OUTSIDE AIR VENTILATION TO FCUS AND
BUILDING EXHAUST - INCLUDES A HEAT EXCHANGER

OPEN OFFICE FCU
VRF FAN COIL UNIT - NORMALLY OFF UNLESS
SPACE TEMP IS ABOVE 76F OR BELOW 70F

8' CEILING FANS
9 BMS AND THERMOSTAT CONTROLLED
UNITS REDUCE COMPRESSOR COOLING

LED LIGHTING
LOW WATTS/SF DESIGN. DAYLIGHTING
AND OCCUPANCY CONTROLLED

WHITE PAINTED CEILING
BRIGHTER CEILING AND WALLS
REDUCE LIGHTING DEMAND

5' CEILING FANS
4 OCCUPANCY CONTROLLED FANS PROVIDE
ENHANCED AIR MOVEMENT IN THE
KITCHEN/BREAKROOM AND FITNESS ROOM

LIVING WALL
3 LIVING WALLS AND A LIVING WINE BAR
PROVIDE A CONNECTION TO NATURE

BUILDING DASHBOARDS
AN ENERGY DASHBOARD, BMS DASHBOARD
AND LEED DYNAMIC PLAQUE ALLOW FOR
OCCUPANT ENGAGEMENT AND FEEDBACK

FIGURE 3.2 NZEB strategy and equipment for DPR's San Francisco office. Courtesy of Integral Group.

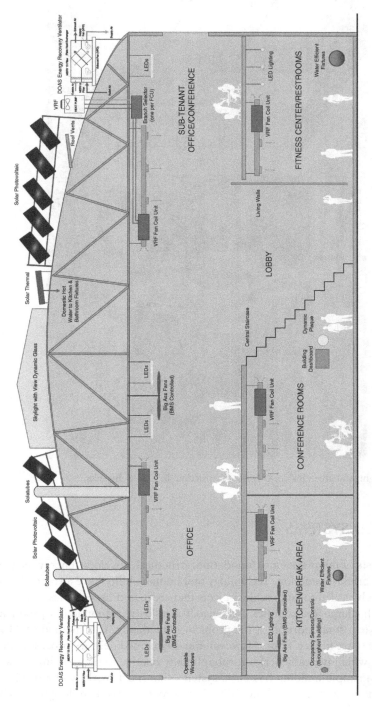

FIGURE 3.3 Section view showing people, space, and light. Courtesy of Integral Group.

FIGURE 3.4   Open, light, and green inside. © DPR Construction, by Lyzz Schwegler.

3.3   INTEGRATED SYSTEMS

Every building consists of many systems (foundation, structural, envelope, energy management, etc.), each with their own primary function. Technology and human ability have advanced exponentially in the past century, which has led to large-scale projects that are incredibly complex. It seems logical that, when faced with such a large and complex challenge, the industry responded by producing specialists who could each tackle a small portion of the problem individually. This method of "divide and conquer" seems to make sense at face value. However, the fragmented delivery process actually results in fragmented systems, which typically perform suboptimally relative to the building's goals.

However, even a simple building has basic systems that must work symbiotically; a waterproof façade and roof alone requires seamless coordination of many subsystem designers, fabricators, and

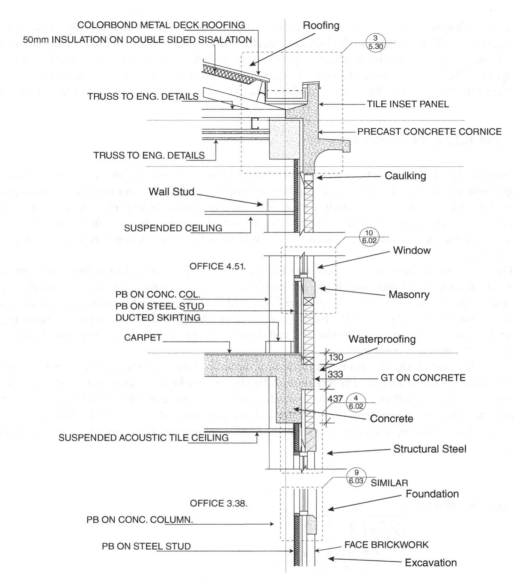

FIGURE 3.5 Buildings leak at the intersection of contracts. Illustration provided to the authors by DPR Construction, Lyzz Schwegler; adapted from Todd Zabelle, Strategic Project Solutions.

builders. Yet, as Figure 3.5 shows, we often break up installation of the façade into separate contracts, which makes achieving a high-performing façade challenging because no one subcontractor is fully responsible for the performance of the whole façade.

The main systems of a high-performance building work in concert, not in conflict with each other. For this reason, no system or element of a high-performing building can be designed in isolation.

Electrical engineers cannot specify the appropriate amount of power without knowing what people and equipment will be working in the building; mechanical engineers must account for the size of the interior spaces specified by the architects and what will be done in them, along with the composition and layout of the exterior skin, and so on. Thus, disciplines must work collaboratively with each other to create a building that truly functions as a whole, not just as an amalgamation of many disparate parts.

3.4   PROCESS INTEGRATION

To create an integrated product, project team members must work together in an integrated way. For example, when designing the façade of a building, a team must take into account energy consumption, natural light, the structure of the building (does it require large cross braces, etc.), and the aesthetics, to name a few. Architects, mechanical engineers, interior designers, and workflow specialists must all give input for the final design; otherwise, the systems will not work together and thus will not perform highly.

There are four main stages of a building's life cycle: define user value, design, build, and operate. The nexus of all of these processes is the design phase (see Figure 3.6). First, owners and designers must work together through many design iterations until both sides are able to clearly articulate and understand the values and goals of the building. Then, as the design is detailed, engineers, subcontractors, and so on must be involved so that the end design is one that still meets established goals and can actually be built and operated.

As the design evolves, its expected performance should be analyzed periodically to ensure that the value the owner seeks and the design stay aligned. Furthermore, a high-performance building generally requires a high level of off-site fabrication, which must be planned carefully. Off-site fabrication has immense benefits to any project—it allows physical components and systems to be produced simultaneously, which is faster, in a more controlled and safer environment, and to be assembled quickly on site.

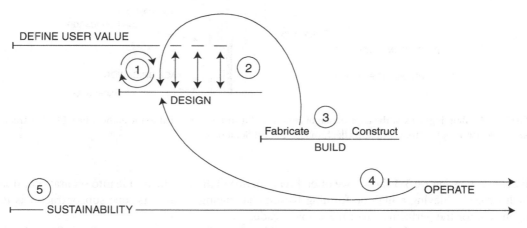

FIGURE 3.6   Integrating knowledge into the project delivery processes when it matters most. Illustration provided to the authors by DPR Construction, Lyzz Schwegler.

Prefabrication also allows for better control of tolerances, which is not only critical for assembly but also for energy performance. However, prefabrication requires the design to be completed as definitely and early as possible; without the knowledge and expertise of the necessary specialists during the design phase, the design will not be reliable enough to put off-site fabrication in motion.

Put more generally, not including construction knowledge in design will likely lengthen the project duration and make it more expensive because effort and time are required for redesign or for inefficient construction. Additionally, the operators' perspectives should be included in the design to ensure that the building can be operated and maintained easily. This process not only applies to new buildings; these phases also apply to the renovation or even decommissioning of a building. Figure 3.6 illustrates the practice of integrating the knowledge of the main disciplines into the project delivery process. Project design strategies and decisions (2) must include upstream user and stakeholder values (1) and downstream construction (3), operations (4), and sustainability (5) knowledge through engagement with people having this knowledge.

Virtual modeling and simulation can and should be used during the design phase. Three-dimensional (3-D) models and related simulations help owners understand how their values might be realized and enable them to make informed decisions over compromises or trade-offs where needed. Detailed 3-D models can also be used for off-site fabrication.

3.5 INTEGRATED ORGANIZATION

Integrated organization means that the right people from every discipline are working together with a clear understanding of their common goals. The "right people" are those with the necessary knowledge and experience and are open to working together as an integrated team. Rather than think of themselves as members of their respective firms, they must see themselves as members of a new organization—the project. Everyone must come to understand that "for the good of the project" means that the project comes first, before the interests of their organization. This is a radical shift for people accustomed to fragmented and adversarial projects, and they must receive strong support from the firms for which they work.

Although integrating the individual managers of each firm in order to manage the project is difficult, creating integrated teams is harder. High-performing teams are cross-functional, multidisciplinary, and integrated on a daily, working level. They also have considerable autonomy; although they receive direction from project leadership, the team members themselves determine how to accomplish goals. Decision making must be enabled at the lowest responsible level by individuals who are closest to the sources of information and best understand the relevant issues. An IPD team is not inhibited by cumbersome "chain of command" processes or by constant second-guessing, which stymies progress and discourages the team. Yet everyone on the team is responsible for their own work and for the work of the team as a whole. IPD works best when team members are empowered and trusted.

But, *empowered* certainly does not mean disorganized. The team should develop challenging goals and protocols that detail exactly what they aim to achieve, by what methods, and what metrics they will use to track progress. By identifying and measuring these "controllable factors," the team is able to correct deficiencies and adjust their system during production, which leads to increased efficiency and effectiveness. Each person on the team must be able to describe their role and, at a deeper level, understand how each small piece of their work contributes to the whole, and ultimately affects achieving the shared project goals.

Leaders cannot expect team members to freely contribute their knowledge if they do not know the state of the project as it evolves. Leaders of IPD teams must find ways to allow team members to "see in": to make management, design, and construction transparent. Every member of the team must understand who the major decision-makers are, what their criteria are, and what information or analysis may be needed or useful for important decisions. The project leadership is also responsible for clearly communicating the owner's values and ensuring that all team members understand what value is and how they will deliver it.

The relative darkness enveloping most project teams is a purposeful and sensible by-product of people not wanting to expose their mistakes in an industry with a long history of finding fault and placing blame.[4] As long as people fear the consequences of openly sharing their thoughts or acknowledging errors as they grapple with amazingly complex problems, they will hold their cards close. That is why leadership must actively create and maintain a culture of transparent sharing.

3.6   INTEGRATED INFORMATION

The importance of integrated information to IPD cannot be overstated; it is the backbone and the source of truth and insight, which allows an integrated team to make the best decisions for the project. There are several main aspects of what we call "integrated information," which includes consolidating fragmented information, extensive use of 3-D models, a robust information technology (IT) infrastructure that allows real-time access to the latest information, and an emphasis on making decisions using all available information.

Sharing information is a linchpin of the IPD organization. Information must remain consistent across all disciplines, and everyone must have access to all current information at any time. A significant, but often overlooked, source of project delay is the time and effort spent locating, re-creating, or transferring fragmented information. One study found that architects and engineers spend 54 percent of their time managing information when they work on fragmented teams (Flager & Haymaker, 2007). Integrated teams are not impeded by a lack of information availability, either due to poor infrastructure or individual disciplines "hoarding" their knowledge.

Using building information modeling (BIM), teams can make decisions after analyzing many options, not just on the basis of a small handful. BIM allows the team to rapidly and consistently explore many design options rapidly and consistently, discuss how different designs will add value (or not), and how they will affect performance targets. BIM allows the team to look at the aesthetics of a design along with the analyses for the salient project objectives. Simulation allows teams to understand the impact of a scenario later down the line, and begin either modifying plans or preparing interventions to mitigate negative impacts and risks. BIM can also help establish an appropriate off-site fabrication strategy and understand the operability and sustainability of an intentional design.

3.7   CONNECTING THE DOTS

Now we will look at the top half of Figure 3.1, which highlights the fundamental methods of IPD: measurable value and simulation/visualization, and production management and collaboration/

co-location. These are the major practical aspects of integrating project delivery—the way a team actually realizes integrated delivery, especially if practitioners actively reflect and improve on these practices as they move through a project.

One of the most important early steps in the IPD process is understanding and defining value. The design team can use simulation and visualization (one of the major uses of integrated information) as a platform to discuss and interpret the values of the owner and the end users of the building, and then use those same 3-D and 4-D models to identify specific performance metrics for the building to meet. By establishing meaningful metrics that align with specific project goals, the team is able to stay on track toward achieving a building that meets and exceeds the owners' needs.

To achieve both an integrated organization and integrated processes, a team must work in the same physical space at least for some significant time or for important decisions, using specific collaborative work practices, which are referred to as production management techniques. One important collaboration method is integrated concurrent engineering (ICE), pioneered by NASA's Jet Propulsion Laboratory (Smith, 1998), where team members from multiple disciplines come together for an intense combined work session, using integrated information to solve complex problems together. A co-located team is able to collaborate fluidly and efficiently, with access to the latest information, and all understanding the implication of outcome metrics. By working together, designing and building virtually, and simulating to test solutions, teams create a "network of knowledge," in which every team member understands where to go for information, rather than leaving it in independent information "silos."

3.7.1 Measurable Value

Currently, most, if not all, projects engage in some sort of value definition or goal setting at a very early stage. Usually, this is a discussion between the owner and the architect, and people working at a later stage of the process rarely hear of it. The goals outlined during this stage are often vague and poorly defined, and there is no mechanism in place to remind the team of goals once work has begun. Effective and efficient building projects require the definition of measurable performance objectives for all goals.

If a team tracks any metrics at all, it is likely the first cost to construct and lost time incidents. Owners are understandably extremely concerned about first cost; however it is not the only value that should be tracked using an established metric. Until this age of powerful and cheap computing, we have never had the ability to simulate and measure alternative designs. Now, we can track not only first cost, but also energy consumption, workflow, natural light, and previously intangible values such as "openness," "connectedness," and so forth. Once a team decides how to measure certain aspects of value that are important to the owner, those values can become design criteria as well, in addition to first cost.

By clearly defining, emphasizing, and tracking the project values, each member of the team is able to make decisions "for the good of the project," because they understand what that means in the context of that specific project. For example, if natural light is a high priority but energy consumption is less important, the team could choose a design with a very high number of windows, even if it resulted in increased energy consumption. Or, if natural light and energy consumption were both of value to an owner, the team could simulate multiple designs highlighting the trade-off between the number, quality, and cost of windows and energy consumption, and the owner could make a highly informed decision between the two (or multiple) designs.

3.7.2   Simulation and Visualization

Visualization and simulation are the main mechanisms to connect integrated information with the design team. By using detailed and accurate 3-D models, the team is able to communicate more clearly and effectively with each other and with the owner. Many owners have little or no experience building anything, and the building they build may be the only one they ever do. They do not have the background of an architect, and are not able to understand complex 2-D shop drawings; 3-D models are vastly easier to comprehend, and ensure that the owner and the design team stay on the same page (Figure 3.7). Simulation allows the team to accurately predict and vividly demonstrate how closely the design comes to desired outcomes and see the consequences of various decisions throughout the process. These models offer rapid feedback, the like of which designers, builders, and clients have not had until very recently.

Simulations also allow the team to carry multiple design options forward for comparison. For example, long- and short-span steel, and precast and timber structural systems could be kept in play along with appropriate mechanical, plumbing, and electrical systems. Distributed and central mechanical approaches, use of space, energy, and natural light can all be calculated and analyzed.

3.7.3   Collaboration and Co-location

To produce an integrated product, a team must learn to work collaboratively and be provided with an empowering environment to do so. Because so many integrated work practices rely on collaboration and consistently getting rapid feedback from teammates, it is extremely important that the team is at least partially or occasionally co-located. On a larger project, team members relocate to a single, large, open office, often referred to as the "Big Room." On smaller projects, they spend two or three days every week or every few weeks working shoulder to shoulder in temporary Big Room or ICE sessions.

Co-location allows all team members to have access to the latest information at any time, which ensures that everyone is working on the "same" building; nobody wastes time working with outdated designs. Without the right environment, an integrated team is not able to most effectively share information or work collaboratively, and thus is not able to create the integrated systems necessary for a high-performance building.

By working in close physical proximity with each other, people from many different disciplines are able to have many quality interactions. Team members will interact frequently when they work in the same space, not only in cross-functional team meetings, but around the water cooler and for activities after work. They come to understand who is responsible for what. They learn who to go to for answers and help, and they begin exchanging information with the "right" people. The frequency and quality of interactions increases dramatically, and problems can be solved faster and more effectively.

Team members learn to view their interactions from a "customer-supplier" point of view. When they are seeking information or work, they are the customer. When they are being asked to produce information or work, they are the supplier. Viewing the relationship this way, team members can understand that when acting as a customer, they are expected to clearly state their needs and expectations. Conversely, when in the supplier role, team members understand they need to know exactly what their customer needs and when, that is, understand their "conditions of satisfaction."

Techniques such as rapid feedback, producing small batches, and sharing work-in-progress early and often allow the team to produce higher-quality work faster. For example, an architect working on

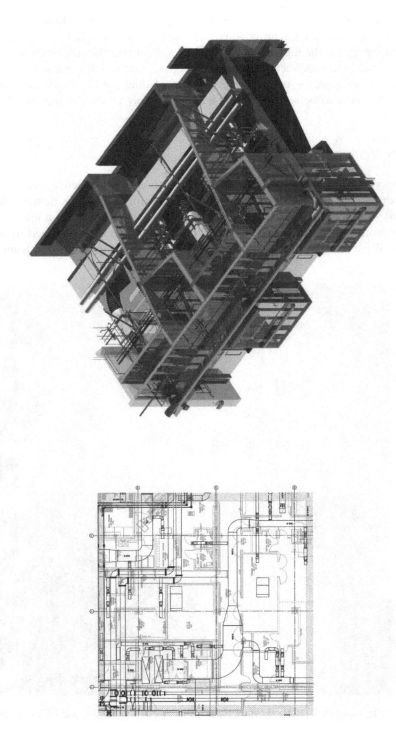

FIGURE 3.7 2-D plan and BIM of framing, ductwork, electrical, plumbing, and fire protection for the UCSF Mission Bay Hospitals project. Licensed by The Regents of the University of California on behalf of its UCSF Medical Center; courtesy of DPR Construction.

window design might take a day to build a partial 3-D model, and then show it to her colleagues at the next table over, who are working on lighting design. The lighting team might immediately see that the windows will be too small for the high-efficiency lighting system that they are designing, and the exterior architect can immediately redirect her window design so that it will work with the lighting system. In this way, the team avoids spending large amounts of time (and money!) on vast bodies of work which will either have to be scrapped completely, or simply cannot be integrated to work seamlessly with each other.

3.7.4 Production Management

Co-locating people does not work miracles by itself. People need to learn and practice specific methods and techniques to produce work as efficiently as possible. First, a team must clearly define milestones, so that individuals and working groups can align their efforts effectively. Next, team members must see and understand each other's workflows and schedules by talking about what each performer needs from others as they pull back from milestones (Figure 3.8). Pull planning allows the team to deal

FIGURE 3.8 Process mapping together. Courtesy of Sutter Health, Ghafari Associates, LLC and DPR Construction, by Peter Lockett.

explicitly with how much work can or should be done at any one time. It also provides an opportunity for team members to understand their relationships as information customers and suppliers, producing the chain of work and network of commitments necessary to achieve milestones, and ultimately fully integrated systems.

Team members quickly discover that producing large batches of work takes more time, introduces lots of difficulty to coordinate work in progress, and requires downstream workers to wait. However, producing in small batches means releasing completed pieces instead of a completed whole, which traditionally has been discouraged out of fear the work will be interpreted as lacking or flawed. Partial, small-batch handoffs require frequent and clear communication, which can be achieved much more effectively when people work closely together in the same space.

Fortunately, process modeling techniques, like value stream mapping, and software tools can be used to analyze workflow, to ensure that all team members are doing the right thing at the right time. Once the batch size and sequence is more or less right, individual work processes can be remapped into "value streams" for delivery of chunks of work. These chunks should match the schedule needs and organizational capacity for production review and approval.

3.7.5 Contractual Framework

The IPD agreement is designed to remove impediments to collaboration, align the interests of the parties, and encourage behaviors that add value to the project. Rather than prescribe specific actions, a well-crafted IPD agreement uses a relational structure with jointly shared risk and reward to create a system that inherently enables and reinforces collaboration. The key parties within this risk/reward structure are bound together through a multiparty agreement that must at least include the owner, designer, and builder. The agreement should also include key consultants and trades, either as signatories to the prime IPD contract or through IPD subagreements.

IPD contracts place authority within the team. Management is a joint activity so that the shared risk is balanced by shared control. This reduces the fear that chills creativity and places decision making in the hands of the individuals with the best knowledge. By requiring major decisions to be made jointly, decisions stay aligned to the project goals and are supported and understood by everyone. IPD contracts also limit liability among team members, which allows them to feel secure sharing information and work that is still in progress. Without liability limitation, fear of litigation quenches vital information exchanges and drives team members back into their "information silos."

In contrast, traditional construction contracts are made between two firms at a time and focus on transferring risk, which often means the incentives of the firms and the project as a whole are not aligned. That in turn leads to firms protecting their individual profits, guarding information, and hampering communication flows.

A traditional approach largely ignores creating structures and relationships that will promote overall project success. Using guarantees, penalties, and risk transfers, these mechanisms attack the symptoms (poor quality, high cost, and excessive duration) without addressing the fundamental causes of poor performance. They excel in assessing liability but do very little to avoid the risks. In fact, the focus on liability assessment and risk transfer reinforces individualistic behavior and exacerbates the dysfunctions that created many risks in the first place.

By itself, the IPD contract accomplishes little. Just as a skeleton creates the potential for motion by providing a structure, an IPD agreement creates the potential for success by providing structures that allow the other elements of IPD to function effectively. And, like a skeleton, if the contract is incomplete or malformed, it can limit project performance. But correctly designed and layered with integrated information, integrated teams, and integrated processes, it becomes a strong and flexible tool for integrated projects.

3.8   APPLYING THE SIMPLE FRAMEWORK

The Simple Framework identifies the elements necessary to integrate project delivery and achieve high-performing buildings. It is not intended to define a set of sequential tasks. Elements may proceed concurrently, and there are different paths available to project teams. We should expect different teams to use the Simple Framework in different ways. And it's likely that they will adjust its application during the project. That's because project teams deal with dynamic complexity, meaning that many variables are interacting and continuously affecting each other. The choice of paths will likely depend on the experience and overall skill of the particular team with integrated project delivery and their level of preparation before undertaking the project. At a particular point in a project, one team might find that the biggest leverage will come from integrating information, another from defining better objectives and metrics, and a third from more co-location.

3.8.1   Nonlinear Flow

We developed the Simple Framework from left to right. We started by asking what an IPD team needs to accomplish and quickly settled on "high-performing building." Then we asked the question: "If that's what we want, what's the best strategy to achieve a high-performing building?" We felt that a building with integrated systems would have the best chance of being high-performing. We continued to ask the question: "How do we best get what we just came up with?" Once we had determined the five main elements of the Simple Framework, we looked for connections between the elements and realized that virtually all of the mechanisms and methods used by leading practitioners connected to the five elements in the Simple Framework.

Instead of looking at how to achieve a high-performing building as a flow diagram, we can also put the high-performing building (HPB) in the center—it is, after all, the main target—as shown in Figure 3.9.

A high-performing building is shaped by its product design (integrated systems [IS]) as well as the organization and process design (integrated organization [IO] and integrated process [IP]) and supported by integrated information (II). Another way to put this is that an HPB is made up of integrated systems, integrated organizations, and integrated processes. It also consists of the physical systems and the digital assets (II) that document the HPB. Considering the main mechanisms that connect these five elements, we noticed that BIM is the primary tool to integrate the information to describe the integrated systems, and simulation and visualization are the primary tools to bring the integrated information to life in support of the integrated organization. Collaboration and co-location are the primary methods with which the integrated organization integrates the processes and carries them out efficiently. Production management methods assure that the integrated processes focus on the most value-adding aspects of creating the integrated systems at all times. Finally, metrics connect the HPB with the other four main elements of the Simple Framework. They are the yardsticks that allow an

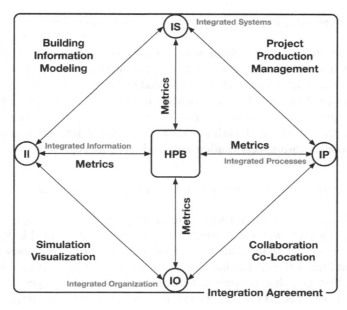

FIGURE 3.9 The Simple Framework diamond.

IPD team to assess the fitness for purpose of the II, IO, IP, and IS. All of this is bounded by an IPD agreement that ties them together and aligns them to the project goals. Represented in this way, the Simple Framework becomes the system architecture of the HPB.

3.8.2 Simple Framework Workflow through Measurable Value

There are three plausible workflows. The first, drawn in Figure 3.10, goes as follows: A team defines a high-performing building through its metrics, then lays out a set (or sets) of integrated systems that should meet the performance, defines the process and organization and the information needed, and produces the high-performing building through visualizing/simulating it in a co-located environment doing things at the last responsible moment and checking whether a particular design meets the performance targets (metrics).

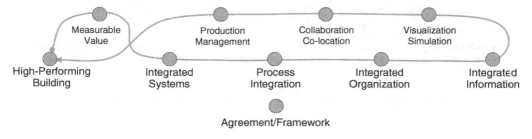

FIGURE 3.10 The Simple Framework workflow through measurable value.

3.8.3   Simple Framework Workflow through Integrated Systems

In the second version, in Figure 3.11, the flow of thinking and work goes from the high-performing building to integrated systems and then to metrics or measurable value. In this case, the team does quick performance tests of sets of integrated systems that make up the building to see whether that set of systems has a chance of meeting the performance requirements, that is, create the value the owner seeks. It then uses simulation and visualization to understand the performance in more detail and to drive integrated information. Once it feels it's on the right track, it defines the process and organization and produces the building through co-location and production management.

3.8.4   Applying the Simple Framework to Product Development and Delivery

Thinking further ahead to when project teams are developing and delivering buildings as products, which we discuss in Chapter 18, the flow might look like the one illustrated in Figure 3.12. The following sequence assumes that an integrated product development team is already in place:

(1) You need an idea of the kind of building you want, (2) the kinds of building systems needed, and (3) the criteria that are going to help you figure out whether a design gives you the performance you want. Then, (4) you need to get into information management right away; if not, you are just

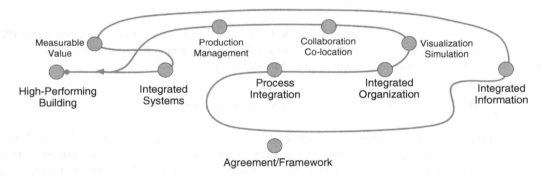

FIGURE 3.11   The Simple Framework workflow through integrated systems.

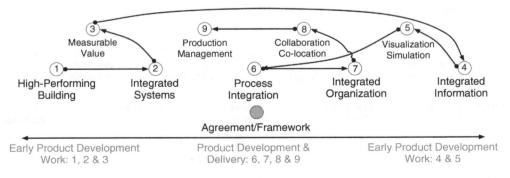

FIGURE 3.12   The Simple Framework for developing and delivering a building as a product.

talking to yourself, with (5) simulation and visualization a big part of information management, that is, integrated information, and, of course, the information needs to help you understand the performance of a design (link between (3) and (4)). From seeing the building in (5) you now can figure out the process (schedule) that'll get you there (6), and you can add the people and companies needed when you need them (7) and really start producing with co-location (8) and production management (9). So (1) to (5) help you make sure you are developing the right building, and (6) to (9) help you do so efficiently.

A further point is that you won't be competitive if you have to start from scratch with the building performance and integrated systems "story" for each project and if you have to invent the information system that supports it a building at a time, so companies will need to invest in these areas outside of projects so that they can hit the ground running when a client asks for a building as a product.

3.9 REFLECTIONS

You might think that a collaborative, data-driven, and performance-focused approach is too difficult to implement. While we have not seen a project that has put all of the Simple Framework into practice, every task that we have recommended has been used in practice. The necessary technology and management methods are available. It is important to start with one or two of the recommended tasks and build your personal and your firm's capabilities from there. We also invite you to consider omitting one of the recommended elements of the Simple Framework and answer honestly whether you still feel strongly that you can design, build, and operate the best facility possible. You might think that your firm and project teams are already working in the way advocated in this book. If so, we invite you to consider the full meaning of the recommended elements and to share your insights with the construction community, in particular in universities, to educate everyone about the possibilities, opportunities, and rewards of working in this way.

You might think that your facilities and the design, construction, and operations processes are just fine as they are today. We invite you to consider that almost half of the energy that needs to be produced in the world—with the detrimental impacts on climate and health—is needed for buildings. We cannot wait until all energy is produced cleanly. The construction industry must quickly find ways to deliver far more sustainable facilities than it does today. We hope that the path outlined in this book helps us do so. As Ross Donaldson, former managing director of the global architecture firm Woods Bagot, said when presenting at Stanford University: "It is our professional and moral obligation to design zero-emissions buildings" (Donaldson, 2010). This is a dramatic vision for the construction industry and requires our utmost attention.

3.10 SUMMARY

The Simple Framework starts with the end in mind—the high-performing building. It then deconstructs project delivery to determine the essential elements to achieve this outcome:

- Integrated systems
- Integrated processes

- Integrated organization
- Integrated information

These elements are supported by the integrated agreement that removes barriers to collaboration and enables the project team to function as a virtual organization. This team manages with metrics, uses visualization and simulation, collaborates and co-locates, and implements the project using production management techniques.

The Simple Framework is robust and continuously applied. It applies from conceptualization through operation. It also applies as an operational plan within a fully integrated organization or as a plan for a virtual organization, such as an IPD project team. And as projects begin to take on characteristics of products, it can provide the predictability needed to execute the high-performing building as a product.

NOTES

1. Ted van der Linden is a green building thought leader. He spent 18 years with DPR Construction. His initial role was as senior preconstruction manager, where he worked with clients on the evaluation of cost, schedule, and strategy for numerous commercial construction projects. In 1998, he led the preconstruction effort on a first of its kind "green" project designed by William McDonough + Partners for Aspect Communications in San Jose, California. Ted was elected to the contractor/builder seat on the U.S. Green Building Council's National Board in 1999 and completed his third term as a national director in 2010. DPR has over 450 LEED APs and typically generates more than 75 percent of DPR's annual volume of new projects focusing on LEED Certifications. Today, he continues to work closely in the industry, driving the market toward darker green projects with a renewed emphasis on process improvement, innovation in green technology, and education. To that end, Ted spearheaded DPR's green commitment on their most recent office in San Francisco, which is targeting both LEED v4 Platinum certification, as well as recently becoming the first NZE Certified commercial office in San Francisco.

2. Mike Humphrey has been with DPR Construction for 24 years and currently serves on the Management Committee. He began his career in 1992 directly after graduating from California Polytechnic State University, San Luis Obispo, with a bachelor of science degree in construction management. He has played a number of roles in operations, preconstruction, and regional leadership. Mike has been on the forefront of sustainable design and construction. He was the project manager for the award-winning Aspect Communications headquarters, which was designed by William McDonough + Partners and Form-4 Architects in 1999 and completed in 2001. This project was well ahead of the LEED rating system and forced a number of important code changes to adopt more energy efficient mechanical ventilation practices. Mike continued to champion sustainable practices in a project executive role on notable projects like the Packard Foundation Headquarters in Los Altos and the Clif Bar headquarters in Emeryville. Later, in his role as regional manager for DPR's new San Francisco office, Mike led the project team in establishing the goals and strategy for achieving a Net Positive Energy building.

3. Mike Messick is a senior project manager at DPR Construction. His passion for sustainable construction began in 2006 while completing one of DPR's first LEED certified projects in San Francisco. Since then,

Mike has completed numerous LEED certified projects, as project manager, including two award-winning net zero energy buildings, each certified by the International Living Future Institute. He graduated from California Polytechnic State University, San Luis Obispo, in 1995 with a bachelor of science degree in construction management. Mike has been in the construction industry for 21 years working in both Australia and the United States.

4. The 9th Annual Fulbright Litigation Trends Survey identifies the engineering and construction industry as spending the most annually on litigation among all industries represented in the survey, with 57 percent of firms spending $5 million or more.

REFERENCES

Donaldson, R. (2010, September 29). *Development of a platform to design zero-emissions buildings.* Presentation at the Center for Integrated Facility Engineering, Stanford University, CA.

Fischer, M., Khanzode, A., Reed, D., & Ashcraft, H. (2012). Benefits of model-based process integration. Invited paper. Fortschritt-Berichte VDI, Reihe 4, Nr. 219. In U. Rickers & S. Krolitzki (Eds.), *Proceedings of the Lake Constance 5D-Conference 2012* (pp. 6–21). Düsseldorf, Germany: VDI Verlag.

Flager, F., & Haymaker, J. (2011). A comparison of multidisciplinary design, analysis and optimization processes in the building construction and aerospace industries. In I. Smith (Ed.), *24th International Conference on Information Technology in Construction* (pp. 625–630). Editor: Danijel Rebolj, Publisher: Department of Civil Engineering, University of Maribor, Slovenia.

CHAPTER 4

Defining High-Performing Buildings

"However beautiful the strategy, you should occasionally look at the results!"

—Sir Winston Churchill

4.1 WHAT IS A HIGH-PERFORMING BUILDING?

There are many definitions for "high-performance building" because what constitutes a high-performance building is different for every owner, user, location, and purpose. *High-performance* is not strictly a synonym for *green* or energy efficient, although a high-performance building should likely include those features. A zero-emissions facility might sound like a high-performance building, but if it is exorbitantly expensive to construct and maintain, it probably doesn't provide much value to its owner. Similarly, an energy-efficient building that is not comfortable and an inviting place to work is not high-performance. The defining characteristic of a high-performance building is that it is the "right" building. This may sound vague (and it is!), precisely because those values are unique to every project.

High-performance is easily understood if it is quantifiable. A process facility, for example, can be categorized as high-performing by comparing its relative efficiency in transforming resources into useful products. Sustainability can be quantified in resource use, CO_2 emission, or other criteria. Metrics are an important tool for predicting and evaluating performance, but not all criteria are quantifiable. The high-performance building, in its fullest sense, is a building that achieves its essential purpose, whether measurable or not.

Functional Aesthetics

People engage with their surroundings on an emotional as well as intellectual level. For some buildings, exciting an emotional response may be its essential purpose. For this reason, a cathedral that helps worshipers deepen their spirituality can be a high-performance building. Memorials may stimulate reflection or may teach through form. The design of a home is similarly high-performing if it accurately reflects the individual preferences of its residents, whether they desire comfort, inspiration or even a quirky aesthetic that reflects their personal sensibilities. Positive engagement between users and a building is a hallmark of good architecture.

The hospital designed and constructed for the Eastern Band of Cherokee is an example of how aesthetic values can blend into, and are an essential aspect of high-performance. The essential purpose of this hospital was to improve health outcomes for the Cherokee, especially with regard to preventative medicine. But to accomplish this result, the Cherokee community needed to feel that the hospital reflected and respected Cherokee values and the natural beauty of the North Carolina mountains. These aesthetic considerations were vital to creating a high-performance project.

Emotional engagement and quantifiable performance are not exclusive alternatives and many buildings are designed to achieve both. A parliament building should be buildable, usable, operable, and sustainable, while simultaneously reflecting the power of government and the rule of law. An airport terminal should provide an efficient flow of travelers and their baggage while enhancing the traveler's experience creating calm in what may be a stressful journey. A university laboratory should relate to its context, inspire collaboration and innovation, and be efficient and sustainable. Too often, aesthetic values have been placed in conflict with cost or efficiency with one being sacrificed for the other. But great architecture can, and should strive to achieve both the measurable and the immeasurable.

The processes described in this book are entirely consistent with great architecture. By beginning with a deep exploration of values, i.e., why the building is being considered, the less quantifiable values can be highlighted at the inception of design and incorporated into the building. Eliminating waste increases the resources available to the designer. In fact, one of the reasons for writing this book is that we saw architecture being sacrificed because inadequate early integration led to later value engineering that compromised function and aesthetics. But by making values explicit from the inception, decisions that affect multiple values can be properly harmonized. The Alfred I. Nemours pediatric hospital in Delaware has an imposing atrium that considered solely on space and mechanical efficiency would seem to be a poor design choice. But considered from the perspective of a parent bringing an ailing child for treatment, the atrium communicates the quality of care the hospital provides and is an excellent design decision. By understanding the fundamental values the project responds to, the team can make the correct design choices.

The essential point is that the high-performance building responds to the users and sponsors needs whether they are measurable or not. Metrics should be used to guide design and implementation and to assess outcomes. But immeasurable values must also be considered and it is the experience, creativity, and wisdom of the team that can optimally balance measurable and immeasurable values.

High-performance is about the whole package. It is the idea that a facility can be resource efficient, environmentally responsible in construction and operation, comfortable and safe for its occupants, easy and cost effective to maintain, and allows its users to perform at their highest level. In short, a truly high-performance building satisfies everyone who designs, constructs, operates, and uses a building as much as possible; it is a building everyone can be proud of.

In this book, we describe the process teams can use to create a high-performance building. Although what constitutes high-performance is project-specific, there are common factors that must be considered on all building projects. We describe these in this chapter and then illustrate how teams have achieved high-performance buildings by optimizing these factors.

4.2   WHAT DOES SUCCESS LOOK LIKE?

We look at building performance through the eyes of the four main stakeholder groups in a building. In chronological order of building delivery, they are the design and build team, the professionals operating the building, the users of the building, and the managers of the building users who need to sustain their business. Hence, we define performance of a building in these four categories: buildability, operability, usability, and sustainability. Because every building faces resource constraints, it is difficult to create a building that is buildable, operable, usable, and sustainable. To optimize a building so that it scores the highest possible marks in all four categories, the project leaders must carefully consider each performance category from the very beginning of the design process. The desired performance for each of these characteristics must be defined and prioritized at the very outset with specific metrics and target values so that broad project goals become measurable objectives. The multidisciplinary design team must then generate design options that perform as highly as possible, using input from all major team members and stakeholders including end users, facility operators, trade contractors, and suppliers.

One particular aspect of building performance deserves early consideration in this discussion. A building needs to not only perform well in the four categories introduced in the previous paragraph. It should also be aesthetically pleasing or beautiful or even inspiring through its appearance. Beauty is the eye of the beholder. A building has many beholders or stakeholders, from the person or company paying for the building (the client or owner) to the people using the building, the building designers, the visitors and neighbors, architecture critics, future buyers, and so on. It is not the purpose of this book focused on project delivery to define building aesthetics for these and other building stakeholders. It is, however, our experience that building aesthetics often suffer because the other project objectives were not considered in an integral, timely, and proactive manner. For example, on an office building, the architect had to sacrifice half of the stone façade to make room in the budget for a mechanical system that allowed temperature control for each office (vs. the original design that allowed temperature control for larger zones only). Because these usability and operability concerns were introduced to the project late, the options to address them within the buildability constraints (which includes schedule milestones and budgetary limits) were very limited, and the only two choices were to sacrifice the aesthetics or to imperil the sustainability of the building by making it more expensive. The owner elected to go with the first of these two options, resulting in a building that is not as beautiful as the original design. We hope that the methods presented in this book lead to fewer such sacrifices and therefore to a more beautiful and more functional or higher-performing built environment.

4.2.1   Buildable

Buildable means that a building is easier to assemble, which means fewer hours are spent in construction, often using less material and requiring less rework. It may also be more energy efficient because elements of the building fit together better (such as a window and an exterior wall). Buildable buildings are also safer to assemble than buildings where buildability was not part of the design objectives. The benefits of a truly buildable building could even extend to workers' wages and career length; more productive workers should get a higher wage, and with an increased need for their expertise during the design phase, workers who can no longer work physically in the field can (and should) build virtually.

A buildable design plans for the best available construction methods and practices, resulting in construction that is as safe and productive as possible. This requires that constructability be considered as part of the design. For example, prefabrication must be decided on as a priority and implemented well before anything goes into production. Modularization and prefabrication are essential to making a design more efficient to build, but the team must design specifically for these production approaches. If constructability is considered only as an afterthought, resources will have to be spent for inefficient fieldwork or for rework, which is inefficient.

4.2.2   Operable

Operable means that the building's systems—the structural, mechanical, electrical, and other systems—work together and are easily maintained and fixed. To create an operable building, the project team must consider all operations and maintenance requirements during design and incorporate building operators as part of the design team. A project leadership team needs to identify operators who become part of the design team. These operators then need to be proactive participants in the design process so that their concerns can be integrated into development of design solutions at the best possible time. As with buildability, it is more efficient to design a building for operability from the start than to fix a design later to make it more operable.

The operators need to provide input to the design and build team about how the building will be commissioned, that is, how it will transition from the construction phase to the use phase. They also need to articulate what information they need from the design-build team to operate the building. It is much more efficient to assemble this information along the way than to search for it once construction is finished. It also will make it much more likely that the operators will be able to operate the building with accurate as-built information.

Operability concerns fall into three main categories. First, an operable building must enable the operators to easily create the building conditions that make occupants productive, which means, for example, that a building's façade, room layout, and mechanical system need to work in unison. Second, an operable building is easily maintained day to day, which means that maintenance can be carried out safely and efficiently given the skill of the available maintenance staff and that they clearly understand which systems or subsystems of a building require a proactive or preventative maintenance approach and which systems or subsystems can be managed using a reactive or "fix it once it's broken" approach. Finally, an operable building must enable easily exchanging major building system components if they break, if a superior technology becomes available, or if the building purpose changes.

4.2.3   Usable

Usable means that the building supports the purposes of the people who live or work in it, or come to it for goods or services. The layout, flexibility, atmosphere, and other practical aspects of the building helps it users achieve their means and ends. For example, in a hospital, a usable design would not require employees to walk from one end of the building to the other to get lab tests or visit patients, and patients would not have to travel 10 floors to receive two different services. Usability is probably the most important of the four performance categories because, after all, a building's purpose is supporting its users in their daily work activities and lives. Without the users, there is no good place to live, make or sell things or provide services, food or entertainment. For a hospital, the main user groups are the patients and the doctors and nurses; for a university building, they are the students, faculty, and staff; and for a retail building, they are the customers and sales staff. Considering user needs is fundamental to achieving usability. However, it can be difficult to obtain information on users' specific objectives, priorities, and feedback on the usability of a design. Later chapters present methods that have been successfully used on projects to define and prioritize usability objectives and validate the usability of a design.

4.2.4 Sustainable

A sustainable building is designed to work in harmony with its natural, social, and economic contexts. A sustainable building allows the client's management team to sustain its business. A sustainable building would, for example, take advantage of its physical surroundings such as "free" cooling for data centers in cold climates or solar panels in sunny climates. In addition to working within their contexts, sustainable buildings should not waste materials during construction or energy during operation. The design of a sustainable building balances the resources for construction and operations appropriately in support of the use of the building. For example, during construction, careful use of materials not only reduces the building's environmental impact, but also reduces landfill and waste disposal costs. Moreover, during operations, owners of energy-efficient buildings pay less to operate and maintain the building, and they should have a more productive workforce—the building users—making them more competitive in their marketplace.

4.3 HOW CAN THIS BE DONE?

4.3.1 Buildable

On a hospital project in California, the project team considered the buildability of patient room headwalls. A headwall is the wall at the head of the patient's bed that provides clinical elements such as multiple medical gases, suction inlets, vacuum slides, and data collection, along with a variety of high- and low-voltage electrical outlets and lighting—all in very limited space. The typical practice is to build the headwall in place and then install the in-wall utilities for electrical, plumbing, medical gases, and the like. This is difficult, slow work that often requires custom fitting and rework. But on this hospital project, the team was organized early during the design phase and included the designers as well as the builders representing all key trades for headwalls (drywall, framing, MEP [mechanical, electrical, plumbing]), so the walls could be designed for prefabrication. Figure 4.1 shows the building information model (BIM) of a prefabricated patient room wall at the head of the bed where all of the monitoring and call systems are located.

Figure 4.2 shows the "spool sheet" view of the BIM. This is printed for drywall carpenters to use in laying out for prefabrication. It's particularly useful for a nonstandard wall such as a headwall.

To plan for fabrication, the team asked many questions about the buildability of the headwall design in terms of product, organization, and process:

Product
- Does it fit in the space?
- Is it safe to work with?
- Can workers lift it?

FIGURE 4.1   Prefabricated headwall BIM. © DPR Construction.

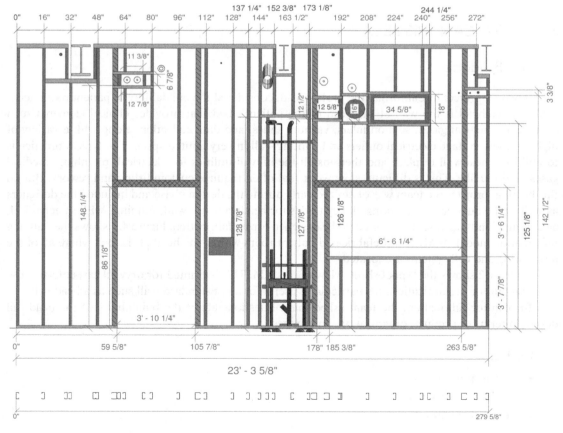

FIGURE 4.2   Wall framing prefabrication spool sheet. © DPR Construction.

Organization

- How should the team be organized?
- Who are the right people to involve?

Process

- Can a fabricated component be delivered to the site without extra handling?
- What is the sequence of installation?
- When and how should the different team members get involved?

The project team consulted foremen who drew from their prior headwall experience to identify important issues to address when constructing a headwall (see Figure 4.3). By incorporating this information, the team was able to prefabricate the wall off site, with utilities already built in and tested, thus avoiding mistakes and rework that often happen when many disciplines must do a lot of work in a tight space. This meant faster and safer on-site installation, with higher-quality work.

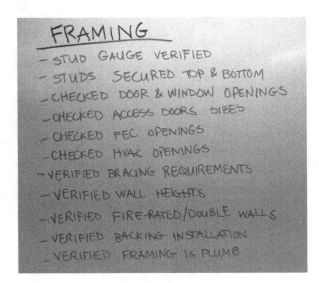

FIGURE 4.3 Checklist for drywall installation, containing lessons learned from experienced foremen. © DPR Construction.

4.3.2 Operable

In DPR Construction's Phoenix Regional Headquarters, building systems included (Figure 4.4) temperature sensors, an air circulation system inside the office space, a heat extraction system (a solar chimney, Figure 4.5), and an air-cooling system composed of passive "shower towers" plus the external louvers, which are actuated based on the air input needed in the space. Figure 4.4 is a cross-section of the office showing the various building systems and their interaction to maintain a comfortable temperature inside the building.

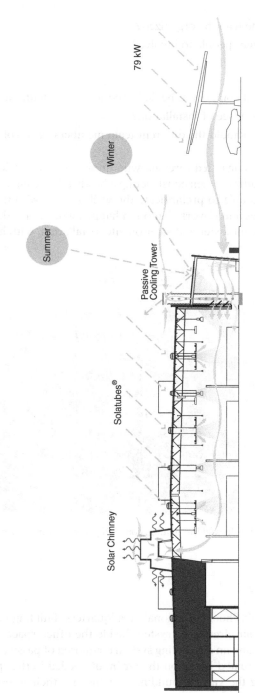

FIGURE 4.4   Bioclimatic strategies. SmithGroupJJR; courtesy of DPR Construction.

FIGURE 4.5 The solar chimney. Gregg Mastorakos; courtesy DPR Construction and SmithGroupJJR.

Figure 4.5 shows the solar chimney, which passively ventilates the building. A conventional design would likely have included a mechanical air-handling unit, which would have consumed energy.

All these systems work together in response to how the building is being used. The temperature is regulated by four passive shower towers, constructed of inexpensive high-density polyethylene (HDPE) piping and sheet metal (Figures 4.6 and 4.7). These work like evaporative coolers, cooling air as it passes through water vapor created by a mister system and water droplets from shower heads within the tower. As the building fills with people and is warmed by the extreme Arizona sun, hot air is released through the louvers into the 87-foot-long, 13-foot-high, zinc-clad solar chimney that sits atop the roof of the office, reportedly the largest of its kind in Arizona. A smart building management system (BMS) operates all the integrated systems automatically as needed.

The shower towers are located on the east façade of the building, which also has full shading and operable windows for fresh air intake (Figure 4.7). The north façade has additional operable windows and full shading as well, while the south and west façades have no openings and R21 insulation. This combination of passive bioclimatic strategies means that the building adapts quickly and effectively to how it is being used.

4.3.3 Usable

At the Sutter Health Eden Medical Center (SHEMC) hospital, the project team was tightly focused on improving usability by incorporating concepts from Lean health care (Chambers, 2011). In this

FIGURE 4.6   Shower towers (black columns), green screening, and operable windows. Gregg Mastorakos; courtesy DPR Construction and SmithGroupJJR.

FIGURE 4.7   Mist (upper left), operable windows, and shower towers. Gregg Mastorakos; courtesy DPR Construction and SmithGroupJJR.

approach, the patient is always at the center of care and the hospital is physically organized to minimize the distance traveled by patients, nurses, and doctors in getting or giving care or services. This patient-centered approach gives priority to optimizing the flow based on the whole outcome delivered to the patient before improving the efficiency of each services step. An early model of this concept developed by David Chambers uses blocks composed of patient care cells, with a universal care unit that morphs based on activity in patient cells using portable technologies to minimize dedicated treatment spaces that can be used for only one thing. The hospital is effective, patient-centric, and flexible. Figure 4.8 shows the model for delivering medical and diagnostic services, organized to maximize value to the patient.

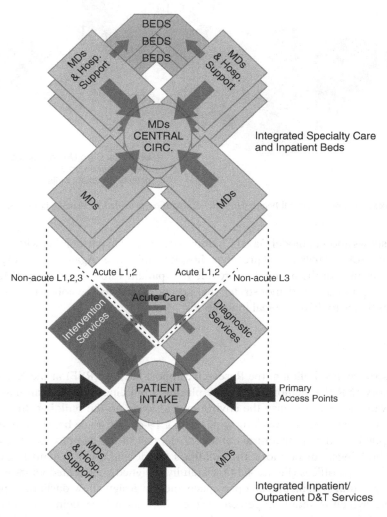

FIGURE 4.8 David Chambers model for medical and diagnostic services. Illustration provided by David Chambers.

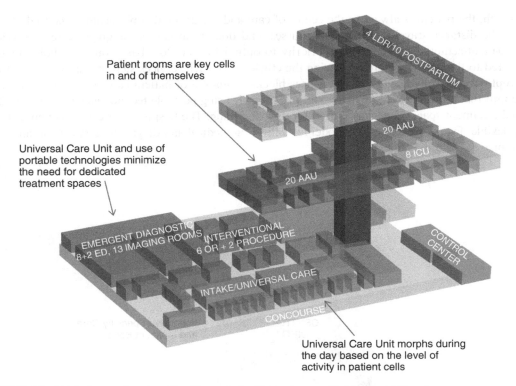

Patient rooms are key cells
in and of themselves

Universal Care Unit and use of
portable technologies minimize
the need for dedicated
treatment spaces

4 LDR/10 POSTPARTUM

20 AAU

8 ICU

20 AAU

CONTROL CENTER

EMERGENT DIAGNOSTIC
8+2 ED, 13 IMAGING ROOMS INTERVENTIONAL
6 OR + 2 PROCEDURE

INTAKE/UNIVERSAL CARE

CONCOURSE

Universal Care Unit morphs during
the day based on the level of
activity in patient cells

FIGURE 4.9 Blocks integrating virtual health care cells. Illustration provided by David Chambers.

Figure 4.9 shows another model David Chambers developed in partnership with health care design and construction professionals for a "prototype hospital" initiative. Again, care is organized into virtual work cells throughout the hospital to provide best patient value. The variety of forms capable of addressing these optimizations demonstrates that no one form need be adhered to, while the idea of optimization continues to drive the real innovation.

4.3.4 Sustainable

Another high-performance building, the Research Support Facility (RSF) at the National Renewable Energy Laboratory (NREL), demonstrates how placing a building on a site can maximize the natural resources available and minimize the environmental loads on the building. In this configuration example, the RSF is composed of two long wings with a 19-degree angle between them (Figure 4.10). This layout responds to the architectural context of the rest of the campus and maximizes the solar income for the buildings. Furthermore, the building's façade and other systems maximize the use of daylight throughout the offices (Figure 4.11). Carefully designed and located windows with light louvers, a filigree structural system, and an open-space interior design allow daylight to penetrate 60 feet into the building, which is about 50 percent farther than normal, reducing the energy required for lighting. Such high energy efficiency also directly supports the business purpose of the buildings' users and owner. Figure 4.10 shows the RSF buildings in the foreground, left of center.

FIGURE 4.10 Aerial view of the RSF buildings. Photograph by Dennis Schroeder, National Renewable Energy Laboratory, NREL.gov.

Figure 4.11 shows the light louver system the NREL-RSF team developed for daylighting. Sunlight reflects to the ceiling, creating an indirect lighting effect. Fixed sunshades limit excess light and glare.

Figure 4.12 shows sunlight redirected through the light louvers deep into the office space, providing ambient lighting during the day most of the year.

4.4 INTERCONNECTIONS

Throughout this book, the term *high-performance building* includes all four characteristics (buildable, operable, usable, and sustainable) described in this chapter. Despite the importance of these characteristics they are rarely translated into specific criteria or metrics to guide design, construction, and operations, leading to divergent priorities and work by project team members. The methods described in this book should allow owners to measure and optimize these four critical characteristics so that they don't have to sacrifice the less tangible, but also important, aspects of a building.

4.5 REFLECTIONS

Truly high-performance buildings require breakthrough performance that is difficult to achieve with today's methods, processes, and organizations because there is often a focus on a few project objectives to the detriment of others. Breakthrough performance depends on integration of building systems and the synchronization of these systems with the objectives of all key stakeholders. Integration entails the

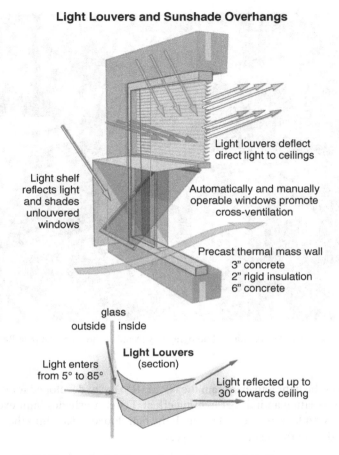

Light Louvers and Sunshade Overhangs

Light louvers deflect direct light to ceilings

Light shelf reflects light and shades unlouvered windows

Automatically and manually operable windows promote cross-ventilation

Precast thermal mass wall
3" concrete
2" rigid insulation
6" concrete

glass
outside | inside

Light Louvers
(section)

Light enters from 5° to 85°

Light reflected up to 30° towards ceiling

A light louver daylighting system reflects sunlight to the ceiling, creating an indirect lighting effect. Fixed sunshades limit excess light and glare. *Illustration from RNL*

FIGURE 4.11   Light louver system. Courtesy of RNL.

timely articulation and rapid resolution of trade-offs. Current practices are too slow and not transparent enough to support integration. But before a project delivery team can create a high-performance facility it has to define—together with the client, users, and operators—what high-performance means. It has to translate their values, wishes, goals, and aspirations into measurable values that guide the development and selection of the best building solutions. It needs frameworks to do so. The main purpose of such a framework is to gain a holistic perspective of building performance and to create specific but mutually aligned performance objectives for a building's systems and stakeholders. Without such a framework, the project delivery team will be unlikely to design and build a high-performance building.

FIGURE 4.12   NREL-RSF open office lit by sunlight. Photograph by Dennis Schroeder, National Renewable Energy Laboratory, NREL.gov.

4.6   SUMMARY

Ordinary buildings are optimized to achieve one or perhaps two goals. As has often been said, you can have cost, schedule, or quality, but not all three. The high-performance building is different. It optimizes all parameters simultaneously: it is buildable, operable, usable, and sustainable. This can be achieved in the high-performance building because the systems are not in competition and the project outcome a compromise among them. Rather, in the high-performance building, the systems work synergistically to assist each other and achieve the project goals.

REFERENCE

Chambers, D. (2011). *Efficient healthcare overcoming broken paradigms: A manifesto* (2nd ed.). Renton, WA: CreateSpace Independent Publishing Platform.

CHAPTER 5

Achieving Highly Valuable Buildings

"Everyone designs who devises courses of action aimed at changing existing situations into preferred ones!"

—Herbert Simon

5.1 WHAT IS A HIGHLY VALUABLE BUILDING?

A highly valuable building is a building that achieves its purpose throughout its life. If it is a research lab, it will facilitate innovation. If it is a school, it will promote learning. If it is a store, it will engage shoppers and encourage sales. It is a physical solution to its owners' and users' needs. Like most solutions to human problems, it has many components. To improve the health characteristics of a community, a hospital must be inviting and comforting to patients and their families so it will be used. It must be a healing environment that enables the health care practitioners to effectively treat their patients. Which means that it must be an efficient, safe, and pleasant environment for the staff. Which means it must be well maintained and maintainable. And it needs to have a high value for its cost to permit it to be initially built and keep operating. If this sounds like the highly valuable building is buildable, usable, operable, and sustainable, it is because it is, or, actually, must be.

5.2 WHAT DOES SUCCESS LOOK LIKE?

The value of a building unfolds over time. When it is completed, we will know whether it met targets for cost and schedule. As it is commissioned, we will gain some idea of its ability to meet energy and operational goals. But other key goals, such as adaptability, improving employee creativity and productivity, or even life cycle and maintenance cost can be determined only after sufficient time has elapsed. Moreover, if the reasons for undertaking a project relate to the community, then it may take considerable time to know whether the project succeeded.

71

The Mosaic Centre in Edmonton, Canada, and the Rocky Mountain Institute's Innovation Center in Boulder, Colorado, both aimed, in different ways, to prove that highly sustainable, attractive, net zero energy use buildings could be built within normal commercial budgets in order to spur sustainable development. One measure of success for these integrated project delivery (IPD) projects (which met their financial and energy goals) is how many others follow the trail they blazed.

There are several characteristics highly valuable buildings share. As noted above, they respond to why the project was undertaken and continue to provide value over time. They are a physical response to a series of problems or opportunities. Highly valuable buildings achieve the principal goals while conserving resources during creation and operation. They operate as a whole building, not just a collection of parts. Their systems are integrated and optimized to the building's purpose.

The upshot is that a highly valuable building superbly addresses the needs of the project sponsors over whatever duration is relevant to those needs. Because these needs vary, the measures for success vary as well.

5.3   HOW CAN THIS BE DONE?

5.3.1   Define Client Goals and Translate Them into Objectives for the Project Team

The core task of a facility development and operations team are: (1) to identify the appropriate goals and objectives that matter to the facility's clients and users to sustain their business or purpose; (2) translate them into the specific performance objectives for the use and operation of the facility and the facility development project; and then (3) through a careful design of the project organization and its work processes and corresponding objectives and metrics, (4) design and build the best possible facility that enables sustainable use and operation (Figure 5.1). This sounds straightforward enough but is, of course, difficult to execute across the many organizational, temporal, and physical boundaries and scales and for the many economic, environmental, and social performance goals that shape each unique facility.

Figure 5.1 shows the two main types of goals and the main types of performance objectives necessary to define the value of a facility. The combination of these goals and objectives defines project value.

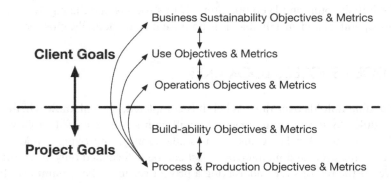

FIGURE 5.1   Client and project goals and objectives.

In our experience, many professionals are anxious to get going on the design and construction of a facility because, after all, that's what they are paid to do. But notice that, without the right design of the project organization, including how the work is going to be done, the experience and the result of the design and construction efforts are unlikely to be satisfying. Also note that without clearly defined performance objectives for the building and without understanding what the users and client value, there is no clear yardstick to distinguish a good design solution from a poor one in terms that really matter to the client.

Why This Is Difficult

Translating the business goals of the client into building-focused client goals and then into specific project goals related to the users, operators, and design-build team is very challenging. Establishing measurable objectives that capture the goals and guide the development of the project is equally difficult. Hence, the formal requirements often inadequately describe the value desired by the users and owner. Because value is not well understood and articulated when the requirements are created and are too often formulated in prescriptive product terms, for example, specifying the number and size of specific rooms in a building and not their performance, the formal requirements fall short of their purpose. Consequently, the connection between the business value of the facility, the facility performance, and the facility design and construction is lost. Furthermore, the facility design often addresses only some of the performance requirements because the designers that join a project team as the work ramps up do not understand all of the requirements and their implications. Consequently, design decisions are rarely made in the context of all the requirements (Kiviniemi, 2005).

"Value engineering" also takes its toll. Cost considerations drive scope reductions that are not aligned with the project's values. Not only is less value delivered, the misalignment changes the balance of values from the balance preferred by the client and users, reducing overall value even more. For example, in a research facility, the information technology (IT) consultant who had joined the project team late was asked for value engineering suggestions. He noticed that the network in the building was designed for a bandwidth that far exceeded anything he had ever seen in other buildings. He recommended reducing bandwidth to more typical levels, saving hundreds of thousands of dollars. Because he did not fully understand the users' needs and values, he did not understand that this reduction in bandwidth would greatly curtail building users' ability to pursue the research goals they had established early in the project. Soon after opening, the owner had to upgrade the network to support the building's research mission, incurring a cost that far exceeded the savings achieved through the IT consultant's "value engineering" suggestion. In another case, budget pressures caused the project team to change the access control system for a building that receives many short-term visitors from a programmable system to a lower-cost key system. While this change helped the project team complete the project on budget, it forced the building users to set up a system (including staff) to issue and collect keys to visitors, reducing the building's operability.

With traditional methods, construction professionals charged with bringing a project in on budget rarely have visibility into what enables users to create income from the facility—or, more generally, what makes the facility valuable to its users—as they make detailed decisions that affect usability. Finally, in spite of the hard work of designers, builders, and operators, many facilities never achieve their potential because the facility handover or commissioning process is short-changed. At the end of the construction phase of many projects, there isn't sufficient time and money for commissioning, key design and construction professionals have moved on to other projects, and the project is tossed to the owner's internal project delivery and facility operations groups who don't fully understand the building from all essential perspectives.

Using Metrics to Make Better Decisions

Metrics allow a project team to see the performance of a facility or a team by comparing two measurements, such as dollars per square foot or fabricatable three-dimensional (3-D) objects created per hour. This enables understanding and ultimately supports making better decisions. A challenge in discussing facility performance is that there are many aspects of a facility for which performance objectives should be set and actual performance measured, but the terminology to describe these performance aspects is not well defined and used consistently across the industry. For example, terms like project goals, team performance, process metrics, outcome objectives, performance targets, leading and lagging indicators, and other combinations of these and similar terms are not clearly defined and mean different things to different people.

A further complication is that some metrics are more important than others to different project participants, that is, the owner representatives, the architect and engineers, the general and specialty contractors, and so on. Some metrics apply to outcomes, usually final outcomes at the end of a sustained effort, for example, the cost of design and construction. Others make the results of work visible as it progresses, for example, the number of tasks actually completed versus the number planned to complete. Some metrics focus on the installation of work, such as lineal feet of wall framing installed by a crew in an hour or day. These are process or production metrics. Stepping back, we can say that we are measuring performance and state that all metrics are "performance" metrics. That is why we have defined the categories of performance objectives shown in Figure 5.1. Throughout the book we have included examples of how these types of metrics are used throughout a building's life cycle to evidence measurable value.

5.3.2 Focus on Value First

A high-performance facility enables its users to create the value they must deliver to thrive in their own business. For example, a bridge allows a certain number of cars to cross each day helping a transportation agency meet its goal of enabling people to go places; a school building allows teachers to inspire, educate, and engage with a certain number and type of students; a home enables affordable and healthy lives of its occupants; and so on. The work of designers, builders, and operators accomplishes this performance and enables this value through the efficient allocation of materials and technical, financial, and human resources. This is a complex endeavor because of the difficulty in predicting many aspects that must be considered when making decisions about a facility. These decisions affect the duration and cost of the design and construction phase or the carbon dioxide (CO_2) footprint during operations, or the expected durability of the facility. In summary, a high-performance facility optimizes its performance across all the cost and income[1] aspects shown in Figure 5.2. As mentioned, this is challenging to accomplish given the unique nature of each facility in its economic, environmental, and social context.

Today's project delivery process often attempts to optimize the design and construction cost and duration (optimization goals 1 and 2 in Figure 5.2). Reducing the time required for design and construction benefits the project delivery team and the client. By finishing early, the project delivery

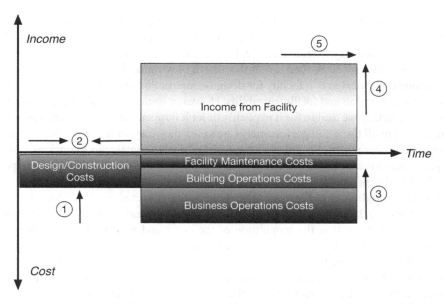

FIGURE 5.2 Cost and income of facilities. Illustration provided by Martin A. Fischer, Stanford University.

team can create more value by starting another project. The client also benefits because it can use the facility sooner, generating more revenue or supporting other organizational goals. The total income that can be generated from a facility not only depends on its opening date, of course, but also on the income that the facility enables (optimization goal 4, the positive segment of the *y*-axis in Figure 5.2), and the duration during which the income can be generated (optimization goal 5, the duration of the use phase shown on the *x*-axis).

A short design and construction duration that imperils the income that the building users can generate because the building isn't well laid-out or reduces the longevity because the building materials or equipment aren't durable will be counterproductive, even though design and construction costs and duration are reduced and income can be generated sooner. Furthermore, owners would also like to keep the operating costs as low as possible (optimization goal 3), including the facility maintenance costs (e.g., the effort required for facility management), the building operations costs (e.g., the energy costs), and the business operations costs (often, the salaries of the staff working in the facility). Finally, in most cases, owners would like to extend the longevity or durability of a facility (optimization goal 5). The design and construction team does not have full control over all aspects of these buckets of costs and income, but it shapes them in many ways. For example, for a casino project in Southern California, Greg Luth, the structural engineer, recognized that the income generated by an earlier opening outweighed the additional design and construction costs of building the casino on top of nine-story columns while simultaneously building a nine-story parking garage underneath the casino.

Trading off these optimization goals is challenging because:

- There is currently no global facility optimization metric or formula;
- Data to inform the key trade-offs are often not available in sufficient quality when needed; and
- The expertise to consider all these project aspects is often brought on board sequentially and too late to devise and maintain a globally optimal design solution.

For example, it is difficult to contrast the value of an additional month of design time against the value of opening the facility a month earlier. With an additional month of design time, could the design team have come up with a way to shorten the construction duration by more than a month or increase the income of the facility by more than a month's worth of revenue, and so forth? It may seem like a tall order to optimize the design of a facility for design-construction and operational costs, income, and design-construction duration and facility durability. However, each facility will have some performance in these respects, whether we design for it explicitly or let it happen. In other words, a project team can let the performance in all these categories emerge however they may, or it can set targets for all of them and collaborate to achieve them. Broadly speaking, since "design signals our intentions" (McDonough, 2011), project teams must become more formal and explicit about considering all these performance aspects of a facility during all life cycle phases, in particular during design.

The value of a building, and the cost to achieve it, are reflected in the different costs and revenue shown in Figure 5.2. The circles with numbers refer to typical optimization goals that are also referenced in the text. Note that the magnitude of the costs and the durations are not to scale and that "costs" and "income" may refer to nonmonetary metrics such as CO_2 emissions/sequestration or other sustainability measurements. Optimizing the values and costs requires the right blend of the four performance types—buildable, operable, usable, and sustainable—defined in the prior chapter.

Buildability affects the design and construction costs (circle 1 in Figure 5.2) and will be determined when the building is completed. Although buildability is determined during the design and construction phase, it would be a major mistake for this to be the only focus of that phase. The remaining performance types—operability, usability, and sustainability—will be determined during the use of the building, but they are largely shaped during design and construction. In fact, it is absolutely essential that the project delivery team considers all building performance goals and objectives during the design and construction phase. In fact, the importance of this point is a primary reason for writing this book.

If they care to create a high-performing building with some certainty, project teams must not only be formal and explicit about these critical performance aspects, they must also become more transparent about them. We have experienced too many projects where project participants held the cards that showed the building performance for which they are responsible close to their chest. Such a lack of transparency rarely, if ever, helps the project team achieve a higher-value building because it leads to rework to adjust the design to improve the performance with respect to the newly shared performance aspects or it leads to a design being built that misses important performance aspects. This transparency must start with the building owner. The building owner must share what she or he expects from the building, otherwise, how can the other project delivery team members know how to best assist the owner in getting there? Put in the words of one of this book's authors, "You must tell the cab driver where you are going if you want him to take you there." Similarly, if an owner is not transparent about the building she wants, for example, the building's target budget, it would be sheer luck if the design-build team created a building design that meets the budget.

Now that we have connected the different costs and income with the main phases of a building's creation and use, we need a framework that connects the levers a project team has to influence building performance, that is, the independent variables, with the performance that is sought, that is, the dependent variables. The next section introduces such a framework.

5.3.3 Apply Design Thinking

Creating a uniquely valuable high-performing building requires what Herbert Simon calls "Design Thinking" (McDonough, 2011). More than four decades ago, both Simon and John Gero developed theories to demystify design. Gero explained design as a process of creating structure or form to produce behaviors that allow people to function in ways they want (Gero, Tham, & Lee, 1992). Put in simple terms: first, one must understand the function of the thing being designed, that is, establish how it needs to perform. Then one has to consider how the thing has to work to meet the needs. Then one should draw on and adapt past experience to conceive the structure of the thing. The so-called "magic of design" involves such cycles of analysis, synthesis, and evaluation to establish the function, structure or form, and behavior of what is being designed.

Specifically, for building projects, in 1896 the American architect Louis Sullivan coined the phrase "form follows function" (Sullivan, 1896). But buildings rarely have only one function or performance requirement. Moreover, there are typically many possible solutions of forms that address the function or performance requirements in differing ways. In the real world, the solution chosen by the project team should optimize the blend of performance types consistent with the project's values. To do so, the team needs a logical and consistent framework that intersects what the team can control (the "levers" or independent variables) and the resultant outcomes (the dependent variables).

Making something buildable, operable, usable, and sustainable are imperative concerns in delivering functionality. Thinking about these four concerns and their application to guide the design of a simple product like a drinking glass and a complex product like a building makes us realize that a building's functions are much more varied and complex than for a glass and involve many more disciplines and stakeholders. For buildings, any decision about a physical component included in a building brings with it a particular mix of organizations and processes. For example, the decision to incorporate automated louvers to provide the right mix of shading and daylight to help optimize the energy performance of a building brings with it a particular set of maintenance activities and organizations and people carrying out these activities. In this way, many decisions made in the design phase have implications for the use phase affecting not only the design and construction costs but also the cost and income during the use phase.

As can be seen from this discussion, the design of buildings should not only address the design of the physical building and its components; it should also address the organizations and processes related to the building's buildability, operability, usability, and sustainability concerns.

To bring a design thinking approach to buildings, applied research carried out at the Center for Integrated Facility Engineering (CIFE) showed that a project organization or delivery team adds value to a building by applying design thinking to three domains: the product, work processes, and the organization. This insight highlights that the design of a high-performing building (or product) depends on the design of the organizations and processes that create the high-performing building. In essence, these are the three levers for affecting project outcomes: A team can change the characteristics of the product (the thing being constructed), it can change what people are doing (the work process), or it can change how people are organizing themselves. The CIFE Product-Organization-Process (POP) matrix can be represented in a 3 × 3 matrix with the three design questions (function, form or structure, and behavior) in the left-most column and the three areas or levers in a row across the top (Table 5.1).

5.3.4   The POP Framework

As mentioned, this framework needs to exhaustively and exclusively consider the independent variables the project team can and must decide on to create a high-value building. If we think about what aspects of a building project the project delivery team can, at least to some extent, control, we realize that they fall into three categories: Product, Organization, and Process (POP). The team can decide on the shape, layout, and makeup of the building itself. We call these decisions broadly product decisions since they refer to the physical components—products—of a building. The team can also decide who to involve and when and how. These are the organization decisions. Finally, it has to decide what the different project participants will do when and in what sequence. These are the process decisions. In addition to defining which objectives to pursue and with what priority, these are the only types of decisions a project team can make. Together, they'll shape the design of the building and how it's going to be designed, built, commissioned, operated, used, and repurposed. The value of the POP framework is that it is a mutually exclusive, collectively exhaustive representation of a building and its stakeholders over time. It not only treats the design of a building as a product design problem, but sees the design of a building holistically as the design of a building's products, organizations, and processes.

The second role of this framework is to connect the aspirations for the building with the POP design decisions and the predicted and observed performance. Table 5.1 shows this framework, with the three design questions on the left and the three levers to influence project outcomes across the top.

The POP framework can be applied at the enterprise/client, user, and facility operations levels. If the organization creates its own products, POP is applied to the enterprise as a whole. If the organization procures the product—like the owner of a building—the focus is on delivering a product meeting the needs of a client. But whether at an enterprise level or a client level, the analysis is essentially the same.

The Enterprise POP model focuses on the client's enterprise to understand the client's business products, its services and organizational stakeholders, the organizations' work, and the aspirations and performance (behavior) to sustain the business.

At the enterprise level, POP is focused on the core products of the enterprise. Consider an automobile manufacturer.

From a product perspective, the *function* of the product is the transportation of people. The *form* of the product is automobiles. The *behavior* is the qualities of the product—for example, mileage, smoothness, and performance—many of which can be measured and predicted.

From an organization perspective, the *functions* of the automobile company are organizational imperatives, such as procurement, management, and marketing. The *form* is the departments within the automobile company that are responsible for these responsibilities. The *behavior* can also be measured in terms of effectiveness in meeting the needs of their internal clients.

TABLE 5.1   CIFE POP Model

	Product	**Organization**	**Process**
Function	What is the purpose/use?		
Structure/Form	What is the structure/form?		
	What does this look like? How is it put together?		
Behavior	How will it/we perform?		

From a process perspective, the *function* is to execute the work. The *form* is the sequence and flow of information and materials among those responsible for creating the product. The *behavior* is the effectiveness in producing work, which can be measured and monitored.

The POP framework is a logical model for discussing the purpose of the enterprise, what opportunities exist ("levers") that can be used to affect the outcome at the enterprise, user, or operational level.

The POP framework is also consistent with the "triple bottom line," the simultaneous achievement of economic, social, and sustainability goals. Communities provide the basic requirements for projects in skilled labor, an infrastructure that supports commerce and the natural resources that are incorporated into the project. In turn, the communities need the projects to be successful—providing revenue through taxes and wages. But they also need the process to repeat indefinitely, that is, to be truly sustainable. The POP framework enables the product, organization, and process to be aligned to these multiple goals and optimized, rather than being bartered against each other in a zero-sum exchange.

The POP model is also effective in providing a project view based on the users' needs. People produce value for their firms and the firm's business customers by processing information and materials. From their perspective of a building, the "product" becomes the place and space in which they will work. As with the enterprise/client model, the product in the user's POP model has function, form, and behavior, as shown in Table 5.2.

The usability or user POP applies the same organization and process, but the product is now the building the company's organization uses to produce its products and services and that the IPD team needs to design. Table 5.3 reveals how the facility (the product) must support the users in achieving the organization and process objectives of the enterprise. It connects the mission of the design and construction team directly to the client's enterprise and users. This perspective is mostly missed in current practice.

TABLE 5.2   Client Enterprise POP Model

	Product	**Organization**	**Process**
Function	Obtain *x*% market share Most innovative product	Be one of the 10 best companies to work for Attract top university graduates	Improve internal process efficiency by 2% each year as measured in $/work hour sold Automate 5% of the workflows each year
Form	The "things"—products or services—the company makes and sells	R&D Department Engineering Sales and Marketing Customer Support	Develop technology options Select technologies for implementation Reengineer the product to incorporate the new technologies Revise the marketing strategy
Behavior	Achieved *y*% market share First to market with an innovative feature	Ranked number 18 in Fortune's best companies to work for Attracted 22 of 28 top graduates who were offered a job	+2.5% process efficiency last year 4% of workflows automated

TABLE 5.3   Usability/Users POP Model

	Product (Building)	Organization	Process
Function	Enhance the company's image Space utilization around 70% 50% better energy performance than code	Be one of the 10 best companies to work for Attract top university graduates	Improve internal process efficiency by 2% each year as measured in $/work hour sold Automate 5% of the workflows each year
Form	Clean room for R&D Product showroom Call center Heating, ventilating, and air-conditioning (HVAC) system Building automation system (BAS) IT system Structural system Façade	R&D Department Engineering Sales and Marketing Customer Support	Develop technology options Select technologies for implementation Reengineer the product to incorporate the new technologies Revise the marketing strategy
Behavior	75% of new employees mention the building as one of the attractions of the company 65% space utilization 62.4% better energy performance than code	Ranked number 18 in Fortune's best companies to work for Attracted 22 of 28 top graduates who were offered a job	+2.5% process efficiency last year 4% of workflows automated

Notice how the facility must support how people organize to do their work safely and productively. If the building doesn't do this well, people will not contribute as much as they could to provide what their customers need when they need it. By understanding how the users will organize their activities and process work, the design and construction team can develop solutions that create the appropriate product to make users healthy, satisfied, and effective. The user POP model shows how the organization and process dimensions drive the design of the facility.

The facility operations POP model, shown in Table 5.4, shows the work of building operators relative to the enterprise, the physical plant, and the work of the users. Notice how focused the function and form of the organization and process are on the facility that has been acquired by or built for the enterprise. Operations are accounting cost centers dedicated to making users more productive in the workplace, which puts them at the nexus of the three sustainability imperatives of the triple bottom line.

POP can also be applied at the operability level. Note that the building is the same as for the users, but the organization and processes are focused on the operation and maintenance of the building.

Aligning Stakeholder Interests for the High-Performance Building

The three client POP models present a picture of stakeholder concerns for the sustainability, usability, and operability criteria of high-performing buildings. This is the starting point for translating

TABLE 5.4 Operability/Operators POP Model

	Product (Building)	**Organization**	**Process**
Function	Enhance the company's image Space utilization around 70% 50% better energy performance than code	Retain staff for 5+ years Zero health and safety issues in building No business impact from cleaning	Respond to critical failures in 30 minutes 100% of the time Recycle 80% of the waste produced by the users
Form	Clean room for R&D Product showroom Call center Heating, ventilating, and air-conditioning (HVAC) system Building automation system (BAS) IT system Structural system Façade	Maintenance staff Energy managers Data analysts Cleaning staff Sustainability coordinator	Analyze building systems data to identify components for preventive maintenance Prioritize maintenance tasks Set up waste and recycling system
Behavior	75% of new employees mention the building as one of the attractions of the company 65% space utilization 62.4% better energy performance than code	Average staff tenure = 6.4 years One health and safety issue last year Occupancy survey showed no business impact from cleaning	98% of critical failures addressed in 30 minutes or less All critical failures addressed in 45 minutes or less 75% of building waste recycled

aspirational goals of client stakeholders into tangible project delivery objectives that can be achieved by the project team. While the specific concerns of each client and their POP models will be different and become more so as the granularity of the models increases, they must be used by the delivery team to establish objectives and metrics to determine success.

Project delivery teams can use Design Thinking and the CIFE POP model to understand client goals relative to the four high-performing building criteria. This in turn allows client and delivery team leaders to translate aspirational goals into objectives and metrics for the project. For example, even high-level POP models for the client enterprise, building users, and operators reveal who the customers and owners are for success criteria or "conditions of satisfaction."

Figure 5.3 puts the different POP perspectives together. The figure shows the POP matrices for the client's enterprise (upper left), the users (upper right), the client's operations and facility management (lower right), and the project delivery team (lower left).

Note the connections between these matrices. The user organization and the work it does should be the same between the enterprise sustainability and user perspectives, arrow (1). The building should support what the user organization needs to do in the building, arrow (2). The product (the building) and its requirements or functions and its performance or behavior should be the same for the users,

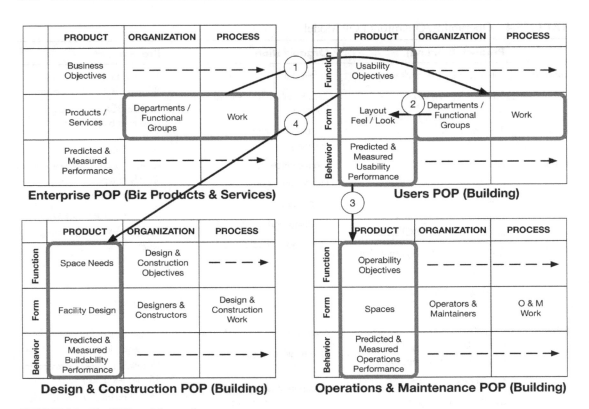

FIGURE 5.3   The POP models together.

operators, and designers and builders, arrows (3) and (4). Of course, the work these organizations carry out and the people will be very different, but they must all focus on the same building.

With the objectives visible, the team can now put its attention to developing the strategies for achieving them (Figure 5.3). It now needs to define how it will meet the objectives through a combination of tasks (processes), building systems (products), and people (organizations).

These strategies can best be summarized with POP matrices that show that it is the ingenious combination of products, organizations, and processes—the three main levers every project team has at its disposal—that shapes the success of any project. The building needs of the client's organization (the building users) that emerge from the sustainability and usability POP matrices must inspire the objectives or functions of the building operators (operability POP matrix) and ultimately the building's design and build—or project delivery—team (buildability POP matrix). The POP matrices highlight the opportunity to address these objectives through an integrated and holistic approach to the resulting design challenge, that is, by considering the design of the product, organization, and process so that the value sought by the users and the client are best met.

Surprisingly few owner representatives and design and construction professionals are taught or encouraged to look for points of leverage other than in the product, that is, the building features. Many owners are frustrated that the first and default choice for cutting costs is to reduce the scope of the product. This strategy cannot produce a high-performing building. In fact, it results in less of a building. POP shows that there are two other points of leverage—organization and process—that can be used to achieve the high-performance goal.

Leaders of every integrated project team can use the POP model to design what they will deliver and how. Answering the questions in Table 5.5 allows leaders to understand the product they need to create and why. When they do, leaders will see the kind of integrated organization and processes they need to create.

TABLE 5.5 Project Delivery Team POP Model Questions

	Product	**Organization**	**Process**
Function	What value-creating activities will the high-performance building support?	What are our objectives? How will we achieve them? What must we control? What do we expect to accomplish?	What will we produce (scope/quality)?
Structure/Form	What spaces and components and systems will make up the building? How will they be arranged?	Who will make decisions about value? How will we organize ourselves?	What methods will the teams use? What steps will we take? How will we communicate?
Behavior	What predictions will we make? What metrics will we use?	What are the measurable outcomes for the team as a whole?	What are our production and outcome metrics?

5.3.5 Ways of Looking at Building Performance

Before giving specific examples of commonly used metrics, we want to connect the POP matrix concept with the four types of performance objectives. The POP matrices offer an additional way to organize and align the many performance objectives, metrics, and targets by organizing a project into its product (what is it, what is it made of), organization (who is involved, who moves the project forward), and process (what is everyone doing) perspectives.

1. Product performance metrics relate the usability, buildability, operability, and sustainability objectives to the product. They describe how the facility should function or perform and how it is performing or behaving. These metrics are the means for predicting outcomes and making decisions during design. Measured over the first year of operation, these metrics become product outcome metrics.
2. Project organization metrics give insight into how well the project organization is able to manage toward the cost, schedule, quality, and safety objectives.
3. Process metrics measure the results of the processes team members are using to achieve project goals. These are typically leading indicators of project organization/team performance. Based on the process metrics, team leaders and members can learn and improve their practices, methods, and tools, and thus project outcomes. Production metrics fall into this category. They are metrics for work processes that contribute directly to building the physical product itself, i.e., what will be delivered to the client.

TABLE 5.6   Categories and Examples of Product, Organization, and Process Performance

High-Performing Building Criteria Types	Product Organization Process (POP) Model for Changing Outcomes		
	Product Performance	Organization Performance	Process Performance
Usable	Healthiness and sustainability of materials, daylight, comfort, acoustics	Stakeholder engagement, response and decision latency, building users' productivity	Testing of product performance, efficiency of users' work processes, flexibility to adjust to changing business purposes
Buildable	Availability, affordability, installability of materials and systems, safety	Cost, schedule, quality, safety	Risk and opportunity costs, information processing and approvals for changes, RFIs and submittals, plan percent complete (PPC), speed of resolution of spatial and temporal clashes with BIM coordination and 4-D simulation
Operable	Energy and water consumption, Accessibility for maintenance	Life cycle cost	Maintenance effort per hour of non-conforming facility performance
Sustainable	CO_2 footprint, occupant safety, durability	Design to WELL Building Standard, American Society of Heating, Refrigerating and Air-Conditioning Engineers (ASHRAE), LEED, and Living Building Challenge standards	Minimized resource consumption, harvesting and reuse of renewable energy, air, and water

Table 5.6 provides an example of how the building's characteristics of buildable, usable, operable and sustainable map to the POP rubric.

Note that it can sometimes be difficult to decide whether a metric measures the performance of an organization or a process, since the two are linked. For example, is the response time to requests for information (RFIs) a measure of the person or organization responding or of the RFI response process? We find that identifying the critical metrics is key and whether to classify them as organization or process metrics is less important.

Currently, team members come to projects intent on managing what is important to their disciplines and the business interests of their company. There is little or no agreement on how to evaluate the project delivery team except for the standard project performance metrics for cost, schedule, quality, and safety. Each entity measures its own outcomes for itself and rarely shares this information. Product performance metrics are left to the design team although they are often not understood and expressed that way and there is rarely feedback from as-used and as-operated buildings to the as-designed and as-built buildings. Organization and process performance is typically left to the construction team, and project teams often find it difficult to connect these broad categories of metrics, let alone align or optimize them. Integrated teams must shed this baggage very deliberately and design a set of measurements they can use to make better predictions for how the product will perform and to improve the way they deliver it.

An IPD project is fundamentally different because of the decisions that will be made and who will make them. Designers are not just designing to an owner's program document. Builders are not just executing the design they are provided. Instead, the team is jointly making decisions with the client regarding how to create the product the client requires. In deciding between alternative systems, materials, or spaces, the team needs to understand why the client is undertaking the project and how its decisions enable the product to meet these needs. The team must design an organization and process that will effectively deliver the product. POP creates a framework for aligning decisions to the project's ultimate goals and values.

Seeing the Big Picture

POP matrices can also help clients and project teams see the impact of design and construction on the big picture of facility income and cost rather than in isolation (Figure 5.4). When viewing building design and construction simply as a cost to be reduced, owners and teams tend to make short-sighted decisions that often lead to much greater costs during operation and use (the brief examples given in this chapter so far illustrate this point). The income is, of course, shaped by the sustained performance of the particular business products in the market, and the business operations costs come largely from the work of the business organization. The design and construction organization is responsible for the design and construction cost and duration. The operations and maintenance organization shapes the building operations and maintenance costs and influences the longevity of the building as well. In addition to these major responsibilities for income and cost over the life of a building, there are, of course, several additional connections. For example, the quality of building operation will likely impact the productivity of the building users and therefore the business operations costs and the income. The early or at least timely completion of a building also drives business costs, income, and the like. Note that the income and cost should not be measured only in monetary terms, since a building requires more than money to run (e.g., energy) and might produce more than just monetary income (e.g., it could create some or all the energy it needs or clean the water it uses).

Figure 5.4 connects the POP matrices relative to facility income and cost. Design and construction can enable increased income and reductions in business operations, building operations, and facility maintenance costs. Note that, for simplicity and to focus on the delivery and use of a building, the client business and building user POP matrices are combined.

Design makes value possible or not. Construction makes value real or not. Value as determined by the customer must be connected and flow from aspirations to the finished product through many hands. Goals must be made into objectives with performance metrics so predictions can be made so performance can be measured and improved every day of the project, leading to the outcomes the customer expects. A building can be seen as high-performing if all of the four of the criteria types are met and in balance, as illustrated simply in Figure 5.5.

5.3.6 Ownership of Objectives

The only way for objectives to be established and validated so they can be used to guide design and construction decisions is to identify an "owner" for each of them. Each owner must be very knowledgeable about the objective and must have a significant stake in ensuring that the facility meets

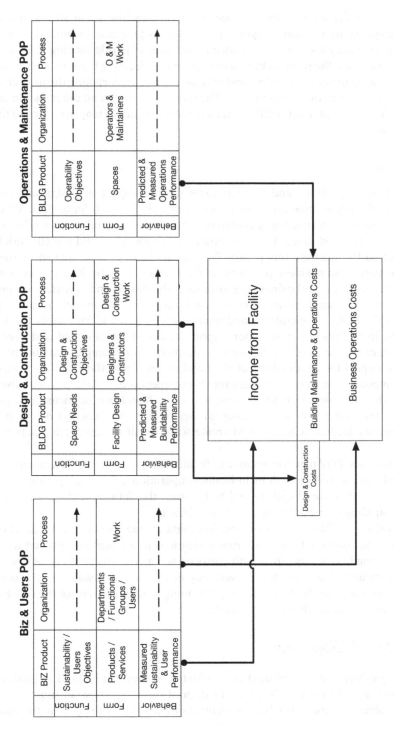

FIGURE 5.4 POP models connected to cost and income.

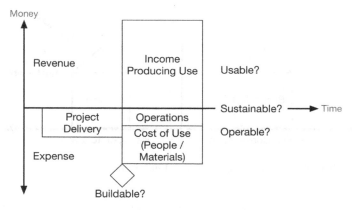

FIGURE 5.5 Delivering a high-performing building.

the particular performance target. Otherwise, they cannot be respected advocates for an objective and cannot be involved in setting, prioritizing, and proactively managing a project with these four types of objectives. Figure 5.4 shows the organizations that typically own each of these four types of objectives.

The client's business organization should be primarily responsible for the sustainability objectives, as the new building should contribute to the prosperity of the client's business in its economic, environmental, and social context. The users of the building (e.g., patients, nurses, and doctors for a hospital, students, professors, and staff for a university building) need to articulate usability objectives. The operators (e.g., facility maintenance staff), should own the operability objectives, and the design and construction team has to bring the buildability perspective to the table.

The Right Advocate

Although an owner may have critical goals for operational efficiency and sustainability, we have seen instances where the owner's project manager—who is likely being reviewed based on meeting a budget or schedule—will not defend the owner's legitimate other goals against a cost or schedule threat. In these instances, the project manager is not the right advocate for the owner's usable, operational, or sustainability objectives.

It is critical that these objectives are considered as early as possible to avoid suboptimization of the building for a subset of these objectives. It is also critical that these objectives are considered in all major decisions, which requires continuity of staffing and definition of the methods for predicting and measuring each objective. Anything short of that—for example, assigning a proxy owner for some of the objectives such as the owner's project manager—will likely dilute the objectives that don't have a formal owner as other stakeholders and team members advocate for design solutions that best meet their needs. A final key point is that the performance of design solutions under consideration needs to be predicted and eventually measured so that the team can guide the project toward the highest possible value and can demonstrate this value through measurements.

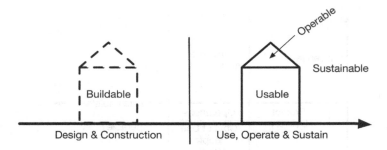

FIGURE 5.6   Realizing buildability, usability, operability, and sustainability.

5.3.7   Measuring the Project Team's Delivery of Value

Many things can be measured while creating and using a high-performance building, shown in Figure 5.6. The buildability of a building can be known only at the time of handing it over to the client because all the measurements of the quality, schedule, cost, and safety performance during the design and construction phases are finally determined. The built facility must, of course, meet the usability, operability, and sustainability criteria established as constraints for the design and construction of the facility because a buildable facility that is not as usable, operable, and sustainable as necessary will have missed the mark. Again, usability refers to what the facility's users do in the facility; sustainability metrics assess how well the facility fits into or even enhances its economic, environmental, and social contexts; and operability refers to the ease of operating the facility to enable its use in its contexts.

Unfortunately, every experienced construction professional we know has numerous stories where usability, operability, or sustainability targets were sacrificed or overlooked to meet buildability objectives, such as design and construction budget and schedule targets. For example, when designing an industrial facility, the project team did not have the budget to consult with the facilities operations staff. As a result, it overlooked operations and maintenance requirements for large pumps installed in the facility's basement. While the pumps could be installed easily, they were very difficult to maintain and even more difficult to replace. All concerns must, of course, be articulated early in the design phase and incorporated into the project and operations and maintenance strategies for a facility.

5.3.8   What You Can Do

If you find yourself on a project that is fragmented and fuzzy in terms of definition of its value, you might say, "Yes, I would like to work on a project that is set up as the book recommends so that everyone's work contributes to the value the building users seek, but my project is not set up like this and I can only control a small part of the picture." It indeed requires sustained management attention to move the project delivery system from a siloed approach to the holistic approach of IPD. So what can you do? The first thing to do is to define the metrics for your work and its impact on the project

TABLE 5.7 Ownership of Performance Targets and Measurements

Owner	Criteria	Metrics	Design		Construction		Use & Operate	
Is	For	Test	Predicted	Validated	Predicted	Validated	Predicted	Measured
Business organization (owner, client)	Sustainability	Building performance	√	√	√	√	√	√
Users	Usability	Function/Form	√	√	√	√	√	√
Building operations	Operability	Systems performance	√	√	√	√	√	√
Design and construction	Buildability	Safety, quality, cost, schedule performance	√	√	√	√ Measured	NA	NA

clearly and align them—as much as possible—with others. Then you need to collect the data for these metrics and actually use what you learn from the metrics in guiding your work. You also need to make these metrics transparent to others, in particular disciplines and professionals who depend on your work and on whom you depend. This is almost fully under your control and creates the foundation for others to make the value of their work visible as well. Before too long, you might find yourself on a project that is well under way toward a more integrated and holistic way of working.

As shown in Table 5.7, different groups within an organization are responsible for providing the performance targets and measurements of the four types of performance objectives and for predicting and measuring usability, buildability, operability, and sustainability in a timely manner.

It is the main responsibility of the business organization to predict the sustainability of the building for its business purpose in its environmental, social, and economic context during design. The predictions can be made by consultants, but the owner needs to organize these predictions and set up a way to validate them. The owner needs to predict the sustainability and validate it as design and construction develop and finally measure it during the use and operations phase. The users and operators need to do the same for their concerns. The design and construction organizations need to predict and validate the buildability and complete the measurement by the end of construction.

5.4 REAL-LIFE EXAMPLES

5.4.1 Sustainability: Intersection of Value and Building Operations

The David and Lucile Packard Foundation established measurable goals and committed to achieve them to demonstrate how to design, build, and operate their new headquarters to be socially, economically, and environmentally sustainable. "343 Second Street," the other name for the Packard

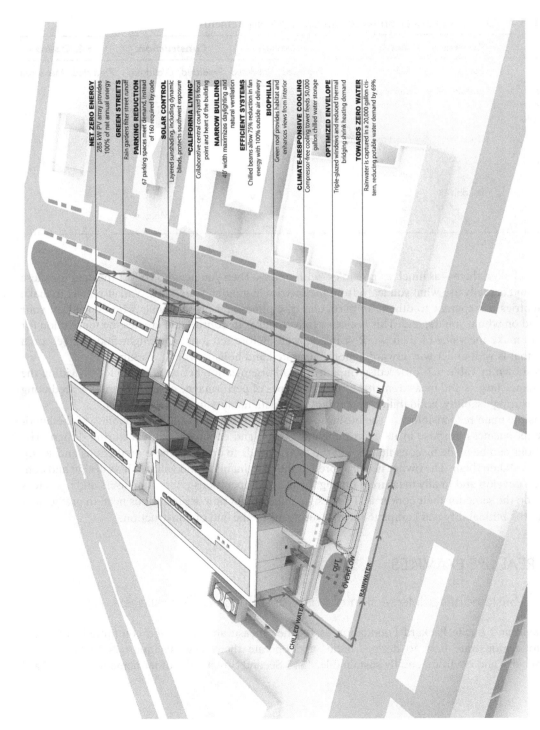

NET ZERO ENERGY
285 kW PV array provides
100% of net annual energy

GREEN STREETS
Rain gardens filter street runoff

PARKING REDUCTION
67 parking spaces meet demand, instead
of 160 required by code

SOLAR CONTROL
Layered sunshading, including dynamic
blinds, protects southwest exposure

"CALIFORNIA LIVING"
Collaborative central courtyard is focal
point and heart of the building

NARROW BUILDING
40' width maximizes daylighting and
natural ventilation

EFFICIENT SYSTEMS
Chilled beams allow 75% reduction in fan
energy with 100% outside air delivery

BIOPHILIA
Green roof provides habitat and
enhances views from interior

CLIMATE-RESPONSIVE COOLING
Compressor-free cooling tower feeds 50,000
gallon chilled water storage

OPTIMIZED ENVELOPE
Triple-glazed windows and reduced thermal
bridging shrink heating demand

TOWARDS ZERO WATER
Rainwater is captured in a 20,000 gallon cis-
tern, reducing potable water demand by 69%

IN

OUT

OVERFLOW

RAINWATER

CHILLED WATER

FIGURE 5.7 David and Lucille Packard Foundation headquarters. EHDD: courtesy of the David & Lucile Packard Foundation.

90

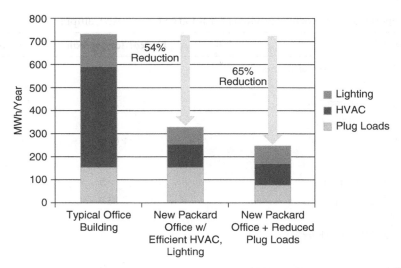

FIGURE 5.8 Energy reduction goals for the David and Lucile Packard Foundation headquarters. Courtesy of Integral Group; courtesy of the David & Lucile Packard Foundation.

Foundation Headquarters, earned LEED® Platinum and Net Zero Energy Building certifications based on data collected in its first year of use, from July 2012 to July 2103. The fact that 97 percent of its inhabitants reported general satisfaction with the building, putting it in the top 5 percent of a national database of building occupant surveys, is impressive considering that everyone has to function differently in their new workplace. This, in turn, was made possible by thoughtful design and the use of technology (Figure 5.7).

The David and Lucille Packard Foundation headquarters, shown in Figure 5.7 demonstrates building operations merged with social, economic and environmental sustainability in the design, construction, and operation.

The first step for the Foundation was to measure how resources were consumed in its workplaces scattered throughout the City of Los Altos, California. The next step was to establish a design strategy based on the Foundation's goals (Figure 5.8) and its time and money constraints. The design team, led by EHDD Architecture and supported strongly by an alliance between Peter Rumsey and the Integral Group, worked with users, developed concepts, and modeled performance. DPR Construction as the builder estimated cost and forecast schedule. Success, from Packard Board funding through design, construction, and operation, had to be predicted, measured, and verified (Knapp, 2013).

Figure 5.8 demonstrates how user value and building operations connect with economic and environmental sustainability. Goals for comfort, indoor air quality, natural light, and energy consumption must be established and modeled, then balanced with construction and life cycle costs. Table 5.8 shows the specific objectives with metrics and targets for building operation the Integral Group created to evaluate the design of a large corporate campus and guide its progress.

TABLE 5.8   Operational Performance Goals for a Large Corporate Campus

Performance Metric	Typical Target Value	Project Target
Energy kBtu/sf/yr (gas and electric) source-based noncompliant T24-2005	250	125
Energy kBtu/sf/yr (gas and electric) site	90	40
CO_2 production (lbs CO_2/sf/ year)	40	10
CO_2 per person (lbs CO_2/person/yr)	1,000	500
Fan system efficiency (W/cfm)	1.5	0.75
Circular pump efficiency (W/gpm)	15	10
Compressor(s) size (sf/ton) (all sf/all compressor tons) (does not include future HPs)	350	1,000
Typical areas cooling (kW/ton)—chiller + tower with dehumidification	1.2	0.3
IT Rooms cooling efficiency (kW/ton—all components	2	0.5
Cooling towers (kW/ton)	0.1	0.01
Boiler efficiency (high/low fire)	75%	95%
Heating water distribution efficiency (W/btu)	5	2
Domestic hot water annual Kbtu/sf/yr (source only 3 to 1)	5	3
Office W/sf connected	1.1	0.9
Office W/sf with daylight	1.1	0.3
Lab W/sf connected	?	1.5
Conference room W/sf connected	?	1
Plug loads (W/sf)	0.7 to 1.5	0.5
Potable water use (gal/yr/sf)—internal uses only	9	5

Courtesy of Integral Group.

5.5   INTERCONNECTIONS

Understanding value for the people who want and need new or renovated buildings has never been easy. It seems that it may have gotten more difficult as the number and diversity of stakeholders, complexity, and cost have increased and project schedules have shrunk. Until very recently, designers and builders could do no more than deliver prototypes as their finished product. That has changed. Now it's possible to simulate and test possible solutions to give those many stakeholders, within and outside of the owner organization, a much better idea of building appearance, functionality, and performance, to truly prototype.

The increase in cheap computing power that makes modeling and simulation possible has occurred almost simultaneously with the realization that Lean, once thought of as a better production system, is essentially a new business system, requiring the creation of a new work culture (Byrne & Womack, 2012). The Lean system and culture supports a disciplined application of process fueled by continuous measurement of performance made visible to those doing the work so they can see problems themselves and make adjustments.

This synergy of predictions based on models and simulations and a culture of performance measurement to expose problems makes integrating project delivery both possible and essential for delivering high-performing buildings that truly deliver the unique value owner stakeholders need.

5.6 REFLECTIONS

Since facility owners and users would rather get a high-performance facility than a low-performance facility, they expect the design and construction professionals they hire to employ a fundamental strategy that makes it likely that the design and construction process delivers a high-performance facility. Reflecting briefly on the definition of a high-performance facility, the expectations of facility owners and users have been changing. They are no longer satisfied with a building for which the whole space program is delivered safely, on time, and on budget. They also expect a building that maximizes their income and that has a strong environmental and social performance without health hazards for its users.

Focusing mostly on scheduling and executing design work discipline by discipline, as is often the case today, is not a strategy that will make it likely that such results are achieved. Because each discipline needs to fend for itself, such a strategy virtually guarantees conflicts between disciplines. These conflicts then require resolution, which puts the attention of the project team members on avoiding really bad outcomes instead of achieving really great outcomes. Rather, the strategy must center on rapid feedback and learning, which includes the rapid generation of 3-D models, 4-D models, and many model-based analyses and simulations to validate predictions and support understanding and decision making.

We see much use of these models by individual disciplines today, but the focus seems to be mostly on documenting design and basic coordination of the geometry. But it is an illusion that this is all it takes to create a great building. We see much less use of these models by the whole team to create the integrated information basis—or single source of truth—and value-focused simulations and predictions that are essential for advancing the project towards the client's objectives with confidence. Seen from the perspective of the project goals and the imperative to integrate the systems that make up the building, the definition of what high-performance means, the strategy of achieving that performance, and the work to achieve it require the engagement of all the key members of the project delivery team. Seen from the perspective of the team members, such a focus motivates and elevates such a full engagement.

With such aspirations, "getting work done" includes how to learn from the other disciplines and how to learn about the combined effect of each discipline's work on the project performance. Without such frequent and rapid learning, the team will learn too late that it veered off course and will have to embark on painful and costly efforts to abandon some disciplines' ideas and to integrate the different building systems as best as possible to bring the project back on course towards its goals.

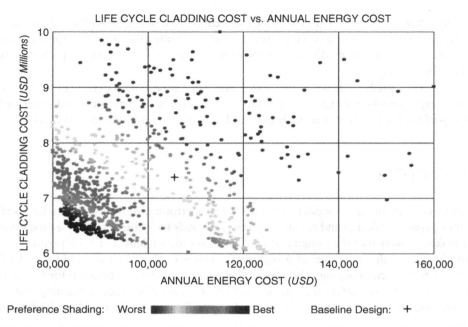

FIGURE 5.9   Pareto plot showing the trade-off between annual energy costs and life cycle cost. Courtesy of Forest Flager and John Basbagill, developed in collaboration with the Beck Group, Dallas, TX.

Figure 5.9 is an example of computationally derived design optimization. Based on work by Forest Flager and John Basbagill, with guidance from Mike Lepech, and support from the Beck Group, this Pareto plot shows the dramatically different performance of different design solutions with respect to life cycle costs and annual energy costs. Each dot represents an automatically generated design solution. Each solution was based on a clear specification of the level of development of the design in the conceptual design stage of the project, the criteria for evaluation, and the methods of analysis. The crosses show the Pareto-optimal designs. The larger black dot is the design solution chosen by the team. The priorities of the client in terms of these two criteria determine which design is superior to other designs. The figure also shows that model-based analysis and simulation are essential to show everyone on the team the many trade-offs that need to be made in a timely and consistent manner. Using this information, the team can select an optimal design, perhaps in conjunction with additional performance criteria.

5.7   SUMMARY

The highly valuable building is a physical solution to owner and stakeholder needs. The first step in achieving a highly valuable building is to deeply understand these needs and translate them into specific goals. The Product, Organization, and Process methodology provides a methodology for identifying the opportunities (levers) the team can use to achieve those goals. Continuous measurement of key metrics—prominently displayed and organized in dashboards—allows project management to assess progress toward creating a highly valuable building and to modify the POP to steer toward a highly valuable outcome.

NOTE

1. Cost and income can be understood in three basic ways: economic or monetary terms; in terms of environmental performance, such as CO_2 emissions; or in terms of human or social impact, such as jobs or quality of life created or affected by a particular facility. For example, as shown by William McDonough, a building might provide a breeding habitat for butterflies to reflect the values of the building owner (Walker, 2011). Optimization of cost and income should consider performance from all of these perspectives.

REFERENCES

Byrne, A., & Womack, J. P. (2012). *The Lean turnaround: How business leaders use Lean principles to create value and transform their company.* New York, NY: McGraw-Hill Professional.

Gero, J. S., Tham, K. W., & Lee, H. S. (1992). Behavior: A link between function and structure in design. In D. C. Brown et al. (Eds.), *Intelligent computer-aided design* (pp. 193–225). Amsterdam, Netherlands: Elsevier.

Kiviniemi, A. (2005). *Product model based requirements management.* PhD thesis, Department of Civil and Environmental Engineering, Stanford University, CA.

Knapp, R. H. (2013). Sustainability in practice building and running 343 Second Street. *David & Lucile Packard Foundation.* Retrieved February 14, 2015, from http://www.packard.org/wp-content/uploads/2013/10/Sustainability-in-Practice-Case-Study.pdf.

McDonough, W. (2011, June 1). *A celebration of abundance: long term goals short term decisions.* Lecture to CEE100 Managing Sustainable Building Projects, Stanford University.

Sullivan, L. H. (1896, March). The tall office building artistically considered. *Lippincott's Magazine, 57,* 403–409.

Walker, A. (2011). Interview with Bill McDonough, Dwell, June 2. Retrieved October 16, 2016, from https://www.dwell.com/article/interview-with-bill-mcdonough-e1d42751.

CHAPTER 6

Integrating the Building's Systems

"Any fool can write a book and most of them are doing it; but it takes brains to build a house."

—Charles F. Lummis

6.1 WHAT ARE INTEGRATED SYSTEMS?

A building can perform at its highest possible level only if all of its systems work together in harmony. *Integration* is not simply putting many systems in the same space; it means designing systems that work in positive synergy; each informs the others, and none work against, or in spite of, each other.

A high-performing building is:

Buildable: Can be built efficiently;
Operable: Helps the team maintaining and operating the building achieve the appropriate level of comfort in an efficient manner;
Usable: Helps the building occupants perform functions they are supposed to perform in the most efficient manner; and
Sustainable: Helps sustain the occupants' business or other purpose for the building and makes a positive contribution to the building's economic, environmental, and social contexts.

Thus, truly integrated systems cannot be designed in isolation. Many systems, such as air conditioning and lighting, are intimately tied to the interior and exterior design of the building. For example, more exterior windows on a building may mean fewer light fixtures, or a different kind of lighting solution is necessary; the type of glazing or the presence of operable windows may change the air conditioning specifications as well. Furthermore, the specialists and tradesmen (usually subcontractors), who have the best knowledge and expertise about the buildability of specific systems and operators who are intimately familiar with operational issues must be involved in the design process of systems to be truly integrated.

This chapter is short because the focus of this book is on integrating project delivery, that is, integrating the organization, process, and information to create high-performing buildings. Hence, we just want to illustrate why integrated systems are the key output of an integrated delivery process, what integrated systems look like, and how they make a building high-performing. There are so many different building systems that together shape the performance of a building that the topic of integrated systems merits its own book.

6.2   WHAT DOES SUCCESS LOOK LIKE?

Why are integrated systems necessary for a high-performance building? Although not all integrated technical systems will lead directly to a high-performance facility, under most circumstances, it is difficult to imagine that a facility with single-purpose systems could perform better than a facility with integrated systems that work in concert with each other. Without integration early in the design phase, systems will inevitably be installed which either do not add value to the building, do not take advantage of potential synergies, or worst of all, work directly in conflict with another system. Crucially, lack of integrated systems ultimately leads to increased cost, and often a loss of productivity—not just during construction but for the entire use phase of the building.

Buildings are used for a variety of purposes, which requires them to respond to the demand of its users in different ways. Systems used in modern facilities have grown exceedingly complex and specialized, such as light-emitting diode (LED) lighting, chilled beams in heating, ventilating, and air-conditioning (HVAC) systems, medical gas systems for hospitals, and a host of exterior skin options. Furthermore, many owners are no longer satisfied with a building that is just delivered on time, on budget, and with no injuries. Owners are increasingly interested in buildings with strong environmental and social performance, no health hazards, ease of operation, and maximization of the business value the users can generate.

Integrating Design through Cost Transfer

In our firm Integral Group, we use the term *cost transfer*. The idea is that you get a better building—a high-performance building or a green building—for the same cost as a normal building. In the simplest example, you take money that would normally be in the mechanical budget and spend it on the performance of the envelope of the building. The owner gets better value because the envelope is going to be longer lasting with lower operational costs, less energy use, less maintenance, and so on, than a mechanical system. And if you understand how to do that well and can take advantage of it, you can get this much higher performing building for the same cost. Figure 6.1 illustrates cost transfer.

A great example of this concept that really influenced me is a project in Seattle the firm Keen Engineering designed many years ago, just before I joined them to run their San Francisco office. The project was a skilled nursing facility for the State of Washington Veterans Administration. They really embraced the idea of cost transfer and said, "we can eliminate all the air conditioning in this building, if we really get the skin of the building right." At the end of the day, after some very complex modeling and simulation work, the answer was yes. As a result, the State of Washington Veteran's

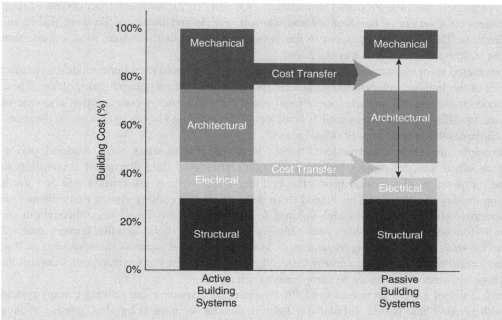

FIGURE 6.1 Cost transfer. Courtesy of Integral Group.

Administration rewrote their administrative code, which said that every space, every patient room had to have air conditioning, to state that patient rooms and other spaces can be naturally cooled if you can prove that it works. They did prove that it worked and that design integration allowed for a much higher-performing building at the same cost through integrated architecture and engineering. They added some thermal mass to the inside of the building, which holds heat and releases it slowly and stabilizes the indoor temperature. They also added more insulation and very aggressive active shading on the outside of the building, that completely controlled the sun on the skin of the building. In order to have a space that requires no cooling system, you have to make sure that it is not getting overheated by the sun—job number one. Those architectural strategies just so happened to cost almost the same as an air-conditioning system for the building, resulting in incredible energy savings. The project won an American Society of Heating, Refrigerating and Air-Conditioning Engineers (ASHRAE) Technology Award, and it was one of the first buildings of its kind to be something like 50 percent better than the Energy Code. It was really a great success. I have been talking to clients about integrated design through "cost transfer" ever since.

John Andary, PE, Principal at Integral Group[1]

6.3   HOW CAN THIS BE DONE?

What does it take to integrate systems effectively? To effectively integrate systems a team must first identify the user values and tie them explicitly to the features that are being considered from a systems perspective for the building. For example, for the Living Laboratory project at Stanford University a team of researchers used the Product-Organization-Process (POP) framework to interview a group of stakeholders on a variety of building use and comfort criteria and these criteria were tied to the building features. This provided the input to the design team to explicitly include those criteria that enhance the building use (Haymaker et al., 2006).

The integrated team of owner, designers, and subcontractors should then build models to predict the behavior of the building relevant to characteristics valuable to the owner and stakeholders. These might include environmental impacts, operational parameters like energy consumption and ease of maintenance, usability considerations such as flexibility of the building's layout to changes in the users' business, constructability concerns, or other factors.

For example, the DPR Construction Phoenix Regional Headquarters team modeled performance for its office remodel project and used CFD (computational fluid dynamics) simulations to predict the temperature distribution inside the building under various use criteria and at various times during the day and the year. This helped them design a passive cooling system that utilizes four direct-evaporative shower towers, 14 high-volume, low-velocity Big Ass Fans® hung throughout the ceiling, an 87-foot-long by 13-foot-high solar chimney, 82 Solatube® lighting units, louvers attached on the exterior walls, and a Building Management System (BMS) that controls these systems so that they respond together to maintain comfort inside the building given the environmental conditions outside. These elements were chosen to complement each other.

Figure 6.2 shows a cross-section of the DPR Phoenix office systems: the parking canopy system covered with photovoltaic panels, tubular day lighting devices for natural lighting, passive cooling tower, and solar chimney for heat exhaust. This conceptual model was developed to demonstrate how the various systems work together.

Figure 6.3 shows a screenshot of the Computational Fluid Dynamics (CFD) simulation performed using a 3-D model of the DPR Phoenix office by KEMA Energy Services, now DNV GL Energy Services, the energy design consultant for the project. This model was used to determine the passive cooling strategy.

Figure 6.4 shows a sketch of the solar chimney that was built. It is metal stud framing with no insulation and clad in zinc. When heated by the intense sun rays of the Southwest, it creates negative pressure in the space, which draws cool outside air through the window louvers into the open office below. Heat is exhausted up and out through louvers in the chimney wall. In that way, the solar chimney acts as the air handler for the passive cooling system.

The plan view of the DPR Phoenix office building in Figure 6.5 illustrates the bioclimatic architectural strategy. The design takes into account the desert climate and environmental conditions, such as no prevailing wind, to help achieve optimal thermal comfort inside. It avoids complete dependence on mechanical systems, which are almost universal in the Southwest. These operate only as support when outside air is too hot to use.

The DPR Phoenix Regional headquarters is a great example of how integrating systems lead to a high-performing building. The systems in the building work together to respond to user needs and environmental conditions. As the user activities generate heat and the building goes into occupied mode, the building management system (BMS) compares the indoor and outdoor temperatures. If the outdoor temperature is cooler than the indoor temperature, the building is put into passive cooling mode. The windows on the north and east façade open, the shower tower dampers open, and the water begins circulating, the Big Ass Fans® are activated, and the louvers in the solar chimney open. All of

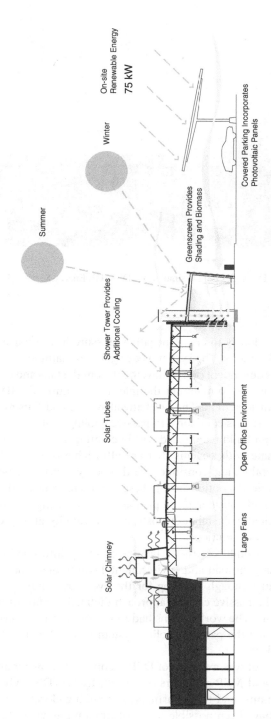

FIGURE 6.2 Cross-section of integrated systems. SmithGroupJJR; courtesy of DPR Construction.

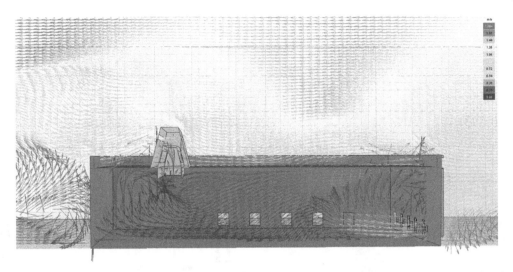

FIGURE 6.3   Computational Fluid Dynamics (CFD) simulation. DNV GL Energy Services USA, In.; courtesy of DPR Construction.

these components work harmoniously to create negative pressure in the space to draw cool outside air across the open office workstations. The system is designed to maintain a max temperature of 84 degrees. Comfort in the space is dependent on air movement from the fans and cross-ventilation. When the passive cooling system is unable to maintain the max temperature, the BMS system transitions cooling to a traditional mechanical system. The mechanical system and fans maintain comfort in the space until the outdoor temperature is conducive to passive cooling again.

The key challenge the design-build team faced was in creating a comfortable (usable) workspace given the very hot desert climate with systems that are affordable (buildable), function well in all foreseeable circumstances (operable) and consume as little energy as possible (sustainable). The team had to think through a range of design options using these criteria to evaluate the performance of each option. For example, the team had considered the pros and cons of using the shower towers versus a mechanical cooling system, evaluating installation costs, maintainability, ability to create a comfortable work environment, and energy consumption.

The industry standard is to build excess or factors of safety into cooling systems to account for worst-case scenarios of occupant load and environmental conditions. The integrated team rallied around the idea of building "just enough" cooling for the everyday conditions that account for the majority of the building life. The passive cooling approach dictated the layout in lieu of the program dictating the mechanical systems. The workstations had to sit between the operable windows/shower towers and the solar chimney for the passive cooling system to be effective. Switching this way of thinking about design was critical.

The team that figured this out was made up of DPR, acting as the owner and general contractor, SmithGroupJJR, the architect and MEP engineers of record, KEMA/DNV GL Energy Services, the energy and mechanical consultants, and trade partners, all working closely from the outset. Without this integration, it would not have been possible to transform a marginal 1972 retail building into a

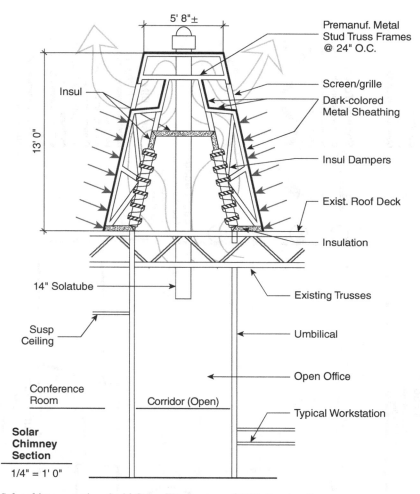

5' 8"±

Premanuf. Metal
Stud Truss Frames
@ 24" O.C.

Insul

Screen/grille

Dark-colored
Metal Sheathing

13' 0"

Insul Dampers

Exist. Roof Deck

Insulation

14" Solatube

Existing Trusses

Susp
Ceiling

Umbilical

Conference
Room

Corridor (Open)

Open Office

Typical Workstation

**Solar
Chimney
Section**

1/4" = 1' 0"

FIGURE 6.4 Solar chimney section. SmithGroupJJR; courtesy of DPR Construction.

certified net zero energy showcase for a cost that could be paid back within 10 years. Life cycle cost, not first cost, was their focus from the beginning.

The criteria to evaluate design options and the expertise needed to address all criteria had to be available simultaneously, not sequentially, in the energy design phase. If any of these systems were designed independently of one another, then each system would not function as well, requiring more expensive solutions to install and operate that in the end may not have created a comfortable working environment. It took the integrated team to identify the specific objectives and targets that are prerequisites for developing the simulation models to test and understand how the systems interact with one another. These findings enabled the team to design the building systems to make a net zero energy building (NZEB) happen in Arizona.

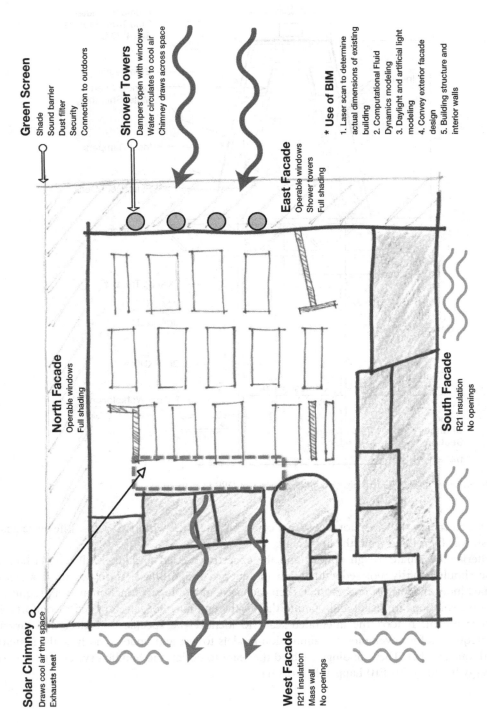

Green Screen
Shade
Sound barrier
Dust filter
Security
Connection to outdoors

Shower Towers
Dampers open with windows
Water circulates to cool air
Chimney draws across space

East Facade
Operable windows
Shower towers
Full shading

* Use of BIM
1. Laser scan to determine actual dimensions of existing building
2. Computational Fluid Dynamics modeling
3. Daylight and artificial light modeling
4. Convey exterior facade design
5. Building structure and interior walls

North Facade
Operable windows
Full shading

Solar Chimney
Draws cool air thru space
Exhausts heat

West Facade
R21 insulation
Mass wall
No openings

South Facade
R21 insulation
No openings

FIGURE 6.5   Plan view of interrelated systems. DNV GL Energy Services USA, In.; courtesy of DPR Construction.

A Practical Approach to Integrated Systems

How do you know if your process and project are integrated? Integrated systems are often the marker of success, but there are important points of integration that can guide the way for design teams. When practicing integrated design, I focus on stages of integration to ensure success throughout the project—each involves different project aspects and varying levels of team member engagement and occurs throughout the project timeline. I was pleased to have been able to introduce and utilize this approach on the DPR San Diego project.

The four guideposts that I will often use include:

1. Integrating the program to performance criteria.
2. Integrating the building to its climate and site.
3. Integrating the team members to each other.
4. Integrating the systems.

This discussion will focus on practical steps of the first guidepost, which occur in predesign work. I find that the most successful projects will include, at a minimum, the following two actions aimed at early integration of systems. The first action is to establish performance metrics for each space in the building and site program.

To accomplish this, the team should work with the owner during predesign to add performance criteria to the space planning tables and adjacency diagrams. This can be done simply by tracking performance-based definitions with the architectural criteria that can be used to inform the engineering basis of design later. Performance criteria may include:

1. Visual comfort: quantity, quality, color, and sources of light.
2. Thermal comfort: define the range of acceptable temperatures and include consideration for the six variables of human thermal comfort.
3. Acoustic comfort: define the activities and understand the desired communication requirements (signal) through the noise.
4. Ventilation options: natural ventilation options, volume of air desired, special filtration, or air quality issues.
5. Energy performance: what uses energy, what generates heat, how the occupants interact with this information.

This simple task connects the overall project goals to the earliest spatial layouts. This approach can improve design quality by aiding the space planner's thinking to introduce natural lighting and ventilation options to the design concepts. Examples of this include integration of day lighting to inform a room's aspect ratio to accommodate natural light from more than one side—which makes for visually pleasing spaces.

The performance criteria tend to not change over the life of the project; this step brings these criteria into the concepts at the earliest possible stage, which makes the integration of systems less costly than at a later point in the project when options are limited and performance is compromised.

Of course, as the design progresses so will the need for the team to verify performance and move from rules of thumb to integrated engineering analysis. If the critical areas of thermal comfort and ventilation performance are determined up-front and tracked throughout, then the later-stage engineering analysis using computational fluid dynamics (CFD) can be more efficient and is less likely to produce surprise results that have major cost and time implications to the project. Using integrated building information modeling (BIM) tools is also very helpful to keep the performance criteria and spatial data in one model that can be assessed for performance at every stage of decision making.

Douglas Kot, AIA, LEED Faculty, Head of Section, Sustainable Buildings and Communities, DNV GL.[2]

6.4   REAL-LIFE EXAMPLES

6.4.1   DPR San Diego Net Zero Energy Building

The San Diego headquarters of general contractor DPR Construction is a converted single story tilt-up office building, built in 1984 and was nearing obsolescence. In 2009, it was completely renovated to achieve LEED Platinum certification and net zero energy performance. From the user's perspective, the goal of the project was to create a more comfortable workspace that would enhance productivity while also reducing the environmental impact. In order to realize those goals, DPR and architect Callison (now Callison/RTKL) had to ensure that the building's systems would take advantage of the natural environment, which in San Diego is mild and dry. It also had to find systems that could be operated and maintained with a reasonable effort and that fit within the design and construction budget and schedule.

Early, during conceptual design, they engaged KEMA, now DNV GL, to develop a passive bioclimatic strategy. This began with a study of climate and operational conditions for the building, as well as creating a definition of comfort for the occupants. The KEMA/DNV GL team developed a CFD model, shown in Figure 6.6, to compare passive strategies, natural ventilation, and thermal comfort. Performance models were developed in parallel to the predictive energy model and the day lighting performance models to ensure system integration.

KEMA/DNV GL worked out a design with architect Callison to use cross-ventilation to naturally ventilate and cool by installing operable windows to the northwest and roof monitors to the south. The drawing in Figure 6.7 shows the third evolution of team thinking.

The bioclimatic strategy encompassed the entire site. This led to reducing the number of parking spaces and replacing the asphalt with landscape. Permeable surface increased, and open space was maximized, which reduced the heat radiating from the building, which had been covered with a black roof and surrounding asphalt and concrete paving. This heat island effect was further reduced by replacing the roof with a high SRI Energy Star product, which contributes to the energy efficiency of the building envelope. Providing both preferred parking for hybrid electric vehicles, and shower facilities for employees has encouraged alternative methods for commuting in an area where most

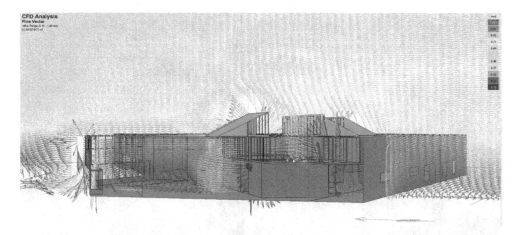

FIGURE 6.6   Computational fluid dynamic (CFD) analysis. CFD analysis.

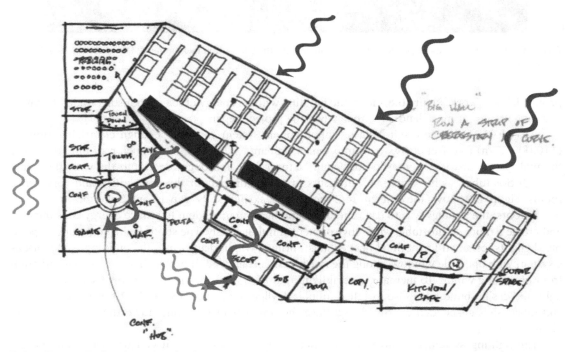

FIGURE 6.7   Bioclimatic strategy for the 1984 tilt-up building. DNV GL Energy Services USA, In.; courtesy of DPR Construction.

FIGURE 6.8   A greener, more permeable site and an energy-efficient rooftop membrane. © DPR Construction, by David Cox.

people drive to work. Figure 6.8 shows the building with its energy-efficient roof surrounded by much less asphalt and far more landscape than when the project began.

A 64 kW-AC Kyocera self-ballasted photovoltaic system generates enough electricity annually (110,000 kWh) to offset energy consumption. A Vaillant 100-gallon, four-panel, roof-mounted solar thermal system provides domestic hot water. Natural light is brought deep into the wide floorplate through Solatubes® with Optiview® diffusers installed throughout the open office and conference rooms. Figure 6.9 shows the efficient roof membrane, skylights, PV panels, and Solatubes®.

Indoors, the strategy was to use ventilation and lighting in tandem to reduce energy consumption and create a more comfortable work environment simultaneously. The south-facing automated operable skylights, oriented for optimal heat exhaust, bring daylight deeper into the building, and work as a system with operable windows along the northwest curtain wall to provide natural cross-ventilation from prevailing winds. To increase natural lighting further, and thus reduce energy used for artificial light, the design-build team also installed tubular daylighting devices (Solatubes®) over workstations and fabric sails suspended overhead to diffuse the bright light reflecting down the tubes, shown in Figure 6.10.

The lighting system is controlled by timers, daylight sensors, or it can be turned on and off manually. The building automation system (BAS) controls both the HVAC system and the operable windows; when criteria for inside air and outside air temperatures are met, the BAS opens the operable windows and turns the HVAC system off. The operable windows are interlocked with the HVAC

FIGURE 6.9   Rooftop equipment to reduce use and produce energy. © DPR Construction, by David Cox.

FIGURE 6.10   Natural lighting and sails with the electric lights off. © DPR Construction.

FIGURE 6.11   BIM for predictive analysis. © DPR Construction.

system, so if an employee chooses to manually open a window, the HVAC system will still automatically turn off.

The team used predictive analysis, including BIM, to determine accurate energy use annually. Solatubes®, ventilation, sails, and ceiling elements were modeled along with the structure, as shown in Figure 6.11. That allowed designers to calculate how many photovoltaic (PV) panels would be necessary to offset annual energy consumption.

The open office culture at DPR and data supported a strategy of greater openness and access to views for all employees. The result was few walls to impede access to perimeter glazing to the north and northwest. Figure 6.12 shows the café area open to the outdoors through glazed roll-up doors.

"Making the best with what we have" guided the design. Over 95 percent of the structural walls, slab, and roof deck were kept intact as well as the majority of the existing curtain wall. Instead of installing new flooring systems, the existing concrete floors were sealed in high-traffic areas. A number of elements were refurbished and reused, such as door openings, made from leftover site materials, casework made from agricultural by-products, and workstations that were relocated from the previous location. Reused materials, such as lumber from pallets and demolition, were incorporated into finish and rough carpentry, creating an integrated systems approach to the material selection as well. Figure 6.13 shows the wine bar in the center of the space, which serves as a central integration point for people.

The BMS manages the mixed-mode ventilation strategy. This ventilation strategy reduces the number of hours the HVAC system runs by 79 percent on an annual basis. Employees control their comfort

FIGURE 6.12 Open office for cross-ventilation and outside connection. © David Hewitt/Anne Garrison Architectural Photography; courtesy of DPR Construction.

with windows, high-volume, low-velocity fans, and thermostats. Energy performance is 42 percent better than prevailing code and as operated uses only 20 percent of the average for commercial office buildings in San Diego.

6.5   INTERCONNECTIONS

Integrating building systems is the critical step in delivering a high-performing building that is buildable, usable, operable, and sustainable. It is tied to integrated organization and integrated process, since it represents the result of the work that an organization does and the processes the organization uses. In order to truly integrate the building systems, the team needs to consider the perspectives of the various experts involved in the project delivery process, and do so in a timely manner. Integrating systems requires the team to develop predictive models by integrating information about the product and use, operation, and building processes and developing visualizations and simulations as illustrated in this chapter.

FIGURE 6.13   Material reuse lowers overall project embodied energy content. © David Hewitt/Anne Garrison Architectural Photography; courtesy of DPR Construction.

6.6   REFLECTIONS

Integrating building systems so they respond efficiently to the user needs in the context of the economic, natural, and social environment is a key to developing a high-performing building. The key to integrating building systems is to understand the user values, map the user values to building features, get an integrated team to build integrated simulation models to predict how the building systems interact with each other to respond to the users' needs, and then choose the optimal solution, which in reality will respond to the user needs to make the building efficient and usable.

6.7   SUMMARY

Conventional buildings are designed as a series of discrete systems that are combined into a building. Whether these systems will interact positively or negatively is left somewhat to chance because the designers and fabricators are focused on their specific disciplines, not the whole building. Because the systems may actually be resisting each other, the resultant building may even be less than the sum of the systems. When building systems are integrated, however, the entire building, all of its component systems, and its relationship to the environment are considered together. As the DPR Phoenix and San Diego offices demonstrate, and John Andary and Douglas Kot explain, considering the systems together and how they could complement each other creates opportunities for exceptional performance. Such buildings, with integrated systems, are greater than the sum of their parts.

NOTES

1. John Andary, PE, LEED AP, Integral Group Principal, Bioclimatic Design Leader, is a mechanical engineer and principal at Integral Group, where he brings over 30 years of energy-focused consulting experience to the firm. John's work at Integral focuses on passive and climate-based architectural and engineering design solutions to improve occupant health, thermal comfort, and energy efficiency in the built environment. Mr. Andary believes that sustainable design is an engineer's social responsibility and has served as principal-in-charge on numerous LEED Platinum and net zero energy projects. Prior to joining Integral Group, John led the MEP Engineering and Energy Consulting teams for the Research Support Facility project at the National Renewable Energy Laboratory (NREL). At 360,000 square feet, this LEED Platinum Certified facility is considered to be the largest measured and verified net zero energy building in the world. Dedicated to widespread adoption of net zero energy, John has recently been working with developers and contractors to implement net zero in the commercial building sector.

2. Douglas Kot, AIA, LEED Faculty, DNV GL–Energy, is Head of Section for DNV GL's Sustainable Buildings and Communities team, where he leads a team of sustainability experts and engineers. Doug has been involved with green building and sustainable planning since 1997; he has led the technical development of dozens of high-performance and zero net energy buildings advising on energy systems, water efficiency and reuse, material impacts and healthy interiors. Doug has worked in all phases and scales of project development—from writing general plans to building detailing—and from concept design through post-occupancy evaluation. Doug also teaches extensively on building energy use, ecological urban design, and sustainable building technologies. He is U.S. Green Building Council LEED Faculty and LEED Accredited Professional with specialty in Neighborhood Development, Buildings Design & Construction, Homes, and Existing Buildings Operations & Maintenance. Doug has three professional degrees: a bachelor of architecture from the Pennsylvania State University, and master of landscape architecture and master of city planning, both from UC Berkeley.

REFERENCE

Haymaker, J., Ayaz, E., Fischer, M., Kam, C., Kunz, J., Ramsey, M., ... Toledo, M. (2006, July). Managing and communicating information on the Stanford Living Laboratory Feasibility Study. *ITcon, 11,* 607–626.

CHAPTER 7

Integrating Process Knowledge

"The whole is greater than the sum of its parts."

—Aristotle

7.1 WHAT IS INTEGRATING PROCESS KNOWLEDGE?

In defining the high-performance building, we have proposed that a project should be buildable, operable, usable, and sustainable. This necessarily requires knowledge of how a building will be used and maintained after construction is complete. To be buildable, it also requires knowledge of how it is most effectively constructed before it is designed. But how can a team integrate this downstream knowledge to satisfy the owner's upstream values? The answer lies in integrating such process knowledge in an integrated organization supported by integrated information.

7.2 WHAT DOES SUCCESS LOOK LIKE?

Users love their new space because of the way they feel in it. Everything they need for their work is at hand. They can collaborate and they can get away when they really need to be by themselves. It's open and light and feels like their home away from home. They can dream of things that are not, but could be, and think of ways to make them so. The group managers can't believe how good their spaces are and how easily they can change things around.

Operators are thrilled with how easy it is to service equipment and reconfigure it when new needs arise. They tell everyone who will listen: "They actually listened to us!" Best of all is that they aren't getting any of the usual complaints from users about temperature, airflow, glare, lack of daylight, and all the rest.

The owner's project manager agrees and tells her peers that she knew when the project would be complete and that the IPD team would deliver it early. She was confident because she was involved in every major decision, knew their schedule impacts, and was aware that the team had maximized off-site prefabrication and use of modular assemblies.

Company leaders realize how much better their and other workplaces could be as they walk through the new buildings. Each whispers to their lieutenant, "Make sure we're in the next building like this." The chief financial officer is really happy about getting a great building without exceeding the approved budget. The CEO says, "Yes, this is what I wanted." The Director of Marketing says, "They got it! This is us."

7.3   HOW CAN THIS BE DONE?

High-performance buildings aren't possible unless all of the systems are integrated. We can't just design each aspect of a building by itself; we must design the systems together, all at once. Take lighting, for example. If you're going to design lighting, you must consider the factors affecting daylight such as building width and floor-to-floor height. Will there be shear walls and/or cross-braces? What is the exterior aesthetic, and how will that affect the amount of glazing? Is the interior layout open, private offices, or a combination? Will there be labs? How wide will the corridors be? How large will the services core be? Each factor affects the next. If the people responsible for each aspect are working in silos, it would be foolish to assume that the systems they design will somehow make the best whole building possible.

The output of the design phase must be the design of a facility that is valuable for its users, can be built, and can be operated. It follows that there are five main process integration needs, listed below and shown in Figure 7.1.

1. User value is translated into design solutions.
2. Design informs and enriches user value and is checked against user value to ensure that user value does not get compromised as the design progresses.
3. Builder's knowledge informs and shapes design.
4. Operator's knowledge informs and shapes design.
5. Sustainability concerns and knowledge informs and shapes design.

A valuable high-performing building requires iterative interaction among client stakeholders and the building's designers and constructors. As described below, this occurs from inception, through detail design, and into completion of construction documents.

1. There must be interplay between the value definition and design processes because users won't be able to fully articulate what they want until they see some possible designs and what they cost and offer.

2. As the design gets detailed, it must be validated against all user values over and over again and the value or the design adjusted when they get out of sync.

3. The building perspective must be brought into the design process because the creation and documentation of a buildable design in the construction documents is a key output of the design process. This cannot happen without the input of those who know how to build.

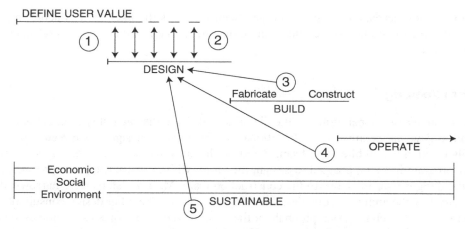

FIGURE 7.1 The five main types of process integration to achieve high-performing facilities (Fischer, Khanzode, Reed, & Ashcraft, 2012). Illustration provided to the authors by DPR Construction, Lyzz Schwegler.

A key aspect in this integration is the off-site fabrication strategy the project wants to pursue. Off-site fabrication allows the parallel production of the physical components and systems, which is much faster than the sequential construction of these components and systems on site. But prefabrication must realistically consider the owner's ability to make key design decisions in a timely manner and without reconsideration. Since the off-site fabrication strategies that really make a difference in schedule duration are multidisciplinary and must be determined early in the design phase, the off-site fabrication strategy must be developed as early as possible.

4. Bringing operational knowledge, that is, how to run the facility during the use phase, to the design phase is also critical because creating a design that can be operated is, of course, also a key task for the integrated design team. Because few designers and contractors have ever operated a facility, this operational knowledge must almost always be obtained from the building's operators.

5. The integrated project team must consider knowledge about the sustainability of a building through its entire life cycle within its economic, social, and environmental context (Elkington, 1998). Even if a building were "perfect" for its users and easily buildable and operable, it needs to have a positive triple bottom line (social, environmental, and economical) to make it sustainable. Given the breadth of knowledge needed, it is unlikely that the team will possess all the sustainability knowledge necessary. Additional consultants may be brought in to advise on social issues like wages and employment practices at the locations where the building parts will be produced or the specific environmental conditions at a site. Trade contractors often have good insights into market, social, and environmental conditions in their fields.

The MaineGeneral Regional Hospital project exemplifies combining economic, social, and sustainability goals. It was delivered 10 months early, under budget and with added value, achieved LEED Gold instead of the anticipated LEED Silver and kept over 90 percent of trade contractor dollars concentrated in the Kennebec Valley to support MaineGeneral's community.

These process integration needs focus on design because that's when a facility really takes shape in response to ideas, needs, and wishes of the facility owner and users. With a clear strategy for addressing the five integration needs, an integrated project team increases the likelihood of achieving a design that's valuable for the users, can be built, can be operated, and is sustainable economically, socially, and environmentally.

Problem Seeking

Collaboration versus cooperation—in the digital age we have the capability to collaborate in real time and in new and creative ways. Traditional collaboration in design was directed and filtered primarily through the architectural team. Based on this information, the architect would create a building program in a silo, proceed to design, then meet with the owner to get approval, and then throw the designs over the fence to the construction team. Most decisions were narrowly defined and managed by the architect, sometimes working closely with the other design consultants. Little consideration was given to the supply chain or the means and methods of construction used to build a design. These were kept separate because the architect and engineers were by contract, excluded from working with the actual way a project was bought and built. The only input from the supply chain came after a project was bid. This was cooperation versus real collaboration.

Over the years, design professionals have implemented many different processes and approaches to programming prior to designing. Due to the "artistic" nature of design culture, many of these processes have been left to the personal approach of the individual designer. More often than not, the designer is not informed by research and knowledge but by common sense and personal experiences and professional judgment. This approach has been inconsistent and often requires rework deep in the design development process or, even worse, in the construction phase of a project. As the designs progress, the cost to change becomes increasingly more expensive.

Now, with the advent of new technologies like building information modeling (BIM), cloud, mobile, etc. and new management approaches—Lean, agile and integrated project delivery (IPD)—we are realizing the opportunity and the need to expand input and use data and information to make the best possible decisions. This has come to be known as evidence-based design (EBD).

What precedents do we have to guide the EBD process? I have for years used a predesign programming approach that was developed by Willie Peña, a leader of Caudill, Rowlett and Scott Architects (CRS). Peña described his rigorous methodology for addressing the key determinants of a building's design in the book *Problem Seeking*, which he published in 1977. These are:

1. Establish goals.
2. Collect and analyze facts.
3. Uncover and test concepts.
4. Determine needs.
5. State the problem.

The design team gathers these through a combination of workshops, interviews, and research. The intent of this approach is to create transparency for the project's designers, users, and other key stakeholders to participate in confirming the goals and objectives of a project.

This approach has proven itself through 50 years of experience on real projects by architects and building program specialists. The details of the approach have evolved over time and have been so popular that Peña's book is now in its fifth edition (Peña & Parshall, 2012).

Initially, problem seeking was used by CRS to design schools. Clients liked the way key stakeholders were engaged and thought this contributed to the success of their projects. The approach accounted for the wide range of factors that designers must consider to create a "good" building design. As the first sentence of the book states: "Good buildings don't just happen. They are planned to look good and perform well, and come about when good architects and good clients join in thoughtful cooperative effort."

The problem-seeking methodology begins with problem definition. Peña believed that there are four basic categories of information that determine design: Function, Form, Economy, and Time. Table 7.1 shows his schema. All four of the design determinants interact together. The project team must optimize the whole before starting to solve any of the parts.

TABLE 7.1 Four Categories of Information That Determine Design

Function	1. People
	2. Activities
	3. Relationships
Form	4. Size
	5. Environment
	6. Quality
Economy	7. Initial budget
	8. Operating cost
	9. Life cycle cost
Time	10. Past
	11. Present
	12. Future

Courtesy of Bruce Cousins.

Peña recommended that these key design determinants be defined simultaneously and that the majority of programming be done prior to engaging design or problem solving. Peña made a strong argument that there should be "a distinct separation between programming and design."

Problem seeking is well suited for IPD. All key stakeholders participate in an IPD process and can contribute to establish a solid building program. An IPD team collaborating on defining the problem significantly increases the possibilities for getting target value design right, the way Haahtela does in Scandinavia, where they create the project budget and "BIM prior design" based on the building program, before the formal start of design. This way, project teams escape the trap of estimating the cost of designs after the design is done.

Small graphic design studies investigating specific issues and systems, especially possibilities for modular construction and extensive prefabrication off and on site should also be done during programing. Social and environmental factors that are often neglected in the rush to design can also be explored and clarified.

We could use "integrated" along with problem seeking to make clear that this is what integrated design and construction teams can and should do. Integrated problem seeking can establish the boundaries and therefore the possibilities for the design phase to follow. The project team, particularly the architect and consulting engineers, would have a solid foundation. Many more alternatives can be explored, and design professionals would have the time they need to iterate positively rather than negatively, as they often do now when the contractors or cost consultants catch up enough with their designs to tell them that they cost too much.

Willie Peña's ideas are every bit as relevant today as they were when he wrote *Problem Seeking*. In fact, the advent of IPD, BIM, and particularly target value design, make them even more so. Every project should begin with integrated problem seeking.

Bruce Cousins, AIA, Principal, SWORD Integrated Building Solutions.[1]

7.4   REAL-LIFE EXAMPLES

A technology company chose to use IPD to realize their vision of creating a workplace to inspire innovation. Given the wide range of knowledge required to design such a workplace, they felt that organizing the project as an IPD project offered the best chance for innovation. They articulated goals and values for enhancing rather than displacing the environment, creating spaces to support their culture of creativity and experimentation, building a workplace free of toxic materials that support healthy living, and realizing a flexible workplace that could evolve for future needs and preferences. The client trusted the IPD team to follow a set of strategies they articulated for each goal in arriving at solutions. They challenged the IPD team to meet these goals within a target cost and delivery schedule. This approach illustrates the advantage of IPD because it allows the project team to design the whole project, in other words, design the best organization and process to create a high-performing building that can be achieved with the owner's resources. Figure 7.2 is an example of an owner's expressed goals.

For an integrated process to unfold, the team leaders need to frankly assess the project's challenges. For example, on a recent project, the IPD team leaders, including the owner's representatives, realized that they faced several big challenges, such as:

1. Understanding the business drivers and stakeholder values underlying the expressed economic, environmental, and social goals of the project.
2. Translating the goals and strategies into tangible objectives with specific performance targets to make these goals measurable for building function, form, and behavior.
3. Describing what decisions had to be made and how they would impact construction.
4. Thoroughly evaluating many more alternatives in far less time than they ever had before.
5. Making grounded predictions of performance compared to desired outcomes.
6. Presenting what was learned so that stakeholders could understand and communicate to others in the client's organization so that decisions could be made.

Creating a Positive Sense of COMMUNITY	Reinforce the Focus on WELLNESS
Create an end-user experience that focuses on *connectivity, employee engagement* and the objective of being *"A Great Place to Work."*	A building to support the *Five Pillars of Well-being*: the physical, social, environmental, physiological, and financial.
Creating Positive Outcomes through INNOVATION	**Positive Environmental Impacts through SUSTAINABILITY**
Utilize *cutting-edge technologies* to support industry-leading *creativity and forward-thinking* innovation.	Support *positive stewardship* focusing on the environment, water, energy, efficiency, and renewables.

FIGURE 7.2 Owner's project mission and goals.

One of the most challenging tasks a team faces is translating the owner's values into measureable and tangible objectives. These objectives guide the design of the building, help the IPD team allocate its resources along the way, and maintain the focus of the project team throughout the project. Tangible and measureable objectives are an extremely important first step in achieving the owner's goals. Note that developing tangible goals is difficult to do even with an integrated team; with a fragmented team that sequentially addresses building performance issues, it is practically impossible.

In this project example, many of the owner's goals had not been codified as objectives, nor were they already modeled in simulation software. The IPD team combined professional expertise with computer simulations and visualizations to support the rapid design iterations that were essential to achieve the level of innovation sought by this owner. In the chapter on simulation and visualization, techniques and methods to predict and communicate expected project performance are discussed further.

7.4.1 Organization, Communication, and New Practices

As will be suggested in Chapter 8 on integrated organization, the technology company's managers and their representatives embraced their responsibilities in the IPD project and senior management teams. They asked their in-house experts in healthy materials, sustainability, and building operations to engage with the design professionals and builders working on the site; sustainability; structure; envelope; mechanical, electrical, and plumbing (MEP); and interiors teams. All of the teams developed designs concurrently and regularly coordinated their work in design coordination and target value. Contrast the concurrent development of designs with the traditional sequential development. To make the concurrent development of the building systems that work together (i.e., that are integrated) possible, the IPD team must develop a strategy to integrate the work of the design teams working concurrently so that one design team does not get too far ahead of the other team. For example, it wouldn't make sense for the MEP team to detail its work while the envelope team is still deciding on the best fenestration and daylighting strategy. The leaders of the integrated process must balance the direct work of each design team with integration work, and everyone on the IPD team must flag misalignments as soon as they see them.

7.4.2   Process Knowledge Integration for the Campus

Integration 1: User Values and Design

The first challenge was to understand and articulate what the goals and related value statements meant for design. What would enhancing rather than displacing the environment look like? Given the site, what should be constructed where? What should remain untouched on the site? What kind of spaces would enable innovation? What should be constructed to increase connections to healthy living, working, and creating? What kind of building elements and systems would be flexible enough to allow for undetermined future needs and preferences?

The only way to answer these questions was for design professionals, builders, operators, and client representatives to define and describe in sketches, models, and words what they thought the future could be for each goal using the value statements as a guide. The team both benefited from and contributed to a parallel effort within the client organization to more precisely identify objectives and metrics in an operational requirements document to guide future decisions about building systems, healthy materials, and so on. Meeting these material and performance requirements was only the starting point. The client expected that the team would be limited only by its imagination and would not be satisfied with a "good enough" design, but would be seeking the best possible alternative by evaluating every possible alternative and recommending which were "good," "better," and "best."

Integration 2: Design and User Value

The architect was at the center of design, taking the lead in proposing alternatives for the form and use of both the outdoor and indoor spaces. They worked closely with the landscape architect, other consultants, and engineering firms, and operated as the clearinghouse for design information and decisions.

Scores of studies were done to decide on the massing and location of the buildings on the site. Figure 7.3, from another project, shows the extent and quality of this work. The architectural team collaborated closely with the site, structure, and MEP teams; cost estimators; and construction planners to forecast the construction cost and duration for each design alternative. This work took time, which pushed the date for the team to present their "Validation Study" consisting of a basis of design and target cost for the project.

The structure team, composed of the consulting engineer, steel fabricator, and concrete contractor, all IPD partners, evaluated 16 possible systems using a method called Choosing by Advantages (CBA). This particular decision-making method requires users to describe important factors in choosing among alternatives and then focus their attention on the differences among alternatives that were relevant to the owner's goals and values. Common practices such as weighting or ranking of factors, and listing pluses and minuses of each alternative are not allowed. Users work through a process that enables them to choose based on the importance of the advantages they see, allowing for both quantitative data and qualitative assessments. The CBA process is transparent, allowing those reviewing the decision to understand the reasoning behind the decision or recommendation (Suhr, 1999).

FIGURE 7.3 Digital bird's-eye view of a new corporate campus. Courtesy of Lend Lease and Perkins Eastman.

Figure 7.4 summarizes the structural team's comparison of three structural system alternatives comparing their CBA scores against the project design goals, which were used as factor categories, with each one having several elements. Ultimately, the steel short span was chosen.

The MEP team included professional engineers, from the MEP consultant, and engineers, operations managers, and BIM specialists from the mechanical and plumbing, fire protection, and electrical contractors. The client's data/telecommunications consultant worked directly with the electrical engineer and contractor. The MEP consulting engineer and MEP team also used the CBA method to evaluate 11 different HVAC systems, some of which required a central utility plant (CUP). The consulting engineer and contractors evaluated six variations of two primary CUP designs. They ultimately presented the client with five alternatives supported by CBA scores, BIM, energy modeling, construction cost estimates, and payback analysis. Figure 7.5 is an example of payback analysis for this type and scale of project. Following the decision protocol, the three top choices were characterized as "good," "better," or "best." In the end, the client chose the combination of "cooling heating water plant" and "chilled ceiling or radiant slab."

The interiors team, composed of architects, a GC project manager and estimator, and a drywall project manager, was the last team to assemble. They had to come up to speed quickly and set about translating goals and strategies for environment, culture, health, and flexibility into forms that would inspire creativity and innovation. The team's language was visual and they became adept at

Summary/Recommendation

	Steel Short Span	Steel Diagrid	Precast Concrete Short Span
Environment	Best	Better	Good
Culture	Good	Better	Best
Health	Best	Best	Best
Flexibility	Best	Best	Good
Project Implementation	Best	Better	Good
CBA Score	955	910	785

TEAM RECOMMENDATION

FIGURE 7.4 Structural CBA matrix. Courtesy of KPFF Consulting Engineers.

Plant & Distribution Options	TOTAL COST OF OWNERSHIP (50 yr NPV)	IRR (50 yr)	DISCOUNTED PAYBACK (includes maintenance, repair, depreciation, replacement, taxes and escalation)
GOOD (VRF w/DOAS)	$142,740,000	Base Case	Base Case
BETTER (CW Loop w/Chilled Beams)	$112,050,000	17%	10.0 years
BEST (CHW Plant w/Chilled Ceiling)	$104,130,000	20%	8.4 years
BEST ALTERNATIVE (CHW Plant w/Chilled Slab)	$85,850,000	45%	5.7 years
	$73,800,000	N/A	Less Expensive than VRF

Negative TCO indicates cost not income.

FIGURE 7.5 MEP systems payback analysis. Courtesy of Integral Group.

FIGURE 7.6   Interior space: light and open. Courtesy of Lend Lease and Perkins Eastman.

proposing concepts and responding to client feedback. In collaboration with client representatives, the interiors team identified metrics for natural light, space within sight of the outdoors, and travel distances between offices, amenities, parking, and so on, and used these as guides. Figure 7.6, also from the digital rendering of the corporate campus, is an example of this work done very well.

Integration 3. Build and Design

Builders took the lead in advancing the project from "Design Intent" to "Virtual Design Construct" in which they would move from routing systems in BIM to detailed models for fabrication. They were also immersed in screening for healthy materials and planning for sustainable, safe, and high-quality construction practices. The builders analyzed schedule and cost impacts for every major design alternative. To make this possible, each team had design and construction professionals working together in multidisciplinary, cross-functional teams. In addition, a majority of these professionals were co-located in an open office on the project site and could ask and answer questions quickly. This enhanced communication and feedback accelerated decision cycles. An example of this collaboration is illustrated in Figure 7.7 showing BIM transitioning to reality for the ConXtech modular steel frame system that enables shorter cycle times for shop drawings, fabrication, and erection.

FIGURE 7.7   From BIM to reality using the ConXtech modular system. © Yong Chan Kim, ConXtech, Inc.

Integration 4: Operations and Design

Client user representatives were embedded in the IPD team and building operations managers were made readily available. They spent many hours working with the architecture, site, sustainability, MEP, envelope, and interiors teams working on designs they believed would be operable, maintainable, and sustainable while meeting the performance targets spelled out in the project requirements. Nowhere was this more evident than in the meetings to decide on five-year performance guarantees the IPD team would give to the client. That is when questions about use and system performance came squarely together.

Integration 5: Sustainability and Design

The client challenged the IPD team to ensure a minimal footprint of the project on the environment, to work with and enhance if possible its physical and social surroundings; ensure high air quality and access to light and views for the building occupants; create a design that offered flexibility of reconfiguring the buildings to maximize life cycle value; and stimulate environmentally sustainable behavior, such as transportation, for the future building users. The project created a "Green Team" that crossed over all boundaries and provided sustainability support to all.

7.4.3   Process Knowledge Integration for Building Systems

Shortly after the other teams were organized, the modular team was established to develop modular solutions within the context of the overall design. Like the others, it was multidisciplinary and cross-functional, drawing on designers and managers from the general, mechanical, and electrical contractors, architect, structural and MEP engineers, steel fabricator, and modular design-builder.

Rack Integration 1: User Values and Design

The modular team's purpose was to explore how modular solutions could support the client's vision and goals for the project while reducing construction and operating costs and scheduled delivery of the project. The designs already chosen or under development were necessarily the starting point. In determining whether to modularize, the modular team had to demonstrate that the modular solution met the client's conditions of satisfaction:

- Be code compliant, including separation of the electrical and data systems.
- Be cost neutral. Pay for the rack structure from savings compared to conventional fabrication and assembly.
- Contribute to the architectural aesthetic.
- Not reduce daylight and visual connection to nature.
- Control reflection, that is, not contribute to glare from reflected light.
- Size utilities to provide flexibility for future unknown business purposes or other changes that require building reconfiguration.
- Provide space for additional services and system reconfiguration to support changes in the type, size, and location of the client's own work teams.
- Locate service connections in nonrated wall spaces to make them as easy to access, operate, and maintain as possible.
- Create typical and repeatable modules.
- Put in place a standard procedure for assembly, inspection and testing, handling, transportation, rigging, and installation prior to commencement of any of these activities.
- Pre-test to performance specifications prior to installation.

The imperative to work with the designs being developed by the other teams made the modular team dependent on them. This was certainly true for a system to carry utilities horizontally through each level of the several buildings. The capacities, materials, location, size, and width of what became known as the "modular horizontal common rack" (MHCR) would be determined in large part by the designs for architectural and MEP systems. Experience with modular design on previous projects had taught design and construction managers that the relationship was in reality an interdependent one. Modular elements influenced architectural elements and systems for an entire building. Core elements such as elevators and stairs, restrooms, and the lobby on each level shouldn't be located and sized without regard for something as large as a common utility rack. There would need to be give and take, which meant that modular design couldn't wait.

Over time, the modular team developed a style that allowed them to both provoke and respond quickly to decisions made within the other cross-functional teams. The team met regularly and spoke whenever they needed input within the co-location space. Team members made and followed up on commitments in their meetings, during their conversations, and through e-mail. They pull-planned their work as best they could, in both long-term and short intervals in spite of the fact that the modular team was not in full control of its own destiny. They needed answers from other teams who were not always able to provide them. They knew the systems that needed to go on the rack, they knew the MEP system's required capacities, they knew the dimensions of the structural bays, but they depended on other teams to determine the placement (e.g., in the corridor or in the offices) and could therefore not address some of the performance objectives, such as the rack's impact on available daylight.

This example illustrates the opportunities and challenges of creating an integrated system through an integrated process and organization. On one hand, the modular rack integrated a number of building systems, offered a better aesthetic, and made fabrication, installation, testing, and maintenance easier. On the other hand, locating a large building element like a rack within the building was not easy. It illustrates the importance of setting up a strategy for building systems integration considering fabrication, construction, and operations and maintenance concerns early in design and to design a process that integrates the work of the various design teams, recognizes the interdependence between these systems at different levels of detail, and advances the design of the individual systems without letting the design of one system getting too far ahead of other systems.

Modular team members realized early that they had to help others visualize what they were thinking. This led them to request help from the mechanical contractor's BIM specialist and from one of the steel fabricators participating in the IPD team. Both of them were also experienced builders. The mechanical contractor's design engineer and modeler became the MHCR designers because their firm was providing most of the utilities on the rack. Altogether, these were supply and return air along with heating, cooling, and domestic water, power and data/communications cable trays. They and other team members were able to go from sketching a design or modification in response to feedback to a virtual model very quickly. As a result, project participants and client stakeholders were able to see and understand possibilities quickly. Figure 7.8 shows a concept similar to the one the team developed.

Rack Integration 2: Design and User Value

The modular team identified several assessments and metrics they needed to make or calculate and report to other project participants and to client stakeholders, especially building operators. They were:

1. Systems performance was a "deal-breaking" set of metrics. The mechanical engineering team worked a few feet away from the mechanical contractor team and was always available to check

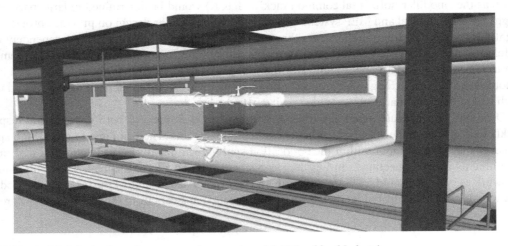

FIGURE 7.8   Modular horizontal common rack concept model. © Southland Industries.

engineering concepts and calculations. This was necessary because so many aspects of the overall design were in play for so long, which made it difficult for the engineers to say, "We're done and we've got it right."

2. Weight was important for the design of the rack itself and was required for the structural engineer's gravity and lateral load calculations.

3. Geometry: overall height, width, and length of the modules; dimensions of each component and distances between them were critical. One of the first decisions the team made was that the modules could not be longer than 32 feet to avoid the additional cost and time required for trucking oversized loads on crowded urban roadways. But other decisions could not be made so quickly. Team members knew that the modules would most likely run in a central corridor, but this was unconfirmed for a long period because the floor plans had not been finalized. The size and location of the modules, both approximate for a long period, were critically important to achieve the client's goal of connecting users with the outdoors. Views and especially natural light could not be blocked.

4. The architectural aesthetic was very important. The rack should contribute to, not detract from, the openness so important in the client's culture. This consideration led the team to an early and significant decision. They decided to leverage the talent and resources available in the metropolitan area to design and fabricate the rack system rather than assembling ready-made components from a manufacturer's catalog. Recognizing the importance of aesthetics, the team began to speak of the common rack as a piece of "mechanical art." Not only would it be functional and perform well, its form had to look like it belonged in an open office environment. It had to look good.

5. The cost and schedule impact of fabricating and assembling off site were also critically important factors. Neither could be greater than conventional installations. In fact, the client expected that there would be time and cost savings, higher quality, and improved safety because so many hours of on-site labor would be shifted to a cleaner, safer, and more productive environment off site.

The team settled on a simple process, shown in Figure 7.9, which they repeated for new iterations and modifications to designs they had previously created. They would start with HVAC, power, and data/communications loads, then calculate system capacities. This enabled them to check structural calculations and building service requirements. At this point, the teams could design or adjust the rack frame before proceeding to fit or adjust the services on the frame.

BIM was the mechanism for rapid prototyping and feedback for planning fabrication, assembly, and installation. The mechanical contractor's BIM specialist followed a process like the one shown in Figure 7.10 to model the many different configurations very quickly so they could be understood and evaluated. The structural engineers could then review and informally approve the new or revised design. At that point, the estimators could revise their budget and check with the project superintendents and scheduler to make sure they were within the allotted installation window.

After creating the first series of designs to support the solutions developed for the building shell and cores, MEP and interiors, the modular team could respond to new inputs from the other design teams and client within a few days because they could get answers very quickly. All of the considerable expertise required to design something this new, to deal with trade-offs and the evolving design of the campus, was within the co-location space or close by. Phone calls or emails were rarely used, and decision latency within the modular team was very short. While the modular team was nimble, it was still limited by the progress being made by other design teams.

For quite some time, the common rack was seen just as another study with similar chances of being adopted into the project as other proposals. The modular team's focus was proving that modular

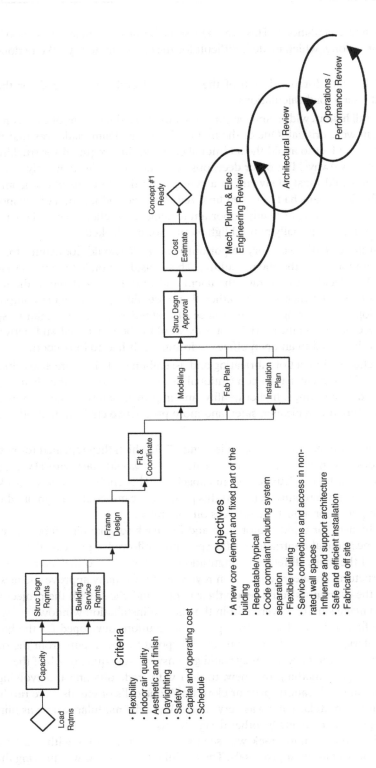

FIGURE 7.9 Common rack design process.

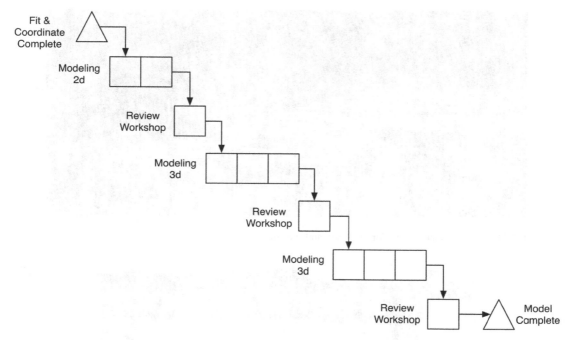

FIGURE 7.10   Common rack modeling process.

assemblies in general and the horizontal common rack in particular could bring more value to the client than conventional construction. The focus changed when the client became convinced of the common rack's benefits and asked the IPD team to incorporate it into the design of every building.

The interiors team, which, unlike the architects laying out the building shell and cores, and the structure and MEP teams, hadn't been working in the co-location space, had to learn how the rack would work with the schemes they had in mind. They came to see how the common rack could work as they listened to the HVAC modeler explain and show what the spaces would look like with ductwork, piping, and conduit installed conventionally and exposed, versus the common rack. Figures 7.11 and 7.12 illustrate this difference. It was truly an example of "a picture is worth a thousand words." They asked the modeler to come back the next day to help integrate the rack model with their concept BIM so they could be confident that the rack would work with the sound baffles they wanted to hang between the exposed steel beams.

With the architectural team on board, the client asked the modular team to move forward with constructing a full-scale physical mockup to prove not only the rack as a product, but that the processes for fabricating, assembling, and transporting the rack would work.

Rack Integration 3: Build and Design

Builders took the lead in organizing the modular team so that it included both the design and fabrication expertise required. Every bit of knowledge was needed to prove the case, especially for the client's expectation that modular construction should save time and money. They first had to design a

FIGURE 7.11   Conventional installation. © Southland Industries.

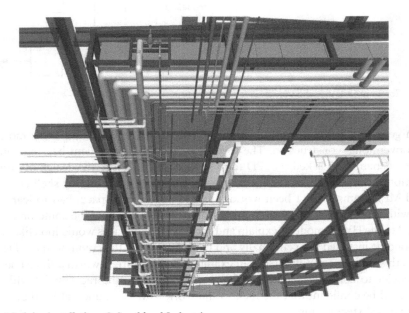

FIGURE 7.12   Modular installation. © Southland Industries.

Lean production system to make and deliver the racks on time. A number of things had to be done, pulling back from installation of the modules to determine the time and cost of fabrication, assembly, transportation, and installation. The modular team worked down their list to determine:

1. Time, equipment, and crew needed to install each module and the duration for each building.
2. Minimum inventory level to support installation within the construction schedule.
3. Location and size of the on-site laydown area for the required inventory.

4. Sizes, exact locations, and fabrication "spool sheets" for the duct, pipe, variable air volume (VAV) boxes, electrical conduit, and cable tray.

5. Methods for joining and anchoring components on each rack and for connecting the modules together end to end.

6. Optimal sequence of assembly and quality assurance testing, that is, who would do what, when.

7. Space, equipment, and skills required for assembly.

8. Number of modules that could be assembled each day.

9. Number of modules that could be transported to the job site given the distance away from the assembly facility.

As the concept BIM was being developed, the structural steel detailer modeled the rack and simulated installation of a module during steel erection. Figure 7.13 shows a typical rack being set in place during erection of a ConXtech modular system. Seeing the simulation gave the modular team enough confidence to calculate costs and time for this scenario. The savings were considerable, proving the case for the modular common rack assuming that an alternative to spray-on fireproofing would be accepted by the local building authority. In case this was not possible, the team asked the fabricator to simulate installation after steel erection and fireproofing. Even though the savings were not as dramatic in this "plan B" scenario, the modular team and project superintendents could see a safe and efficient process. The team created an alternate, conservative budget based on the "plan B" scenario proving that the rack would at worst be cost neutral.

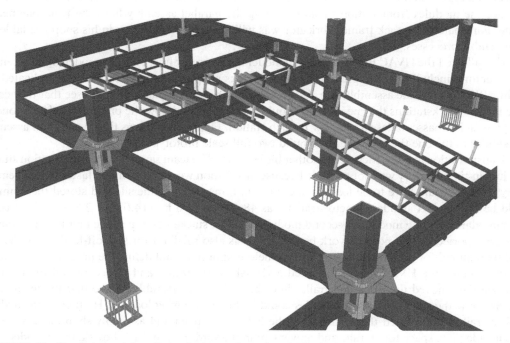

FIGURE 7.13 Simulating rack installation during steel erection. © Elizabeth Fortune, ConXtech, Inc.

The team identified six different conditions for the rack and modeled all of them with BIM. Within each of these conditions, there were different configurations. It turned out that the common rack was six families of racks. Two modules of the same type, each with its own configuration, were fabricated and assembled to test the assembly process. Before production began, modelers placed components exactly in the BIM so that fabrication "spool" information could be transferred directly into computer numeric control (CNC) cutting, bending, and welding machines to fabricate the sections of duct and VAV boxes and cut pipe. Figure 7.14 shows a typical common rack spool sheet.

The modular team could not wait until they had an approved design to plan how they would make and install the rack. This planning was a prerequisite for estimating cost so IPD team leaders could decide whether to recommend building a full-scale prototype to enable building operators to evaluate functionality. The modular team decided to begin planning fabrication soon after making significant changes in the height and width of the rack at the request of space planners.

The mechanical contractor team took the lead in planning the assembly and asked their shop managers to help plan not only duct and piping fabrication but also the assembly of all components. Fortunately, these managers had studied Lean production design prior to relocating and expanding their fabrication shop three years earlier. These shop managers applied their knowledge to design "level flow" so that no workstation would be overburdened while others stood idle. Figure 7.15 shows a production study similar to the one developed by the mechanical contractor's production managers together with counterparts for the other trades installing materials and equipment on the common rack.

After it became clear that the steel fabricator's shop would not be available, the modular team's project manager began an urgent search for a facility with sufficient space and with gantry cranes to move the modules from station to station along the parallel assembly lines the team planned to create. Fortunately, the rack frame fabricator was willing to make two bays in his shop available for the assembly processes.

This allowed the HVAC BIM specialist to model the assembly line. First, he modeled the assembly hall structure, including above-grade foundation blocks supporting the steel. He then replicated the module BIM to show the assembly line. As a result, the production planners could see that they could have the five workstations they needed, and could also understand how the prefabricated components and pipe rack subassemblies would flow to each station. Figure 7.16 shows the layout of an assembly line similar to the one the team set up to build two full-scale prototypes.

Transportation and logistics were another big concern. The team saw from their production studies that just-in-time delivery was not possible because installation would be so much faster than assembly. Every module would have to be loaded onto a truck, transported to the site and stored there until it could be installed. A standard tractor-trailer was 48 ft. × 8 ft. 6 in. (14.6 m × 2.6 m), which could accommodate only one module. A second module could be stacked on top of the first for the modules to be transported during normal work hours. This was also the limit for safe off-loading by the two forklifts required for each module. To confirm their assumptions and define the processes for loading and off-loading, the HVAC modeler created a virtual tractor-trailer and placed two of the modules on it. He then moved the truck and trailer into the end of the assembly hall so that the team could visualize how the shrink-wrapped modules could be handled during loading. He then replicated the stacked trailers and placed them outside of the hall so the team and assembly shop manager could allocate adequate space for storage and movement in the yard. Figure 7.17 shows a truck loaded with four rack sections similar to one created by the HVAC modeler.

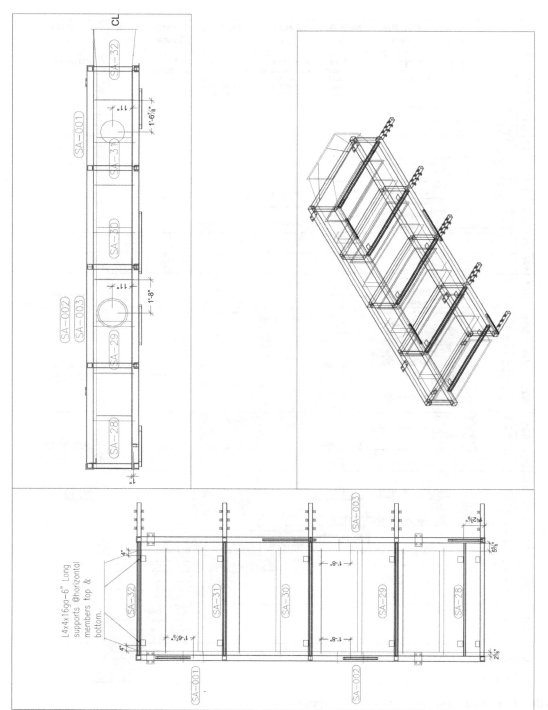

FIGURE 7.14 Fabrication and assembly spool sheet. © Southland Industries.

135

DATE		Fit Up Bay	Cop Pre Fab	Weld Bay 1	Weld Bay 2	Weld Bay 3	Outside Stage	Truck-ing		Weld Bay 1
		WELDER 1	WELDER 2 PIPE INSTALLER 1 PIPE INSTALLER 2	WELDER 3 PIPE INSTALLER 3 PIPE INSTALLER 4 DUCT INSTALLER 1 DUCT INSTALLER 2	WELDER 4 PIPE INSTALLER 5 PIPE INSTALLER 6 DUCT INSTALLER 1 DUCT INSTALLER 2	WELDER 5 PIPE INSTALLER 1 PIPE INSTALLER 2 DUCT INSTALLER 1 DUCT INSTALLER 2	MAT. HANDLERS			WELDER 1
8/11	6-8 8-10 10-12 12:30-2:30	Fitup Frame - MODULE 6								
8/12	6-8 8-10 10-12 12:30-2:30	Fitup Frame - MODULE 8		Weld Frame - MODULE 6						Weld Frame - MODULE 6
8/14	6-8 8-10 10-12 12:30-2:30	Fitup Frame - MODULE 5 Fitup Frame - MODULE 4	Single Pipe Prefab - MODULE 5 Single Pipe Prefab - MODULE 5	Duct Install - MODULE 6	Weld Frame - MODULE 5					
8/14	6-8 8-10 10-12 12:30-2:30	Fitup Frame - MODULE 7 Fitup Frame - MODULE 3	Single Pipe Prefab - MODULE 5 Single Pipe Prefab - MODULE 4 Single Pipe Prefab - MODULE 7	Piping Install - MODULE 6	Weld Frame - MODULE 5 Duct Install - MODULE 6	Weld Frame - MODULE 4				
###										
###										
8/17	6-8 8-10 10-12 12:30-2:30	Fitup Frame - MODULE 3 Fitup Frame - MODULE 1	Single Pipe Prefab - MODULE 1 Single Pipe Prefab - MODULE 1 Single Pipe Prefab - MODULE 1 Single Pipe Prefab - MODULE 1	Weld Frame - MODULE 7	Piping Install - MODULE 5	Weld Frame - MODULE 4 Duct Install - MODULE 4 Piping Install - MODULE 4				Weld Frame - MODULE 7
8/18	6-8 8-10 10-12 12:30-2:30	Fitup Frame - MODULE 1 Fitup Frame - MODULE 2	Single Pipe Prefab - MODULE 1 Single Pipe Prefab - MODULE 2 Single Pipe Prefab - MODULE 6	Duct Install - MODULE 7 Piping Install - MODULE 7	Weld Frame - MODULE 3 Duct Install - MODULE 3	Piping Install - MODULE 4 Weld Frame - MODULE 1				
8/19	6-8 8-10 10-12 12:30-2:30	Fitup Frame - MODULE 8 Fitup Frame - MODULE 9	Single Pipe Prefab - MODULE 8 Single Pipe Prefab - MODULE 8	Weld Frame - MODULE 2	Duct Install - MODULE 3 Piping Install - MODULE 3	Weld Frame - MODULE 1 Duct Install - MODULE 1				Weld Frame - MODULE 2
8/20	6-8 8-10 10-12 12:30-2:30	Fitup Frame - MODULE 9 Fitup Frame - MODULE 10	Single Pipe Prefab - MODULE 9	Weld Frame - MODULE 2 Duct Install - MODULE 2 Piping Install - MODULE 2	Weld Frame - MODULE 8	Piping Install - MODULE 1				Weld Frame - MODULE 2
8/21	6-8 8-10 10-12 12:30-2:30	Fitup Frame - MODULE 10 Fitup Frame - MODULE 11	Single Pipe Prefab - MODULE 10	Piping Install - MODULE 2 Weld Frame - MODULE 10	Duct Install - MODULE 8 Piping Install - MODULE 8	Weld Frame - MODULE 9 Duct Install - MODULE 9				Weld Frame - MODULE 10
###										
###										
8/24	6-8 8-10 10-12 12:30-2:30	Fitup Frame - MODULE 11 Fitup Frame - MODULE 12	Single Pipe Prefab - MODULE 11 Single Pipe Prefab - MODULE 11	Weld Frame - MODULE 10 Duct Install - MODULE 10	Weld Frame - MODULE 11	Duct Install - MODULE 9 Piping Install - MODULE 9				Weld Frame - MODULE 10
8/25	6-8 8-10 10-12 12:30-2:30		Single Pipe Prefab - MODULE 12	Piping Install - MODULE 10	Weld Frame - MODULE 11 Duct Install - MODULE 11 Piping Install - MODULE 11	Weld Frame - MODULE 12	Insulator - MODULE 6 Insulator - MODULE 5 Insulator - MODULE 4 Insulator - MODULE 7			
8/26	6-8 8-10 10-12 12:30-2:30		Single Pipe Prefab - MODULE 12		Piping Install - MODULE 11	Duct Install - MODULE 12 Piping Install - MODULE 12	Insulator - MODULE 3 Insulator - MODULE 1 Insulator - MODULE 2 Insulator - MODULE 8	Load - MODULE 6 Load - MODULE 5 Load - MODULE 4 Load - MODULE 7		
8/27	6-8 8-10 10-12 12:30-2:30						Insulator - MODULE 9 Insulator - MODULE 10 Insulator - MODULE 10 Insulator - MODULE 12	Load - MODULE 3 Load - MODULE 1 Load - MODULE 2 Load - MODULE 8		
8/28	6-8 8-10 10-12 12:30-2:30							Load - MODULE 9 Load - MODULE 10 Load - MODULE 11 Load - MODULE 12		

FIGURE 7.15   Production study. © Southland Industries.

Weld Bay 2	Weld Bay 3	Weld Bay 1	Weld Bay 2	Weld Bay 3	Weld Bay 1	Weld Bay 2	Weld Bay 3
WELDER 2	WELDER 3	DUCT INSTALLER 1 DUCT INSTALLER 2	DUCT INSTALLER 1 DUCT INSTALLER 2	DUCT INSTALLER 1 DUCT INSTALLER 2	PIPE INSTALLER 3 PIPE INSTALLER 4	PIPE INSTALLER 5 PIPE INSTALLER 6	PIPE INSTALLER 3 PIPE INSTALLER 4
		Duct Install - MODULE 6					
Weld Frame - MODULE 6							
Weld Frame - MODULE 5					Piping Install - MODULE 6		
	Weld Frame - MODULE 4		Duct Install - MODULE 5				
	Weld Frame - MODULE 4					Piping Install - MODULE 5	
				Duct Install - MODULE 4			Piping Install - MODULE 4
Weld Frame - MODULE 3		Duct Install - MODULE 7			Piping Install - MODULE 7		Piping Install - MODULE 4
	Weld Frame - MODULE 1		Duct Install - MODULE 3				
	Weld Frame - MODULE 1		Duct Install - MODULE 3			Piping Install - MODULE 3	
				Duct Install - MODULE 1			
Weld Frame - MODULE 8		Duct Install - MODULE 2					Piping Install - MODULE 1
	Weld Frame - MODULE 9		Duct Install - MODULE 8		Piping Install - MODULE 2		
				Duct Install - MODULE 9	Piping Install - MODULE 2	Piping Install - MODULE 8	
				Duct Install - MODULE 9			
Weld Frame - MODULE 11		Duct Install - MODULE 10					Piping Install - MODULE 9
Weld Frame - MODULE 11					Piping Install - MODULE 10		
	Weld Frame - MODULE 12		Duct Install - MODULE 11			Piping Install - MODULE 11	
				Duct Install - MODULE 12		Piping Install - MODULE 11	
							Piping Install - MODULE 12

FIGURE 7.15 (*Continued*)

FIGURE 7.16   Assembly hall model. © Southland Industries.

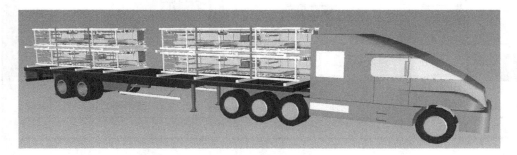

FIGURE 7.17   Transport model. © Southland Industries.

Creating mechanical art required the fabrication team to choose their "cleanest" flange joint for ductwork and to decide on "propress" pipe connections. The rack components would need to be fastened to the frame in a standard way while not detracting from the "art." The frame and components would be painted before performance testing and shrink-wrapping for transport. By the time the client authorized building the physical prototype, everyone on the modular team realized that they had gone far beyond conventional construction and were, in fact, engineering and producing a manufactured product. Figure 7.18 is a photo of a common rack installed on another project by the mechanical contractor. It shows how the prototypes looked when they were lifted to the height at which they would be installed in the assembly hall for inspection by owner representatives and team members.

FIGURE 7.18   Installed rack similar to the prototype sections. © Southland Industries.

Note the breadth and depth of the knowledge applied to develop the best design possible in this example. Also note that this knowledge was applied at a time when every aspect of the design of the common rack could still be changed. Furthermore, the modular team made its work efficient by integrating the information from all participants in building information models of the many design, fabrication, assembly, and installation alternatives (see Chapter 10 for an explanation of how integrated information supports integrated teams and processes).

Rack Integration 4: Operations and Design

Client building operators actively engaged with both the MEP and modular teams to insure that these systems achieved two objectives:

1. Flexibility for changing preferences and needs in the future
2. Access to valves, equipment, and data cabling for operation, maintenance, and reconfiguration

The possibility and relatively lower cost of oversizing the heat and cooling water piping was a significant advantage of a radiant slab supplied from a central utility plant. The modular team's job was to design their system to accommodate the size and loaded weight of the piping along with conduit and cable tray. The team was also responsible for planning how the utilities would feed and exit the rack.

FIGURE 7.19   Electrical and data trays on the lowest level of the rack to simplify reconfiguration. © Southland Industries.

The same system engineers, modelers, and construction managers working with both the MEP and modular teams synchronized the HVAC, plumbing, electrical and data/communications designs. They shared the BIM and spreadsheet models of production cost and time so that everyone was visualizing and reading the same information.

Flexibility for the client meant that data cables could be quickly reconfigured to support ad hoc workgroups. That's why the modular team placed the electrical conduits in a tray on one side and data opposite in its own tray. These were left open on each side, with the cable tray always having the greatest distance from the opposite corridor wall. Figure 7.19 shows a model similar to the ones the HVAC modeler created so building operators could understand the advantages of the modular rack for quick changes of power and data cabling.

The client's operations managers knew the ergonomics of operating valves and inspecting and maintaining utilities. In addition to product choices, these ergonomics dictated how much space their staff needed to work using the equipment they preferred. Valves would not be located on the rack; rather, they were to be placed in walls accessible to operators. Large ceiling access panels would be installed in private offices adjacent to the common rack where access would be required. The VAVs, placed on top of the rack, would be located on one side of the central corridor and would not run over electrical and data rooms. Figure 7.20 shows modular assemblies of supply lines with valves designed for easy access.

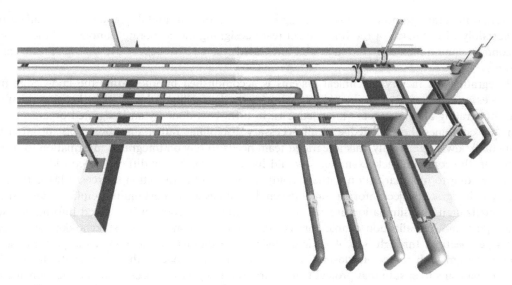

FIGURE 7.20 Modular assemblies of supply lines with valves located for easy access overhead and in wall cabinets. © Southland Industries.

Operations affected architecture and could not be an afterthought or just information gathered during an architectural programming exercise. Because so many alternatives were in play for the site, architecture and engineered systems, including the common rack, the MEP, and modular teams continually engaged the operators. As a result, these engineers and managers were confident that the MEP systems and the common rack would be both usable and operable when the decision to move forward with the mock-ups was made.

Rack Integration 5: Sustainability and Design

In addition to the economic advantages mentioned already, the team embraced the broad and ambitious goals of the project outlined above and decided to leverage local expertise and capacity for production of the rack, use only "healthy" materials for the rack so that indoor air quality would not be impaired, and ensured that the energy efficiency of the buildings was maintained by the rack. In that way, the rack embodied and operationalized all the applicable building sustainability goals.

7.5 INTERCONNECTIONS

Our example demonstrates that the best approach to achieving a high-performance building is to create synergies between the technical systems that make up the facility and between the facility users and the facility. These synergies are best generated through integrating these systems as much as possible, that is, through integrating knowledge of building operations, design, prefabrication, and installation. It was certainly within reach of the IPD project team designing the corporate campus and the modular horizontal common rack to create such well-integrated systems. We believe every project team can do the same.

Integrating the facility's technical systems is accomplished by integrating the knowledge from the processes that conceive, design, build, and operate the facility rather than by keeping them separate. Integrating all this process knowledge is, in turn, best done by an integrated organization or team that combines the knowledge of all necessary disciplines rather than by an organization that fragments knowledge. Finally, this integrated team needs to rely on integrated information to be successful. In the common rack example, it would have been extremely difficult to consider the many factors relating to modularization without ready access to all information sources. BIM is the main strategy to integrate project information between disciplines to support multidisciplinary analysis.

Visualization and simulation are the main mechanisms to connect integrated information with the design team. Visualization enables project team members and project stakeholders to see each other's perspectives through 3-D, 4-D, and other visualizations to understand the performance of a facility design with respect to desired performance metrics, that is, the value of the facility. The main mechanisms to ensure that project teams carry out integrated processes are co-location and collaboration, also known as integrated concurrent engineering (ICE), pioneered by the Jet Propulsion Laboratory (JPL) for the design of space missions (Mark, 2002). Lean production management is the most effective mechanism today to guarantee that each project team member's efforts are contributing to an integrated facility. Finally, performance metrics must become the operational definition of a high-performance facility and allow the comparison of different facility designs so that the best design can be selected.

Note that the integrated team must not only design the product (facility), but also the organization and process (Kunz & Fischer, 2007), and the information generation, sharing, and use mechanisms that deliver the project. A contractual framework is needed to support this work.

7.6   REFLECTIONS

A high-performance facility enables its users to create the value they must deliver to thrive in their own business. Once built, a facility functions as a whole, with all of its technical systems and social organizations working together or working individually and against each other. Given that a facility is conceived and used as a whole, it is surprising that design and construction management concepts—in particular budgeting and scheduling methods—and textbooks, software, and practices focus largely on how to decompose a project into small manageable parts and how to assign contractual responsibility to each part (Barrie & Paulson, 1992).

The expectations of facility owners and users are changing. They now expect a facility with strong environmental and social performance, a facility without health hazards for its users, and a facility that maximizes the business value its users can generate with it, that is, a sustainable facility. Hence, we need a new fundamental strategy because the essentially reductionist or "divide and conquer" approach to facility design and construction depends entirely and only on exceptionally good work of designers, builders, and operators to create a facility that performs well. The "divide and conquer" system works against a facility that performs at a high level as a whole. While occasionally the heroic efforts of the project team members align and a high-performance facility emerges from the traditional, fragmented project delivery process, the traditional project management approach does not scale to create high-performance facilities consistently and globally.

As we have shown in this chapter, processes that focus on the whole are more likely to achieve exceptional results. We foresee that individuals and firms that make this switch will have a more rewarding work experience because it will likely be more productive and effective.

7.7   SUMMARY

The design and construction industry has divided what users and project owners want from the information about how to best achieve it. The knowledge about the construction, operation, and use processes is often introduced sequentially, greatly diminishing its impact on making a building high-performing. Integrating process knowledge makes knowledge about all critical life cycle phases available to the IPD team early and comprehensively. In the common rack example used in this chapter, the needs—financial, operational, and aesthetic—were clearly communicated to the project team. An interdisciplinary team evaluated prefabrication options against these project values to achieve a solution that was buildable, usable, operable, and sustainable—and attractive!

NOTE

1. Bruce C. Cousins, AIA, is a principal of SWORD Integrated Building Solutions. Bruce has over 40 years of diverse experience in architecture and construction project management. As a practicing architect and firm owner, he specialized in the design of higher-density service-intensive mixed-use projects such as resort hotels, town centers, and health care and community facilities. For the past 10 years, he has served as a leader, consultant, and mentor for Lean and integrated project delivery (IPD) for both architects and builders. His experience with implementing an IPD approach to projects began in 2007. He worked with preconstruction managers on several projects using traditional GMP agreements to transition to a collaborative Lean project delivery approach. These included a community services and city hall building in Wellington, Florida, and a research and development laboratory facility for Kansas State University in Olathe, Kansas. Bruce has established and directed BIM—Virtual Design and Lean Construction for Architects and several top 100 general contractors. As a senior leader and coach, he has led project teams to achieve operational excellence, documented savings, and efficiencies by using the tools and techniques of the Lean production and product development principles, understanding of human behavior, and BIM technology.

REFERENCES

Barrie, D., & Paulson, B. (1992). *Professional construction management*. New York, NY: McGraw-Hill.

Elkington, J. (1998). *Cannibals with forks: The triple bottom line of 21st century business*. Gabriola Island, British Columbia, Canada: New Society Publishers.

Martin Fischer, Atul Khanzode, Dean Reed, and Howard Ashcraft (2012). "Benefits of Model-based Process Integration." Invited Paper, *Proceedings of the Lake Constance 5D-Conference 2012*, Fortschritt-Berichte VDI, Reihe 4, Nr. 219, U. Rickers and S. Krolitzki (eds.), VDI Verlag, Düsseldorf, Germany, Pages 6–21.

Kunz, J., & Fischer, M. (2007). *Virtual design and construction: Themes, case studies and implementation suggestions*. Working Paper #97, Center for Integrated Facility Engineering, Stanford, CA.

Mark, G. (2002). Extreme collaboration. *Communications of the ACM, 45*(6), 89–93.

Peña, W., & Parshall, S. A. (2012). *Problem seeking: An architectural programming primer* (5th ed.). Hoboken, NJ: Wiley.

Suhr, J. (1999). *The Choosing by Advantages decision making system*. Westport, CT: Quorum Books.

CHAPTER 8

Integrating the Project Organization

"People need to find their own language for describing the intent of their efforts in ways that work in their own context, as part of developing their own strategies and leadership practices. How we talk about our work matters. But the key lies in our personal journey of reflection, experimentation, and becoming more open, not the words we use."

—Peter Senge

8.1 INTRODUCTION

We propose that there are four tasks every project team must do well to succeed, shown in Figure 8.1. At the end of the day, a team gets paid for value-added work. Doing only or mostly value-adding work requires coordination, leadership, and decisions—otherwise, it's unlikely that each team member and all the subteams will carry out the work and work processes outlined in Chapter 7 that lead to a high-performing facility with integrated systems. Team members with the most experience and responsibility must create a new culture and organization. This includes setting up the coordination scopes and mechanisms and helping team members and owner stakeholders determine which decisions must be made, when and how to advance toward the project goals, and so on. These same leaders help team members decide on their next steps.

8.1.1 "Best Practice" Today

Sometimes it helps to look at things from a different perspective, in this case professional sports. A well-known adage in competitive sports is: "You play like you practice." Don't cut corners during practice because it is foolish to expect that, suddenly, on game day you will be able to complete a play in a way that you never could during practice. Nobody would expect any sports team to perform well if they simply practiced their individual moves in isolation and came together once for the game.

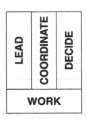

FIGURE 8.1   Four tasks every project organization must do well to succeed.

Imagine the owner of a professional football team who wants to win a championship game. He hires an experienced manager and coaches, who have led teams to victory. The head coach and offensive and defensive coordinators meet together a few times a week for months before the game designing and documenting plays they think will work. Perhaps they walk through some of the plays in their office, though only one of the coaches has played the game professionally. The head coach tells the owner he is confident their strategies can win, but he hasn't yet hired players and they will have never played together when they take the field. This also means he has no statistics to predict their performance, like the offense's red zone efficiency, the quarterback's completion percentage, or the defense's ability to create turnovers.

On game day, the coaches send players onto the field one at a time from the locker room. The players are there because they were the lowest-cost players available. They are meeting each other for the first time, have not seen anything that has happened so far in the game, have never seen this field, and they don't even know who they are playing against.

Although what we've described above sounds ridiculous, it is an accurate description of the design-bid-build (DBB) way of delivering projects. Many owners have opted for other approaches. Design-Build (DB) at least gets the some of the players together before the game. Owners who don't feel comfortable putting the contractor in control prefer construction manager at risk (CM@R), but this doesn't result in deep integration, either. Unfortunately, none of these approaches we've described seem likely to result in a championship. Yet one or the other is the way that most projects in our industry are run. We are trying to win a championship by first selecting the cheapest team, keeping them apart, and then asking them to each play their best individually.

Because team members in non-IPD projects do not win or lose together, but by themselves, individual leaders must first think about how a decision affects their budget, schedule, and bottom line, and maybe, if it doesn't cost anything, the project. Team members make sure their work is internally coordinated and sequenced before thinking about the project. Leaders and team members must make sure their company's interests are protected in making decisions, and then try to do what's best for the project. The imperative to first take care of each participating organization's interest results in having to redo work because it does not meet the needs of other team members and the project as a whole. This rework is accepted as the normal course of business and not seen for the waste that it is. No one really knows how much there is because the rework is invisible. The rework consumes the time of valuable human resources and certainly adds cost to projects. Figure 8.2 shows the result of organizing this way.

Winning every game requires a different kind of organization, one that enables team members to do their best while supporting their teammates. No single player can succeed alone and apart from

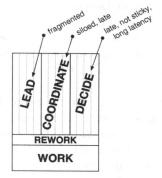

FIGURE 8.2   Fragmented leadership, siloed coordination, late decisions leading to rework.

the others. This winning team is created and operates very differently than the ones we usually form. It is integrated and reflects the way sports teams actually work. Team members work together from early in the process, getting to know each other personally, understanding each other's work processes, learning how to make reliable promises, and building virtually together so that once construction is in full swing, they already know how to solve problems quickly and effectively.

8.2   WHAT IS INTEGRATED ORGANIZATION?

For us in this book, *organization* means a body of people working within a structure, arranged in some order and connected to an external environment. This organization is a community, with shared rules and norms. Boiling this down, we have people with different levels of accountability and responsibility, knowledge and skill carrying out actions so the entity can provide things or services to other people in the wider world. Ford Motors provides cars and trucks. Mass General Hospital provides health care. DPR Construction provides buildings.

Although the subject of teams is closely related to organization, they are not synonymous. For our purposes, we define *team* as a group of people linked together in a common purpose. We ask you to think of organization as structure and arrangement. Teams are people with purpose. Both are necessary and you can't have one without the other to design and construct. There's a lot to say about both, which is why we want to deal with them in separate chapters, starting with organization.

The structure of true integrated project delivery organization is formal, in that it is established by contract or agreement, unlike other forms of project delivery that only establish responsibilities and financial obligations. The IPD agreement defines a structure for the organization that includes a group to provide high-level leadership, another for day-to-day project management, and multidisciplinary/cross-functional teams to implement design and construction. These are composed of people with knowledge of different specialties such as structural concrete, steel framing, mechanical, electrical, and exterior skin systems. Within the various disciplines, teams need managers, estimators, and fabrication and installation experts—that is, people who are responsible for different work processes or functions. Like all project organizations, it is virtual because there are no permanent employees. It is also temporary, existing only for the single purpose of delivering the project.

We believe all this can be summed up in a single sentence, as follows. An integrated organization in IPD is a collection of people organized in an integrated structure that is aligned to the project. Essentially, it is a group of individual organizations—and their employees—that embrace a common set of values and goals and act as if they were one company—a virtual organization we call the "Project." In fact, the question "Would we do it this way if we were a single company?" is a pretty good test for the depth of integration.

We suggest that the organization aligns people to the project in four ways. The integrated project organization first connects people's actions, information, and decisions. Individuals are not left to do this on their own in a hit-and-miss fashion. Second, the integrated project organization is literally built through and on people's use of language to make and keep commitments to do what they believe needs to be done as contributors. Third, the integrated project organization promotes individual and collective learning so that the organization's IQ is greater than any single individual's. Fourth, integrated project organization connects the work that people do through its structure to the unique combination of things that the end customer, the client, has defined as value.

8.3   WHAT DOES SUCCESS LOOK LIKE?

How can you tell if a project organization is integrated? Is it because the parties have an integrated project delivery (IPD) contract and/or share some amount of risk and reward? Is it because the general and trade contractors have been brought into the team early in design? Is it that people are co-located, working in the same space? Is it because they are using building information modeling (BIM)? Is it because they draw lots of diagrams on the whiteboard with arrows connecting boxes to show how they will work together? Is it because they plan with sticky notes on long rolls of newsprint or butcher paper? We don't think that any or all of these things mean that a project organization is integrated even though you are likely to see people in integrated project organizations do all of them. We advise watching and listening to the people in the room. We hope to see and hear the following.

- Decisions are made as if the project participants (including the owner) were a single, vertically integrated organization.
- Leadership is truly shared because people trust each other. People seem to take turns at leading.
- People decide what to do, how to do it, and who should do it based on what's best for the project.
- There's a buzz in the co-location offices. There are lots of conversations—passionate, but positive. A few are wearing headphones while they work. Many are smiling and joking with each other. Everyone is engaged.
- Project leaders take responsibility for building a strong network of commitments. They talk about offers, requests, and promises.
- Team members create space for new possibilities by listening to each other in dialogue rather than first trying to persuade.
- Delivery team members and owner stakeholders thank each other for discovering problems early. Everyone understands that no problem is a problem.
- New and different ideas and approaches are welcomed and heard.

- Team members make sure that they know who their customers are within the organization and what they really need.
- Everyone believes their first responsibility is to learn and improve every day.
- Cluster teams and their leaders and project managers meet briefly every day to discuss progress, questions, roadblocks, and decisions.
- Managers and leaders model rather than talk about behaviors.

8.4 HOW CAN THIS BE DONE?

8.4.1 Sutter Health's Five Big Ideas

The "Five Big Ideas that Are Reshaping the Design and Delivery of Capital Projects," written for Sutter Health primarily by Hal Macomber of Lean Project Consulting, is a very good starting point. In fact, David Pixley and other leaders of Sutter Health Facilities Planning & Development, asked that every organization in their supply chain begin using the Five Big Ideas on Sutter projects starting immediately after the Sutter Health Lean Project Summit on March 23 and 24, 2004 (Macomber, 2004a).

Regardless of what design and build companies did on their other projects, Sutter Health expected project leaders, their own people and others, to educate project team members, so they could work differently than before. We take the Five Big Ideas to mean the following:

1. Make constructors part of the team from day 1, so design can be made buildable, drawing on what experienced builders have learned.
2. Establish relationships based on trust.
3. Realize that the only way things get done is through people making and keeping commitments. Management's job is to model and support these practices so that everyone understands and operates this way.
4. Decide and act based on what is best for the project rather than what is the least cost, easiest way to design or construct one piece or another.
5. Everyone should produce single pieces, or small rather than big batches, to learn whether they are providing the right thing as quickly as possible. Everyone is responsible for getting this feedback from their customer as quickly as possible. The question is, "Am I/are we providing what you, the customer, has asked for?"

Five Big Ideas that Are Reshaping the Design and Delivery of Capital Projects[1]

Declaration

We are setting out to transform how capital projects are designed and delivered. This initiative is noble and necessary. We believe that capital projects cost too much; they take far too much time; they often fall short of our objectives; and they kill or injure too many along the way. It need not be this way.

Five Big Ideas

These big ideas can transform projects. Together, they form the foundation for innovating project delivery systems and approaches. There are solid historical and theoretical foundations for this claim. Companies around the world have adopted one or more of these ideas to improve their practices. These companies report significant gains. We aim to transform the industry by applying these ideas.

1. **Collaborate; really collaborate, throughout design, planning, and execution.**

   Constructable, maintainable, and affordable design requires the participation of the range of project performers and constituencies. Since abandoning the master-builder concept, and separating design from construction, we have been patching a poorly conceived design practice. Value engineering, design assist, and constructability reviews mask an underlying assumption—that design can be successful when separated from engineering and construction. Design is an iterative conversation; the choice of ends affects means, and available means affects ends. Collaborative design and planning maximizes positive iterations and reduces negative iterations.

2. **Increase relatedness among all project participants.**

   People come together on architecture, engineering, and construction (AEC) projects as strangers. Too often, they leave as enemies. Facilities projects today are complex and long lived, requiring ongoing learning, innovation, and collaboration to be successful. The chief impediment to transforming the design and delivery of capital projects is an insufficient relatedness of project participants. Participants need to develop relationships founded on trust if they are to share their mistakes as learning opportunities for their project, and all the other projects. This will not just happen. However, we are learning that relationships can be developed intentionally.

3. **Projects are networks of commitments.**

   Projects are not processes. They are not value streams. The work of management in project environments is the ongoing articulation and activation of unique networks of commitments. The work of leaders is bringing coherence to the networks of commitments in the face of the uncertain future and co-creating the future with project participants. This contrasts with the commonsense understanding that planning is predicting, managing is controlling, and leadership is setting direction.

4. **Optimize the project not the pieces.**

   Project work is messy. Projects get messier and spin out of control when contracts and project practices push every activity manager to press for speed and lowest cost. Pushing for high productivity at the task level may maximize local performance but it reduces the predictable release of work downstream, increases project duration, complicates coordination, and reduces trust. In design, we incur rework and delays. In the field, this means greater danger. We have a significant opportunity and responsibility to reduce workers' exposure to hazards on construction projects. Doing so can bring about greater than 50 percent improvements in the safety on the work site. We are committed to do all that is possible so that the people who build these projects are able to go home each night the way they came to work. The way we understand work and manage planning can increase that messiness or reduce it.

5. **Tightly couple action with learning.**

Continuous improvement of costs, schedule, and overall project value is possible when project performers learn in action. Work can be performed so that the performer gets immediate feedback on how well it matched the intended conditions of satisfaction. Doing work as single-piece flow avoids producing batches that in some way don't meet customer expectations. The current separation of planning, execution, and control contributes to poor project performance and to declining expectations of what is possible.

We are setting out to change project design and delivery forever. Please join us.

Prepared for Sutter Health by Hal Macomber.

8.4.2 Creating an Integrated Organization

We believe there are four pillars supporting the roof of the Integrated Organization house. They are as follows:

1. Connecting people's actions, information and decisions;
2. Using language to coordinate action to create value;
3. Creating a learning organization; and
4. Aligning everyone's effort with value creation for the client.

1. Connect People's Actions, Information, and Decisions

We believe it's essential for integrated project delivery team members to see the networks within which they are operating. This is not a matter of new terminology; it's a fact of life. With that understanding, participants need to understand another reality, which is that there are three types of interdependencies. They also need to be aware of the large effort required to coordinate work within the network, which is affected by these dependencies. These understandings make clear the need for multidisciplinary/cross-functional teams as the fundamental unit of IPD, and also the need for daily tiered communication so that everyone in the network knows what is happening all the time.

Recognize and Account for Interdependency in the Project Network The first step is to understand and account for interdependency. In the 1967 book *Organizations in Action*, sociologist James D. Thompson defined three types of interdependence within an organizational structure: pooled, sequential, and reciprocal. Pooled interdependence is when organizational units, be they groups, teams, departments, or business units complete work independently on their own, and then contribute it to what others have done to produce a whole product or service. The work has been designed so that teams/units can work independently of the other teams. When each team has done its work this part of the project is complete. Management's main role is to define and align the completion criteria for the work of each unit, shown in Figure 8.3.

Sequential interdependence occurs when one unit in the overall process depends on input from a previous unit, that is, the first unit, be it a team or installation crew, produces an output necessary for the next unit's performance. Perhaps the most obvious example of sequential interdependence is an assembly line. In addition to defining and aligning the task deliverables and handover criteria, management needs to focus on the supply of information and materials. Coordination must be very good to support sequential interdependence. Figure 8.4 shows teams working in sequence.

In reciprocal interdependence, discipline A requires the input of discipline B to contribute something (knowledge or work). The challenge is that discipline B also needs the input of discipline A to do its work, that is, their work is reciprocal or interdependent. Consider the interdependencies in the common rack example in Chapter 7. For example, the common rack design team depended on the interior design team and vice versa. Thompson believed that the best way to manage reciprocal interdependence is through constant sharing of information and mutual adjustments among participants. Reciprocal interdependence, shown in Figure 8.5, requires the highest level of interaction, collaboration, and management attention (Thompson, 1967).

All three types of interdependence exist in any project. The greater the need to move forward quickly and the richer the set of performance criteria, the more likely work will be reciprocally interdependent in design.

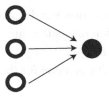

FIGURE 8.3   Pooled interdependence. Illustration provided to the authors by DPR Construction, Lyzz Schwegler.

FIGURE 8.4   Sequential interdependence. Illustration provided to the authors by DPR Construction, Lyzz Schwegler.

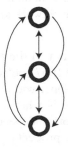

FIGURE 8.5   Reciprocal interdependence. Illustration provided to the authors by DPR Construction, Lyzz Schwegler.

The most important outcome of the preconstruction phase must be a well-specified strategy and approach to the construction phase with only pooled or sequential work needed to construct the building because reciprocal work in the field—work that must be coordinated by the crews installing the various building systems because the design is not sufficiently specified—is slow, costly, unsafe, and often of lower quality. Building virtually using BIM offers builders the opportunity to sequence work in a way they couldn't before. For example, the crew installing cast-iron drain, waste, and vent piping on the Temecula Valley Hospital was able to install their work before the metal-stud wall framing because both systems were modeled in detail to the extent that foremen could see the proximity of pipes and studs. Typically, the plumbers would have worked in the same space as the framing crew or had to follow them. The plumbing crew was much more productive having a clear floor and time by themselves while the framing crew was able to meet their productivity targets.

Figure 8.6 shows what we believe are the curves for pooled, sequential, and reciprocal interdependence in integrated design and construction. A relatively small multidisciplinary/cross-functional team designs the product, organization, and processes through high interaction. Once cluster teams are organized and begin to function, they can work on their own to develop concepts originating from design and build leaders, proceeding for short intervals on their own, aware and empowered to request information from those in other cluster teams. Traditionally, the proportion of sequential interdependence was high in both design and construction, especially when contractors were excluded completely or partially from participation in design. Increasingly during the past few decades, general contractors have resorted to more pooled interdependence to speed up construction work. Although reciprocal work is valuable during design/preconstruction, it can be very costly during construction in terms of

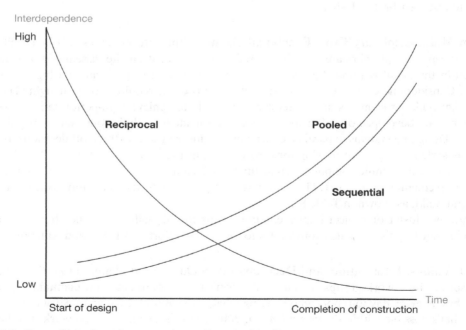

FIGURE 8.6 Types of interdependence over time on integrated projects.

productivity, time, safety, and quality, leading to complaints from their subcontractors about lower productivity and higher labor costs.

Although Thompson didn't characterize these three types of interdependence as networks, it's obvious that they are once you imagine dots inside the subteam circles representing people who must coordinate their work. So we have the networks and interdependence between people and teams described in the Five Big Ideas. Although recognizing networked interdependence is a good beginning, it begs the question of how can an integrated organization's managers make sure the right work is being done when it is needed.

Managers often repeat the adage that if something can't be measured, it can't be managed. Professors Raymond Levitt and John Kunz of Stanford University noticed in the late 1990s a pattern of poor outcomes from projects in which timelines had been dramatically cut to meet market demand, forcing teams to implement "concurrent engineering." Design, engineering, and production work were being done in parallel and not sequentially, as they had been previously. Levitt and Kunz wanted to know what they and project participants could learn by building computer models showing the flow of information through various "actors" within project organizations in different industries, including aerospace and construction. Their SimVision software application consistently exposed a significant amount of what Levitt and Kunz called "hidden effort" that team members needed to do to coordinate between themselves and others, and between project sub-teams. They also found that this work was invisible to team managers and was not accounted for in staffing plans and schedules. They concluded that this hidden effort was a significant factor leading to projects exceeding their budgets and expected duration, along with disappointing quality. Levitt and Kunz recommended in their 2002 Center for Integrated Facility Engineering (CIFE) paper (Levitt and Kunz, 2002) that project organizations be designed by simulating the stresses they will face in coordinating work, just as engineers simulate stresses to test their bridge design.

Work in Multidisciplinary/Cross-Functional Teams   Ultimately, materials will be specified, purchased, transported, prefabricated, and delivered or transformed on site, assembled, and tested for operation in the detail required for the owner's unique product to perform as a high-performance building. Considerable expertise in every step of this process is required to get it right. The experts must design building elements and systems concurrently to achieve high-performance across all of them. If they are developed one after the other, choices made to optimize one system may limit those for others. Designing systems in parallel, concurrent engineering allows trade-off decisions to be made so long as system designs are moving forward in sync, at the same level of detail. This is why it is important to organize multidisciplinary cross-functional teams, often called "clusters" and sometimes "Project Implementation Teams (PIT)," composed of people with the capability required to design, supply, and build, as shown in Table 8.1.

Requiring cluster or project implementation teams to talk daily can reduce both response and decision latency (lag time), which contributes to the hidden effort that Levitt and Kunz revealed.

Connect Actions, Information, and Decisions Every Day   David Mann, in *Creating a Lean Culture,* describes how work groups within Lean organizations coordinate and maintain alignment by meeting briefly every day to report who is doing what, surface problems. and take action to mitigate them, if that's possible within the group (Mann, 2005). This is also how scrum works (Sutherland & Sutherland, 2014). Mann explains that alignment comes from having the team leader participate in

TABLE 8.1   Multidisciplinary/Cross-Functional
Team Knowledge

#	Capability
1	Business Operations and Planning
2	Functionality/Use
3	Facility Operations
4	Project Management
5	Production Management
6	Integrated Design
7	Healthy Materials
8	Energy Design
9	Building Controls and Commissioning
10	Building Form and Structure
11	Engineered Systems
12	Modeling and Simulation
13	Life Cycle Costing
14	Fabrication and Assembly
15	Logistics
16	Building Methods

an identical meeting at the next level, and that leader do the same at the next or top level. In a presentation to the 2010 LCI Congress, Paul Reiser of the Boldt Company, presented an image inspired by Mann's method. Figure 8.7 shows this at three levels. Cluster team leaders participate in the daily stand-up meeting of their cluster, in person or by phone if the team is not co-located every day. They also huddle with all of the other cluster team leaders every day. The leader of that group, who might be the chief designer or constructor, depending on the stage of the project, huddles with the core/project management team each day.

What happens in the daily tiered coordination meetings? We recommend that the standard agenda of the basic work-group meeting be the same as the daily scrum. As with scrum, all check-in meetings should be time-boxed to 15 minutes with every member standing up to ensure that they don't exceed that limit. Each team member should report on what they did yesterday to help the team meet its next goal, what they are doing today, and what impediments stand in their or the team's way. The team leader reports his team's progress, their decisions or the ones that they are working toward making, and the impediments they face in the cluster leads' daily check-in. The cluster team leader, likely the person in the role of chief designer or constructor, reports the progress, decisions, and impediments in the top-level project management meeting each day. Decisions and actions taken by the PMT (project management team) to remove the roadblocks are communicated back to the chief designer or constructor immediately if necessary or in the next day's cluster leads meeting. Cluster leads report them

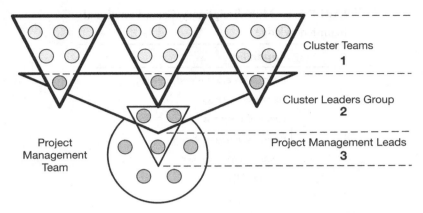

FIGURE 8.7   Three tiers of daily meetings to connect actions, information, and decision. Courtesy of Paul Reiser and Tariq Abdelhamid; illustration adapted by the authors.

immediately or in the next day's cluster team meeting. Everyone is informed about what everyone else is doing in this way, not by e-mail, voice mail, or in a casual one-on-one conversation.

We agree with Mann and Reiser, and do not see a better or more effective way for team members to know what others are doing. We also think this same structure should be used during construction. The teams we've seen who have done this have been successful; the ones who haven't have struggled. We've noticed significant resistance by project leaders who are new to, or not committed to, integration to the idea of meeting, even briefly, twice a day every day. In every case, it is because they do not see coordination as work. (Mann advocated that this be part of "Leader Standard Work" for Lean managers/leaders.) Instead, they see work as being with their own team solving design problems, estimating cost, planning and managing construction, or managing the project. Taking time every day to meet with people outside their company doesn't seem worthwhile and may be uncomfortable.

The result of work groups not talking every day and having their actions, decisions, and obstacles reported and discussed at the next management level is that team members complain that they do not know what others are doing, which causes them to do the same thing two or three times. This increases the hidden effort of coordination exponentially. People get frustrated, particularly the owner who is paying the bill for having many more working on the project early. They legitimately wonder whether they are getting the advertised benefits, and rightly so.

Patrick Lencioni on Daily Check-ins in *The Advantage*

The most powerful impact of having teams meet every day is the quick resolution of minor issues that might otherwise fester and create unnecessary busywork for the team. For instance, when team members don't see each other more than once a week, or even less often, they end up trying to resolve the endless administrative issues that surface every day with an e-mail here and a voice mail there and a hallway conversation in between. That sets off a flurry of more e-mails, voice mails, and hallway stops as the situation changes and more people on the team need or want to be looped in. It would be fascinating to actually track and calculate the amount of time and energy that leaders spend chasing down issues that could be sorted out in a 30-second conversation if everyone were in the same room for a few minutes every day.

A big part of the beauty of the daily check-in is that the leaders know they're going to see their colleagues within 24 hours, so rather than firing off an e-mail or a voice mail or interrupting someone in the course of their day, they simply make a note to bring up a small issue at the next day's meeting. There is something undeniably efficient and liberating about this, which makes the protest I hear from executives all that much more absurd. It's as though they're saying, "Do you realize how busy we are trying to solve problems that result from our lack of communication? We can't possibly spend ten minutes every day preventing them."

Patrick Lencioni (2012)

2. Use Language and Commitments to Coordinate Action

A project organization in which people are taught and expected to use language effectively is far down the road of solving the challenge of exposing, planning, and coordinating work. Rather than being imposed from the top down, control emerges from a self-regulating/governing team. How can this be possible? It begins with people making clear requests for what they need from another. They learn to describe their conditions of satisfaction (CoS) and date when they need what they requested. Knowing the context in which they are making their request because they are immersed in the situation, they may not be surprised when the other person, the "performer," explains what they can provide by that date. They negotiate, perhaps modifying the CoS or delivery date and come to agreement. Now each person can plan for delivery. The requestor/customer has learned not to ask for more than they need and how to describe it much more clearly than they ever did before. The performer/supplier has learned that they can and should say, "No, I can't do that," and make a counteroffer. Everyone has learned that they must understand the CoS and due date, and learned the criteria for a reliable promise (Macomber, Howell, & Reed, 2005).

Five Elements of a Reliable Promise

1. The conditions of satisfaction are clear to both performer and customer.
2. The performer (promissor) is assessed as competent to perform or has access to that competence.
3. The performer has estimated the time to perform the action for completing the promise and has allocated (blocked) that time on the schedule (calendar).
4. The performer is sincere in making the promise. In the moment the promise is made, the performer is not having a private, unspoken conversation, which contradicts fulfillment.
5. Regardless of what the future holds, the performer will make good on the promise—particularly if the promise cannot be performed, taking responsibility for whatever consequences may ensue.

Hal Macomber (2004b)

Very few project managers ask team members to make reliable promises in current practice. Instead they operate on a mental model of directing people and expecting compliance. They do not give people the opportunity to ask whether it's the right thing to do, or can be done in the time requested. Command and control managers, regardless of whether or not they are pleasant or mean-spirited, are building their house of control on sand because they are not providing the opportunity for people to share a different perspective, one that may be better informed because these other people are closer to the actual work.

Fernando Flores began to think about how humans shape the world in language while sitting in a prison camp in Chile after the democratically elected government in which he was serving was overthrown by a military coup (Flores, 2012). After gaining his freedom and earning a PhD at the University of California, Berkeley, he and his business partner, Chauncey Bell, and their associates, including Hal Macomber, helped thousands of business leaders understand a very powerful mental model showing how people coordinate and get things done. Flores called it the "Basic Action Workflow." The diagram, shown in Figure 8.8, shows this "atomic structure of commitment," a term he used to make the point that the workflow was the essential core of commitment making. (Macomber & Howell, 2003).

Three other elements of what Flores called the "Grammar of Action" are critically important for every team member and especially leaders. They are "Assessments," "Assertions," and "Declarations." A manager and leader can confidently express commitment to outcomes by declaring it. "We will deliver every one of our client's conditions of satisfaction within the target cost and make all of the profit which we are entitled." Think of how different this is than "I don't see how we'll ever get to target cost with this wish list." Everyone in an integrated organization must understand their responsibility to state their assessments clearly and without blaming individuals. "Things are all screwed up because Bob doesn't know what he's doing" is an assessment but not a useful one. "I believe we are not making the consequences of indecision clear to client stakeholders" can lead to productive dialogue. The assertion that "we could not complete the soil borings because of the bad weather" is useful, if not good news. Now at least we can replan. Table 8.2 shows the "Grammar of Action," which everyone in an integrated organization should understand.

FIGURE 8.8   Basic Action Workflow showing how people use language to make requests and promises and fulfill them. Courtesy of Fernando Flores.

TABLE 8.2   The Grammar of Action (Macomber & Howell, 2003)

Action	Example	Definition
Declaration	*"We will put a man on the moon and bring him back safely in this decade."*	*Creating a space of action, not to be confused with a promise.*
Request	*"Please deliver the submittal on Thursday."*	*Calling for a statement of commitment.*
Offer	*"I will go to the city for you to get the permit."*	*Performer makes a commitment as the opening of the conversation.*
Promise	*"You can have the crane at noon."*	*Statement of commitment to provide something specific by a specific time.*
Assessment	*"We are making good progress."*	*Offering an opinion with or without any basis for the assessment.*
Assertion	*"All tasks were completed as promised."*	*Statement of fact. Includes an offer to provide evidence.*

Just as important as knowing how to use words is awareness of people's moods. Flores spoke of these not just as feelings or emotions, but as automatic assessments people make about the future based on past experience. He taught company and team leaders that they needed to pay attention to moods assessments so that they could shift them from negative to positive in their organizations. He helped them understand that their levers were building trust in internal and customer relationships and especially cultivating a mood of satisfaction with customers. Without this awareness, people become the victims of their moods and organizations suffer the consequences when they turn sour. These moods, shown in Table 8.3, can be influenced by leaders for better or worse.

The final piece that Flores taught was to listen by establishing a shared space of possibilities for the future. Fernando Flores explained this and the ideas above fully in a collection of his early essays written for clients and colleagues collected and edited by his daughter, titled *Conversations for Action and Collected Essays: Instilling a Culture of Commitment in Working Relationships.*

TABLE 8.3   Positive and Negative Moods (Davis, 1998)

Positive	Negative
Ambition, persistence	Resignation, boredom
Serenity, peacefulness, joy	Despair
Trust, prudence	Distrust, skepticism
Acceptance	Resentment, anger
Wonder	Confusion
Resolution, speculation, urgency	Panic, worry, anxiety
Confidence	Arrogance

3. Create a Learning Organization

We've observed that project organizations that perform exceptionally are ones in which people are encouraged to ask and answer questions together, within working groups, which lead to good decisions that drive the project forward. Conversely, projects that operate based on the thinking of a relatively small group usually run into trouble and underperform. The questions are not only technical, about what and how to build, but also and especially about why team members, including leaders, are doing things a certain way. We've noticed that when people pay lots of attention to decision making, performance in safety, cost, quality and schedule also improve. When score isn't kept and there are no metrics, or only reported up the chain versus within the project, there really isn't a way to focus improvement efforts.

Peter Senge launched a movement for organizational learning with his book, *The Fifth Discipline: The Art and Practice of the Learning Organization*. In 1997, *Harvard Business Review* described it as one of the most seminal management books in the past 75 years, and thousands of people have joined the Society for Organizational Learning (SOL), which he founded, to learn how to implement Senge's theories in their organizations. Senge came out of the Systems Thinking movement and was also inspired by W. Edwards Deming, who taught Japanese industrialists and their engineers how to consistently produce high-quality products by applying the Shewhart Cycle. The Japanese later applied it to improving work processes, which Deming described as the Plan-Do-Study-Act (PDSA) learning cycle. Top managers and engineers from Toyota were among Deming's students and used PDSA to invent and perfect the Toyota Production System (TPS), which became known as Lean production. We encourage readers to study both Deming and Senge to gain a deep enough understanding to sustain continuous improvement and organizational learning.

By the close of the 1980s Peter Senge was convinced that Systems Thinking was not sufficient by itself to change the way people interacted with the natural world, with each other and others in organizations in which they spend most of their time. He had also come to believe that Systems Thinking, thinking of the whole rather than the parts, was a discipline that requires continual practice, just as the martial arts do. This fifth discipline could not be sustained without four supporting disciplines, which we've illustrated in Figure 8.9. He wrote and published *The Fifth Discipline* in 1990 to explain his ideas and founded SOL to help people learn and practice Personal Mastery, Shared Vision, Mental Models, and Team Learning so they could become systems thinkers.

Peter Senge believed organizations had to master learning because of high interdependence, which is our starting point for thinking about integrating project organizations. Senge, building on Deming, talked about how the prevailing system of management had to change. We created Table 8.4 to show how we've seen this system play out on building projects. The left-side list is what Peter Senge believes Deming thought was the prevailing system of management (Senge, 2006). The one on the right is what we have seen on building projects.

The starting point for Deming and Senge was to stop doing things that prevented the people who comprise the organization from learning together. The new system would support what Senge in his 2006 update of *The Fifth Discipline* described as "Core Learning Capabilities for Team," represented as a three-legged stool, shown in Figure 8.10. The legs were Aspiration, Reflective Conversation, and Understanding Complexity. Senge believe they mapped almost directly to three of Deming's four tenets of Profound Knowledge with the exception of theory of variation. The three were "understanding a system," "theory of knowledge" (the importance of mental models), and "intrinsic motivation" (the importance of personal vision and genuine aspiration). We see them reflected in the Five Big Ideas.

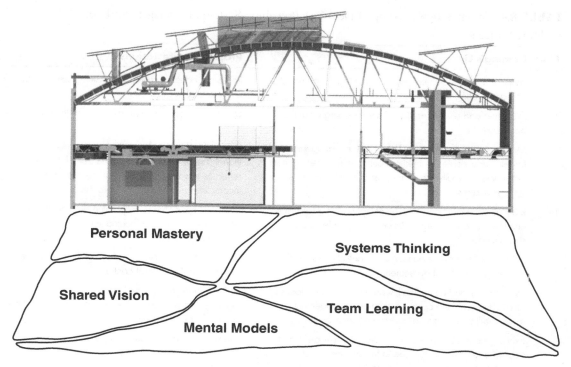

FIGURE 8.9 Peter Senge's Five Disciplines. Courtesy of Integral Group; illustration provided by CDReed.

Senge teaches that people in organizations need to think in "circles of causality," where every influence is both a cause and effect, rather than in straight lines of cause and effect, which leads to blame and brings no lasting change. This is the only way people can understand and deal with complexity and is an axiom of Systems Thinking, the fifth discipline. Senge used the three-legged stool to illustrate that "Understanding Complexity through Systems Thinking" had to be supported by four other disciplines, comprising the other two legs.

The Aspiration leg is composed of "Personal Mastery" and "Shared Vision." As with any discipline, personal awareness and growth is gained through reflection and practice. It starts with each person having a vision for themselves while at the same time being aware of the situation they are in. Senge believes the dichotomy between personal vision and awareness of situation is the source of creative tension, which leads people to draw current reality closer to their personal vision. He also believes there can be no shared vision without people having their own, individual visions. Shared vision is when people hold an idea in common upon which they are willing to act. It is a powerful force, one necessary for achieving almost anything of significance, according to Senge.

The Reflective Conversation leg is composed of "Mental Models" and "Dialogue." Senge teaches that culture and organizations provide us with assumptions of why things are the way they are and how things work. Sets of assumptions form the mental models on which we operate to shape the world. He points out that our mental models often prevent us from seeing a different reality, and especially of understanding things from a systems perspective.

TABLE 8.4   What Senge's Listing of Deming's Prevailing System of Management Looks Like on Building Projects

No.	Deming's List	Project Manifestations
1	Manage by measurement, especially using short-term metrics and devaluating intangibles	Focus on reducing cost for each item, assuming this reduces total project cost.
2	Expecting and rewarding compliance—doing what the boss says.	Do what you're told. Managers do the thinking.
3	Management by outcomes. This is where management sets targets and people are held accountable for meeting them regardless of whether they are possible within the existing systems and resources.	Deliver the design documents on schedule even though the owner decisions have been late. Meet schedule in spite of the fact that the design is incomplete and the permits were late.
4	Right versus wrong answers. Technical problem solving is emphasized and diverging (systematic) problems are discounted.	We don't have time to map value streams or plan because we're already behind schedule.
5	Uniformity. Diversity of opinion and conflict are suppressed in favor of superficial agreement.	Don't raise your concerns in front of the owner, the architect, or the general contractor.
6	Predictability and controllability are the goals and focus of effort. The goal is to control outcomes. Management assumes responsibility for planning, organizing, and controlling.	Push the work through and maintain schedule regardless of problems. Critical path tasks cannot be delayed for any reason.
7	Excessive competitiveness and distrust. People are pitted against each other and made to feel like they must or are in competition with one another, breeding distrust—which is tolerated.	We've got to protect our fee and take care of ourselves before the project.
8	Loss of the whole. Fragmentation is accepted; innovations are not shared.	First, we design in our separate offices; then, we bid and take the contractors with the lowest price and let them figure it out in the field if the design isn't perfect.

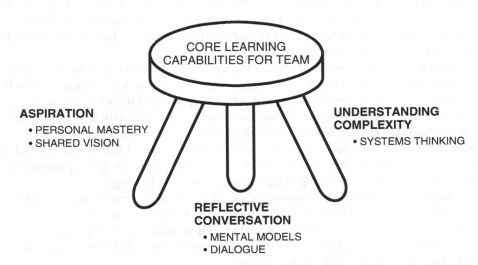

FIGURE 8.10   Peter Senge's three-legged stool of Core Leaning Capabilities for Team. Courtesy of Peter Senge; illustration provided by CDReed.

For Senge, the discipline of mental models is the continual examination of our and others' mental models from a systems viewpoint. Dialogue is the motive force for the discipline of Team Learning. Senge points out that dialogue was practiced by the ancient Greeks and in Native American cultures. It occurs when individuals suspend their assumptions to "think together" by sharing their ideas and perceptions without trying to win over others. Senge believes that teams that practice dialogue become aware of deeply ingrained patterns of defensiveness that undermine team learning. Senge believes that "team learning is vital because teams, not individuals, are the fundamental learning unit in modern organizations" (Senge, 2006).

4. Connect Work through the Project Organization to Customer Value

Use the POP Model to Connect Work through the Organization to What Is Being Built IPD teams must be closely aligned to be effective. Unlike machines on the factory floor or equipment on a job site, the only way humans can align is through conversation, hopefully assisted by powerful visual controls and visualization. To do so, the integrated organization forms multidisciplinary/cross-functional teams to accomplish the work. The organization is the customer of this work and often aggregates information into packages or components into assemblies. These packages are necessary for or are directly value-adding to the building product. We propose the Product-Organization-Process (POP) model (Levitt, Kunz, Luiten, Fischer, & Jin, 1995) described in Chapter 5 as a mental model that can be used to help people connect their work to the project organization as the intermediate customer, and then to the building product itself. We think it's as simple as managers and team leaders asking why and what are we building? What will we do together? How will we do this work? These questions will almost certainly spark a learning dialogue and possibly better ideas before team members implement a process. This can be step 1, useful preparation prior to mapping the value stream for whatever piece of the project team members are taking on. Both POP modeling and value-stream mapping are "learning to see" activities in that they allow people to see and understand the situation. POP models can be constructed in conversations where people are making assessments and assertions about what the project organization should do with the information they are producing so that it contributes the greatest value. Figure 8.11 shows POP's use as a mental model.

These questions can spur dialogue about all sort of things. For example, an estimating team could ask why they are doing periodic estimates? Who are their customers in the project organization? What are these people going to do with the information? How is this contributing value overall? Team

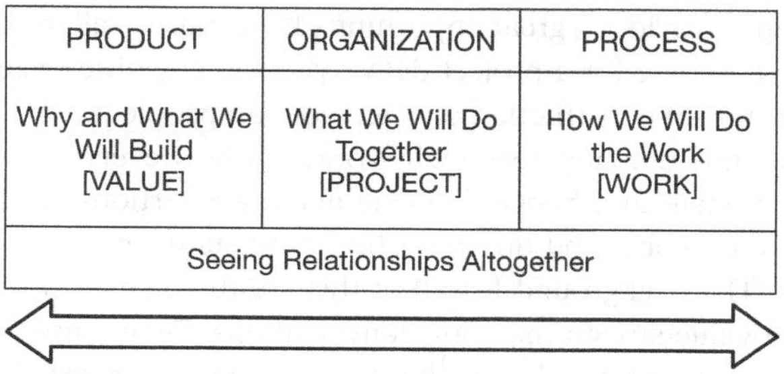

PRODUCT	ORGANIZATION	PROCESS
Why and What We Will Build [VALUE]	What We Will Do Together [PROJECT]	How We Will Do the Work [WORK]
Seeing Relationships Altogether		

FIGURE 8.11 Product-Organization-Process mental model.

managers and leaders could be engaged. Who needs these estimates in the owner's organization and what do they do with them? How often are they needed and for what scopes of work? At the very least, estimators and managers could gain a better understanding of who their customer is and what they really need. Skeptics will say all this is always clear and obvious, and that this is a waste of time. We beg to differ based on our experiences. It seems that what is obvious to an experienced project manager is often not clear to younger people or those who've done things quite differently on other projects. It seems that everyone assumes more than they should. This takes us back to Flores' instructions for being clear on conditions of satisfaction in making requests, offers, and promises, and Senge's counsel on Reflective Conversation. And we can't forget that POP modeling takes place in the context of an interdependent network of people dealing with high complexity.

Map the Flow of Value   Imagine the next step being that managers and the people executing the work would map value streams. Think of the amount of time saved if people began with clarity of purpose and a shared vision on the importance of every process. Unfortunately, we've seen only a handful of teams, all of them on successful IPD and integrated design-build projects, do this kind of work.

Lean practitioners learn that mapping how value is delivered is the essential first step in seeing how to improve any process. That's because people need to understand the current state so they can see opportunities and obstacles to change it for the better. Toyota calls these maps "Information and Material Flow Diagrams" because that is the content. The organization is not mapped, the flow of work products is. In fact, one of the rules is to not ask each area manager to map their piece, but instead ask the people who do the work. The very first step is do decide what a team should map. Most of the time, that is the process used to deliver a family of products, viewed from the customer perspective. The second step is to ask the person with responsibility for understanding a value stream and improving it to act as the Value Stream Manager. This person should report to the top person on the site.

Mike Rother and John Shook, in their highly regarded book, *Learning to See*, describe value-stream mapping as a language for understanding flow. They say that the first step should be gathering the facts on the shop floor, followed by drawing the current state map using special VSM symbols with those most familiar with the work. Those people will think of ways to improve almost immediately, which should be captured in a second, future state map. Mapping the future will often point out current state information that has been overlooked. The final step is to prepare and use an implementation plan that describes on one page how the future state will be achieved. Rother and Shook (1999) discourage allowing the mapping to drag on. They believe that all of this can and should be done in two days.

Mapping value streams provides a great opportunity for people to align themselves and their work processes. The steps are the same for a project delivery team. The difference is that there is no shop floor to visit prior to the start of construction on site. Nor can people go to every home office because they could only see how that company does their work, not how everyone will work together. Team members map the current state in a Senge dialogue making assertions about what should be done, and assessments about what works and doesn't. The future state emerges as requests and offers are made for improvements. The background for all of this are the mental models people carry around, which others question in value-stream mapping conversations. People use language to describe how they currently work, and in the process learn what they can do to improve. That's why the statement, "We don't have time to map processes and value streams" is so wrong.

Work within the Learning Cycle One of the great strengths of Lean organizations and work is "process discipline." That means taking the time and making the effort to develop the best-known way for doing something to produce value for the customer, whether they be next in line or at the end. Value-stream mapping (VSM) is this first step. Toyota's collective genius was to put VSM into the hands of the people doing the work along with their managers, supported by expert facilitation. Japanese engineers, particularly those inside Toyota, redrew the Shewhart Cycle that Deming had taught them for developing quality products, making it into what has become known as Plan-Do-Check-Act (PDCA). Figure 8.12 shows both cycles together. In his writings and four-day lectures, Deming described it as a cycle of learning and substituted "Study" for "Check" to emphasize that this step was not simply checking boxes on a form, but serious analysis that required time, effort, and expertise (Latzko & Saunders 1995). Toyota and other Japanese manufacturers made Plan-Do-Study-Act (PDSA) the cornerstone of continuous improvement efforts by engineers and factory floor people. We see no reason why PDSA cannot be used in design, construction, and commissioning. The prerequisites are education and effort, both of which take time away from people repeating failures and doing rework.

Structure Work to Produce Greater Value Forming and working in cluster teams creates the possibility of asking three critical questions during all phases of design: (1) what will we build? (2) how will we assemble the pieces? and (3) how should we buy or fabricate those pieces? (Ballard, Koskela, Howell, & Zabelle, 2001). BIM in the hands of Virtual Design & Construction practitioners would allow teams to model what will be built, how it will be assembled and how it will be fabricated – as was done for the modular horizontal common rack that we described in chapter 7. Figure 8.13 shows "Work Structuring" at the intersection of Product Design, Process Design, and Supply, emerging through the answers to the three questions. Although they seem simple enough, they require multidisciplinary/cross-functional expertise and participation. IPD is the delivery system that enables

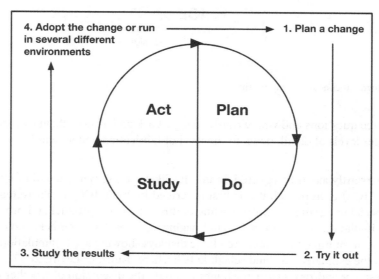

FIGURE 8.12 The Shewhart Cycle on the outside of the Deming/Japanese Learning Cycle on the inside.

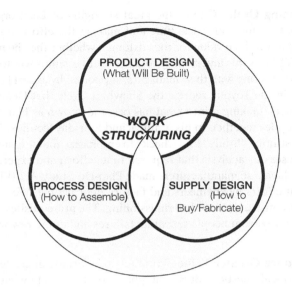

FIGURE 8.13   Work structuring. Courtesy of Glenn Ballard. Illustration recreated by the authors.

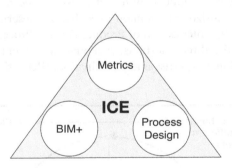

FIGURE 8.14   Integrated concurrent engineering.

this. Ask only three questions and you're done? Yes, just ask and answer them over and over again at progressive greater levels of detail, down to the level of fabrication and assembly.

Engineer Concurrently as an Integrated Team   It's likely you've never heard of integrated concurrent engineering (ICE) even though it was first described in 2003. ICE, as illustrated in Figure 8.14, is a scheduled special integration event consisting of three elements: product and project performance metrics, BIM + simulation, and process design. Imagine delivery team members and owner stakeholders seated in front of multiple interconnected large displays showing a 4-D simulation of construction side by side with the estimated cost and schedule (Chachere, Kunz, & Levitt, 2003).

The chief designer and/or constructor might facilitate the discussion of whether it would be possible to incorporate a new design feature without putting opening day at risk. Every cluster team lead

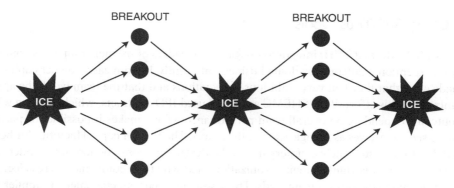

FIGURE 8.15   ICE accordion model. © CIFE/SPS VDC Certificate Program. Illustration recreated by DPR Construction, Lyzz Schwegler.

is there along with project management team leaders. Key owner stakeholders, who do not typically participate in team meetings, will have cleared their calendars to be there. Beforehand, everyone has had the opportunity to read a description of the alternatives at play and knows the agenda for the first of three 4-hour meetings scheduled within a seven workday period. Figure 8.15 shows how the "accordion" meeting schedule will look (John Kunz, Stanford University "Integrated Concurrent Engineering" lecture, created January 12, 2003). Resolving the issue and making the right decision becomes the delivery team's priority. Between the ICE sessions, cluster teams focus their attention on answering the questions that are raised in the previous session and improving possible alternatives.

ICE leverages co-location and multidisciplinary/cross-function teams, cluster lead coordination meetings and the project management/core team. It does not replace them. Incorporating ICE sessions into the design schedule at points where big decisions must be made can significantly accelerate the evaluation of design alternatives and stakeholder decisions.

8.5   REAL-LIFE EXAMPLES

Has anyone mastered a complex building project by creating an integrated organization using the Five Big Ideas themselves or the equivalent? Yes, certainly. We've closely observed this on four projects and are confident there are many more that readers will tell us about. We describe three of them in the pages that follow in this chapter and the fourth in Chapter 13. In this section, we present thumbnails of the Sutter Health Eden Medical Center (SHEMC) and Temecula Valley Hospital (TVH) projects as examples of how integrated project organizations are created and operated. Then we follow up with a case study of the University of California San Francisco (UCSF) Medical Center's Mission Bay Hospitals project.

Integrated organizations and outcomes are formed by disciplined practice. The SHEMC team followed the Five Big Ideas while applying BIM and Lean construction tools such as target value design and the Last Planner® System. The TVH team focused on Lean thinking, the use of BIM, and the same Lean construction tools as SHEMC. Applying Lean construction on projects requires the Five Big Ideas, whether or not they are made explicit. Essentially, the "operating systems" of both IPD projects were very similar.

8.5.1   Clear, Ambitious Goals

In the cases of SHEMC and TVH, integration began with leaders of partner companies committing to delivering an exceptional outcome by doing business differently. In both cases, owner capital facilities leaders made clear that they had very ambitious expectations and that the team would be appreciated and recognized for this effort. Both SHEMC and TVH used IPD contracts, which set up a win or lose together outcome depending on overall team performance, which makes it possible for partner company teams to see themselves as a single team at the outset. The VPs for capital facilities for both Sutter Health and TVH made sure their counterparts in the design and construction firms understood why each project was very important to their organizations and why their companies were selected. These conversations included the project team leads. These were not one way, customer to supplier "orders." The IPD contracting process makes reversion to the "I want/my way" business-as-usual approach very difficult because the owner has agreed to share governance with delivery partners. In both cases, these IPD partners quickly gained confidence that top and project-level owner leaders had everyone's interest at heart, not just their own. Both the owner and IPD team members were committed to the project and each other.

Both owners were clear on what success would look like for their projects, especially for the people who would work in and use their hospitals. Sutter Health made the first exhibit in the integrated form of agreement (IFOA) for the Eden Medical Center a clear statement of expectations for what the project delivery team would deliver. The introduction to the document prefaces the goals as follows: "A project is not considered successful by the owner unless it meets the owner's goals. Often, these goals are unstated, not clear, vary with time, or vary with person. On this project this will not be the case. The goals will be explicitly stated in this document."

Although the "Official Statement of Owner's Goals" did not use the POP framework we discussed above, every statement can be placed within it. It's worth noting that the team discussed these goals over the nearly six-month period in which they validated that the clinical program could be designed and built within Sutter Health's time and money constraints. Stating the goals in the contract emphasized their importance. Seeing them in a POP matrix together with the three classic design questions of form, function, and behavior of the building, as shown in Table 8.5, invites a dialogue about what the project organization and work groups should do.

Bill Seed, then Universal Health Services Staff VP Design & Construction, invited three teams to develop innovative healthcare concepts for a new medical center in Temecula, California. Later, after launching the project and forming an IPD team, Bill made the establishment of Conditions of Satisfaction (CoS) for and by the team the first order of business. Every team member had an opportunity to offer his or her thoughts and decide on what the CoS would be. As with Sutter Health's statement of goals, we've also put those developed by Bill Seed and the Temecula Valley Hospital team into a POP matrix, shown in Table 8.6.

IPD team members on both the SHEMC and TVH projects have said that they believe the early focus on goals and outcomes, especially for the integrated organization, laid the foundation for the success that both teams enjoyed. Making customer goals explicit at the outset provided anchors for team members as they developed ways of working and interacting to solve the considerable technical challenges they faced to make good on their commitments to Sutter Health and Universal Health Services.

TABLE 8.5 Sutter Health Eden Medical Center: Official Statement of Owner's Goals in a POP matrix

Question/ Lever	Product	Organization	Process
Function	**Goal 4: Health Care Delivery Innovation** Cellular concept of health care design to be utilized Control center concept to be utilized Electronic health records system implemented	**Goal 6: Design & Construction Delivery Transformation** The building will significantly transform the delivery model for the design and construction of complex health care facilities	
Structure/Form	Validation Report The Validation Report described the use, capacity, and square footage for each space in the "Clinical Basis of Design."	**Goal 6: Design & Construction Delivery Transformation** Integrated form of agreement contract be utilized Higher number of direct signatories to contract, higher percentage of total budget under IFOA New incentive structure New method of defining project goals	**Goal 6: Design & Construction Delivery Transformation** New methodology for the design process New methodology for planning and tracking commitments New methodology of active engagement with the state regulatory agency Far more extensive usage BIM I Virtual Design and Construction Use of target value design Sophisticated Commissioning & O&M Handover Energy Modeling
Behavior	**Goal 2: Project Cost** Total Project Cost of the project shall not exceed $309,000,000. **Goal 5: Environmental Stewardship** Meet anyone of the following: 1. The standards for certification at the SILVER level per LEED for Health Care (draft version) 2. The standards for certification at the SILVER level per LEED NC v2.2, 3. Achieve CERTIFIED level per LEED for HealthCare (final) 4. Achieve CERTIFIED level per LEED NC v3.0.	**Goal 1: Structural Design Completion** The first incremental package will be submitted to OSHPD for review no later than December 31, 2008. **Goal 3: Project Completion** The replacement hospital shall open fully complete and ready for business no later than January 1, 2013.	

TABLE 8.6   Extracts from the Invitation for the Temecula Valley Hospital the Initial Study and Conditions of Satisfaction Placed in a POP Matrix

Question/ Lever	Product	Organization	Process
Function	Invitation Letter	Invitation Letter	
	"It is absolutely imperative that innovative concepts be employed to optimize the size of the OSHPD compliant facility." "Creativity should stem from multi-use service spaces and consolidation of traditional departments in an attempt to deliver efficient healthcare."	"…self form creating a highly OSHPD experienced, Lean, Integrated Project Delivery team."	
	"An approved site plan showing building pads, parking and boundaries is included. Make every attempt to limit significant alterations to this plan to stay within our current entitlements. The hospital site is limited to 6 stories and the MOB sites to 4 stories."	"teams might consist of planners and builders who can collaborate well at a conceptual level to best define an innovative solution while understanding the economic impacts of that plan."	
Structure/Form			
Behavior	3) Community and Social Responsibility	1) Project Delivery Success	
	• Positive press in the local and regional press	• Maintain Conditional Use Permit by securing major modification approval in November 2010	
	• Physician buy-in as reflected by hiring rates	• Maintain or reduce the Target Value Cost of $144M for 140 beds	
	• Neighborhood satisfaction score of 3.5 (out of 4)—survey to be conducted	• Deliver the Owner's Manual six months prior to opening (approx. 3rd quarter 2012)	
	5) Facility Operational Success	• Certificate of Occupancy by the 1st quarter of 2013	
	• 30% more operationally efficient than the best performing UHS facility	• Construction safety reflected by?	
	• Patient Family Centered Care Delivery and Design reflected by HCAHPS scores of _____.	2) Project Team Participation and Satisfaction	
	• Safe Patient Care Environment by improving/reducing _____.	• Every team member firm finishes this project with a profit	
	• Community endorsement by the use of our facilities versus others in the area	• Secure one new project as a team by the issuance of the Certificate of Occupancy	
		• Two visitors (owners/industry colleagues/ additional team firm employees) in the Corona Big Room per month	
		• Two or more educational presentations in the Corona Big Room per month	

TABLE 8.6 (*continued*)

Question/ Lever	Product	Organization	Process
		• Every team member an active participant in at least one Lean organization	
		• Predictable outcomes as a result of labor efficiency	
		• Reliability and trust as shown by measuring promises made versus promises kept	
		4) Relationships with Regulatory Agencies	
		• Maintain promise of UHS being OSHPD's best customer	
		• Zero defects in all agency submittals	
		• Drawings in OSHPD possession for a time period 15% lower than the lowest established records	
		• "No excuses" surrounding OSHPD, City, etc. for not meeting CoS milestones, etc.	
		• Trade partners considered a business partner of OSHPD at the completion of the project	

The people of the IPD partner companies for the SHEMC and TVH projects became more than business partners. They formed strong professional relationships and aligned their thinking and work practices beyond what any one of them had ever done before on a project. The trust people gained in each other by speaking clearly and acting with good intent every day allowed them to share leadership at every level and collaborate deeply to solve problems that stop many other project teams dead in their tracks. Although the sketches we provide in the following pages are brief, they show why and how this unfolded.

8.5.2   Sutter Health Eden Medical Center Hospital

Eleven companies became IPD partners to design and build the SHEMC replacement hospital on the Eden Medical Center campus in Castro Valley, California. Four years earlier, Sutter Health had stopped the work of another design-construction manager team after the estimated construction cost became too high. This time around, Sutter Health wanted to leverage the Lean health care design concepts of David Chambers, then their Director of Design. Sutter engaged teams of architects and builders through a competitive and cooperative process called the "Coopetition" in order to develop prototype designs that could be used as a starting point for new hospitals. They asked one of the "winning" architects, the Devenney Group, to form a team with the general contractor, DPR Construction, that had successfully delivered a large non-acute medical center using BIM and Lean construction practices. Devenney invited consultants it had worked with on the Sutter Health Prototype Hospital Coopetition, and DPR invited the contractors from the medical center project. Sutter proposed adding Ghafari Associates to join the team as a partner for leveraging BIM for Lean processes.

Sutter Health asked Digby Christian, one of its project managers who had previously managed a much smaller medical center IPD project, to manage the SHEMC project. Digby brought confidence in IPD and a disarming ability to question everything. He believed (and does today) passionately that team intelligence was more important than any single individual or organization. Digby knew that he had to bring back decisions from within the Sutter Health organization, and also work through many others with the IPD partners in the Core team. Including the existing hospital's director of Facility Operations in the project leadership team made that easier. Besides Digby and his operations counterpart, the project executives for the architect, general contractor, one design and one trade partner sat on the Core team. Their practice was to talk until everyone agreed on the decision. No one, including Digby, could decide by himself or herself. The Five Big Ideas were true north for how the team was going to operate.

The replacement hospital had to be delivered on time and for the funds available. Otherwise, there was no project. Starting with a high-level program, the team spent two months developing a concept, estimating the cost, and planning how to construct it within the budget and schedule constraints. The team presented two "Validation" reports in succession to Sutter Health and waited for approval. Once the board of directors accepted the report and authorized the project, the team was faced with a big decision. Should they start designing even though the detailed clinical program was not firmly established or wait for two months? They decided to wait and lose the time rather than develop designs that might not work. The team remained small, but together.

The team searched for office space near the project site and rented a storefront on a main street. Although it was narrow and deep, it worked as home base. Soon the walls were covered with a large value-stream map, concept drawings, schedules, spreadsheets, and other documents. Two 4 × 6 foot computer displays were added when design began in earnest. Design professionals from Phoenix and Los Angeles would meet with Bay Area consultants and contractor staff for a minimum of two days every other week, and often weekly. The Big Room meeting agendas were packed with design reviews, BIM coordination, cluster team breakouts, value-stream mapping, and work planning. Weekly budget updates were also done in the Big Room meetings along with reviews of the project risks and opportunities. Commitments made and kept were tracked. Support staff in the home offices would often participate remotely via web conferencing. Shortly before the start of construction, the Big Room moved to a set of job site trailers configured as a large open office, a big meeting space and four smaller meeting rooms. No one, including Digby Christian, had a private office.

Soon after design commenced, the team saw that the market cost to deliver the detailed program, along with operational and sustainability goals, exceeded the target cost. From that point, the core team asked the cluster teams to develop alternatives to deliver equivalent value for lower cost. They made a rule that no concept or detail could be proposed without first considering its cost and schedule impact. This work had to be done within the design review schedule Sutter Health had negotiated with the Office of Statewide Health Planning and Development (OSHPD), the state agency responsible for safe design and construction of all hospitals in California. Big decisions had to be made quickly so that the structural system could be designed and modeled for submission and review under the existing building code. The project team was able to meet that and all subsequent design package delivery dates because they continued to map and refine work processes and pull work backward from the milestones. Making cost an input for design decisions allowed the team to maintain the project target costs through the completion of fabrication-level design documents.

Value-stream mapping and pull planning[1] sessions were led by whoever was most qualified, which depended on the discipline/scopes in focus. Design and construction leaders were always available in person. Team members continually planned and replanned activities and handoffs for design and BIM with sticky notes on top of the value-stream map on the large wall in the Big Room. Many sessions began with Digby or Samir Emdanat of Ghafari asking team members to map the best way to work for the project as a whole. Team members responded by progressively refining and sharpening their processes, especially when they needed to make up for lost time. The 40-foot-long map was incrementally transformed into a pull plan to achieve near and midterm milestones. The BIM Execution Plan was based on the value-stream map.

Like all California hospitals, every detail of the SHEMC replacement hospital design had to be reviewed and approved by OSHPD before that particular piece of work could be installed. Digby Christian's curiosity led him to ask people why they did what they did, and whether they could think of a better way. Samir Emdanat brought training and experience as a design architect along with a very good understanding of Lean and BIM gained from work on General Motors' "Digital Factory" initiative for constructing auto plants. The project team's collective knowledge and experience with OSHPD informed their value-stream mapping. Across the board, team members were universally interested in discovering better ways to work together. Then there was the reality that designers and major trades had become business partners when they signed the IFOA, putting their fees and profits at risk. All of these factors motivated team members to design systems, then model and coordinate them so they could be efficiently prefabricated and assembled. Everyone came to understand they had to do this before the materials and components were ordered to reduce rework, which Digby described at the end of the project as the "Point of Release" strategy for reducing risk for the owner (Figure 8.16).

Core team members and cluster and IPD partner company leads participated in weekly design coordination, cost review, and planning meetings. They reported status, revisited the work plan, and committed to what their teams would accomplish in the next two weeks. Digby Christian expected every team member, and especially cluster leads, to express their concerns about anything threatening the budget or the schedule. He challenged cluster leads to speak out rather than remain comfortably silent, which designers and contractors sometimes do to avoid being seen as negative and disruptive. By taking a role as an active participant and agitator, Digby helped cluster leaders get "comfortable with being uncomfortable" and start to shift the paradigm to deep collaboration and integration.

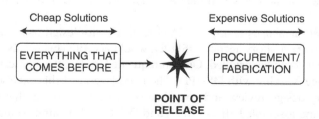

FIGURE 8.16 Digby Christian's "Point of Release" strategy. Courtesy of Digby Christian and Sutter Health. Illustration recreated by DPR Construction, Lyzz Schwegler.

8.5.3   Temecula Valley Hospital

Seven companies came together as partners to design and build the Temecula Valley Hospital. Universal Health Services (UHS) was the owner. HMC Architects was the architect of record, and DPR Construction and Turner Construction formed a joint venture to act as the construction manager/general contractor. Southland Industries was the design builder for heating, ventilating, and air-conditioning (HVAC) and plumbing. Bergelectric was the design builder for electrical. Southwest Fire Project was design builder for fire protection. DPR Drywall, a division of DPR Construction, was brought in as the interior framing and drywall trade partner. All of the delivery team partners believed that Bill Seed, the Universal Services vice president for Capital Facilities, wanted them to succeed financially. In fact, knowing that gave the partners the confidence to sign the IPD contract when the profit pool was well below what they were entitled to earn.

The team's first estimate, based solely on their understanding of the scope, prior to starting design, was several million over the "Allowable Cost" target. In approximately 11 months, the team showed that they could deliver at the target, but with one-third of the profit they were allowed to earn according to the contract. The architects, GC and design-build engineers, and estimators worked diligently together to prevent wasting time and energy carrying forward concepts that would break the budget. It wasn't easy and there were differences that had to be talked through. HMC Architects invited the builders into a process that they had previously used only with other design professionals. They asked all the IPD partners to participate in room-by-room reviews of features and locations drawn on a whiteboard so they could easily be changed. The room designs could not be advanced until every discipline agreed that they were right.

Discipline and construction experts with varied responsibilities came together in six clusters to analyze designs for constructability, cost, schedule, and facility operations. The architect's PM and project superintendent led the cluster dealing with entitlements and regulatory agencies. Site/civil included the architect, GC estimator, and project superintendent. The design-build partners and GC estimators made up the mechanical, electrical, and plumbing cluster. The structural engineer and steel fabricator/erector and GC project executive worked together in the structural cluster. The architect's designer worked together with the drywall trade partner and subcontractors in the skin cluster. The interior design architect worked with a GC estimator in the interiors cluster.

The TVH team leads met two full days each week during design, but not in their own collocation space until they mobilized on site. It wasn't until then that the team jelled. The project leadership team coordinated directly with the cluster leads, who attended the Big Room meetings during design. The cluster teams were kept small while staff members working in their home offices did design work. The cluster leads were responsible to the project leadership team for aligning their design teams with the decisions made in the Big Room meetings. These occurred in different places as space was made available.

One of the biggest challenges facing the TVH team was to design using design details that the California State hospital regulatory agency, OSHPD, would accept. This was because every detail had to be approved before it could be installed. One prerequisite for almost every team member, regardless of level, was experience working with OSHPD. The team leveraged this experience as they detailed. The questions in every design review and coordination session were, "Is this technically correct?"; "Can it be fabricated and assembled efficiently?"; and "Will OSHPD approve it?" Trade general foremen worked with their BIM specialists to improve prefabrication and installation. They also planned construction workflow beginning early in design, as it was being developed.

The Lean construction consultant, internal expert, and coaches educated designers and builders on what is considered waste by Toyota and Lean thinkers. All of the project design and preconstruction, foremen, and crew leads were introduced to the PDSA learning cycle during "on-boarding" training. They also learned how to pull-plan and make reliable commitments. These concepts and practices, coupled with many years of collective frustration with conventional construction management, inspired much deeper collaboration and ambition to eliminate waste and strive for continuous improvement across all trades. This resulted in the deepest implementation of Lean construction we have seen.

Once team members were co-located, informal and structured collaboration increased significantly. Wednesday was meeting day for the project leadership team and the general and trade foremen directing the construction crews. Core team members joined design leads and estimators in Big Room management and planning meetings. These sessions continued each week during construction with Core team members joining project managers; superintendents and foremen in a series of planning and management meetings scheduled one after the other throughout an entire day in the Big Room. The schedule included a "Gemba walk" for the project executives and managers to walk with the superintendents and general foremen to see construction with their own eyes. Project managers, including UHS people, were required to attend the weekly production planning meeting that established next week's schedule based on promises made by the foreman. The managers heard about any issues that needed to be resolved for work to move forward.

The project leader for each partner, including UHS, had authority to make decisions. These people managed the project together in a "core team." There was "no rank in the room" in these meeting; everyone was expected to state what they believed to be true. The core team's practice was that everyone had to agree; when they couldn't, they would call Bill Seed. Core team members jointly managed the project budget, meeting each month to report cost increases and savings to their IPD business partners. They agreed on the project and scope budgets against which they could bill each month.

8.5.4 Outcomes

The Sutter Health Eden Medical Center replacement hospital was delivered for the target cost one week ahead of the milestone for OSHPD sign-off. It opened for business six weeks ahead of the "first patient" milestone. There were almost no compromises to the space program, which is very rare in hospital construction in California. All of the owner's goals were achieved. The building achieved LEED Silver certification, which is extremely difficult for a facility operating 24 hours every day of the week. Ninety-seven percent of inspections passed the first time. The injury and lost-time incident rates were very low. The "time-on-task" for the major trades was 74 percent, far above the industry average of 50 percent.

The IPD partners made somewhat less than their allowed profit because they exceeded the target cost. This was mostly attributable to rework and lower productivity due to out-of-sequence work caused by OSHPD rejecting details for end of wall intersections with the exterior skin. This was a consequence of having to find a way to connect interior walls in an extremely rigid structure to an exterior skin system designed to be flexible. The end-of-wall connection details approved during OSHPD design review were later rejected by the inspectors-of-record on site. The IPD team learned that they had to be certain that design details were "inspectable," meaning that there was precedent or clarity that an exception would be allowed.

Outcomes for the Temecula Valley Hospital met and exceeded the owner's expectations. Bill Seed sent an e-mail message to all team members praising their achievements. Cost to construct was

40 percent below market. The constructed cost per bed was significantly lower than for any other hospital in the last several years. Universal Health Services believed that the facility would meet their goal of 30 percent operational improvement. The 17 lbs./sq. ft. was the lightest structural steel framed hospital built under the new OSHPD requirements. The team had maintained a very good relationship with OSHPD all the way through the project. The IPD trade partners had achieved a 200+ percent increase in labor productivity. The Temecula Valley Hospital was the fastest construction of a similar-size California hospital, built from foundations to substantial completion in 20 months. No contractor lost and almost all made money on the project. The IPD partners earned additional profit from shared savings due mostly to increased crew productivity. To a person, every individual on the delivery team reported that the teamwork was the best of any project in their experience.

8.6   A CASE STUDY: INTEGRATING THE UCSF MEDICAL CENTER MISSION BAY HOSPITALS PROJECT

8.6.1   Forming an Integrated Organization

Any successful integrated organization is heavily dependent on the goals and constraints of the project. The same is true for the UCSF Mission Bay Hospitals, which were opened in 2014 after a 45-month design and construction schedule (Figure 8.17). It currently serves pediatric specialties, the adult surgical oncology program, and a women's birthing program. There are 16 imaging rooms,

FIGURE 8.17   The UCSF Mission Bay Hospitals. Licensed by The Regents of the University of California on behalf of its UCSF Medical Center; courtesy of Stantec Architecture Inc.

20 operating rooms, and 289 patient rooms. Heating, cooling, and power for the hospital and connected outpatient clinic are supplied by an energy center. Altogether, the facility is 878,000 gross square feet and has 60,000 square feet of roof gardens. The total construction budget was $765 million, pared down from an original $965 million cost estimate; the project budget was $1.52 billion. It earned LEED Gold and was opened eight days early.

An Award-Winning Project

- 2016   Engineering News-Record (ENR) Best Projects, Best of the Best, Health Care
- 2015   Structural Engineers Association of California (SEAOC) Excellence in Structural Engineering Awards, Award of Merit—Sustainable Design
- 2015   Shaw Contract Group—Design is ..., Awards, Design Award Market Winner—Health Care
- 2015   Precast/Prestressed Concrete (PCI) Institute Design Awards, Best Health Care/Medical Structure
- 2015   ENR California's Best Projects 2015, Best Project Award—Health Care
- 2011   Celebration of Engineering and Technology Innovation Awards, Scenario-Based Project Planning

Stuart Eckblad, formerly director of Project Administration, Kaiser Permanente, came to UCSF Medical Center as director of Design and Construction in 2006 as a long-time advocate and practitioner of collaboration. Long before IPD came into being, Stuart joined with other industry leaders to cofound and serve as president of the Collaborative Process Institute (CPI) in the mid 1990s. CPI's mission was to educate owners and industry leaders about building collaborative cultures to enable project teams to deliver extraordinary results. As associate director of Kaiser Permanente's National Facilities Services, Stuart worked with other industry leaders in Kaiser's Alliance program to incorporate CPI's ideas and practices in building collaboration on their projects. In 2007, he chaired the AIA California Council committee, which drafted and published the "Working Definition of Integrated Project Delivery."

UCSF Medical Center is one of the leading research hospitals in the country and UCSF leadership and the community could not settle for anything less than a world-class facility. Stuart was determined that the new UCSF Mission Bay Hospitals project would be designed and built by an integrated team of collaborators. Working with system-wide and campus leadership, Stuart was able to adapt university practices, contracts, mindsets and policies to facilitate the signing of Cost Plus Guaranteed Maximum Price (CPGMP) contracts with the general and major specialty contractors, which embraced a target cost and innovations to achieve $200 million in project savings from the original cost estimate.

Because of constraints dictated by statute and university policy and procedures in effect at the time, the architect and engineers initially worked alone with a cost consultant, hospital administrators, department heads, doctors, and support staff to define the program, establish design concepts, and develop the design without the collaboration of the contractors who would eventually construct the facility. The estimated cost continued to rise, while at the same time the world economy crashed, and the U.S. economy sank into recession causing difficulty in raising funds beyond what the State of

California could contribute to the hospital. By the time the build team organizations were brought on board beginning one year after the start of design, the cost of construction had to be significantly reduced. At the same time, the design had to be detailed, modeled, and coordinated to meet the schedule negotiated with the OSHPD for building code review.

Traditionally, the response to such intense cost pressure would have been for the facility owner to take advantage of the depressed construction market by putting the project out to bid for a fixed price, thereby exposing the project to bid busts and cost and schedule overages. However, Stuart had indeed succeeded in convincing university officials that he could produce better results by engaging the general contractor and key trades early and adopting major IPD concepts. Stuart in 2008 authorized the leasing and modifications of 17 job-site trailers that would be joined together to form the Integrated Center for Design and Construction (ICDC)©,[2] a very large open office on site where team members would be co-located to spur team integration. DPR Construction was selected in a competitive process as the GC and joined the project in August of 2008, followed in December of 2008 by eight design-assist MEP, concrete, steel, drywall, and doors-frames-hardware subcontractors chosen via a competitive process. All were contracted to work under CPGMP contracts with target costs and target incentives to reduce overall construction cost by $200 million and to meet schedule and other constraints.

Stuart along with leaders of Cambridge CM, Anshen + Allen (now Stantec Architecture), and DPR chose Stanford University's Center for Integrated Facility Engineering (CIFE) to lead a four-day team building workshop on virtual design and construction (VDC). They hoped to achieve two goals: first and primarily to form an integrated team with the new members, and second, to teach project managers, designers, and BIM specialists how to create and use multidisciplinary models of the facility (the product), and to leverage their organization, and their work processes to deliver the greatest value for every dollar UCSF Medical Center invested. The Mission Bay Hospitals team VDC workshop was held on the Stanford University campus in March 2009, two months before the team began co-locating in its on-site "Big Room" in June 2009.

The VDC Method diagram in Figure 8.18 shows how virtual design and construction unfolds. The facility to be built is represented in building information models. People responsible for installing the work coordinate systems and components. Work is planned and simulated by linking BIM objects to schedule activities. Information flow to build what is represented in BIM is modeled tied to the schedule to make sure the people in the project organization can process it in time. The dotted lines in the shape of a "V" illustrate how VDC project delivery teams can use metrics to predict performance of their designs, work processes and project organization. This is not possible without BIM and experienced designers and builders working together at the same time.

The Mission Bay leadership team decided to bring together a mix of design professionals, construction project managers, engineers, and shop drawing detailers. Including the executives who were invited to drop in periodically, nearly 40 people came together in the CIFE "iRoom Lab," surrounded by big touch-screen displays. Project leaders set clear overall goals for the workshop.

- Develop VDC implementation plans for the Mission Bay Hospitals project.
- Establish a culture of production management and measurement.
- Recognize the importance of short latency for superior project performance.
- Understand the role of VDC for superior project performance and plan for the incorporation of VDC into the daily work processes.

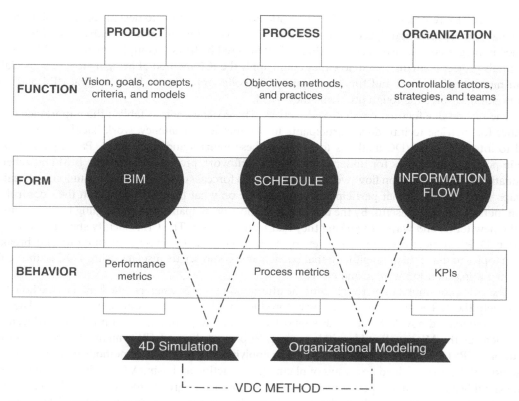

FIGURE 8.18   VDC Method. Illustration provided by Raymond E. Levitt, Stanford University; re-created by DPR Construction, Lyzz Schwegler.

- Learn how to work in and manage with integrated concurrent engineering sessions.
- Learn how to design the product, organization, and process of the project with VDC methods.

Every team member felt a sense of urgency to achieve a high level of collaboration quickly to face the major challenges. Very few, if any, had ever worked in a Big Room. Few had ever heard of VDC. No one had ever been asked to value engineer while detailing design for permit submittals. The challenge for the CIFE facilitators was to clearly communicate new ideas and practices in a very short period of time. Success would be measured in how quickly and well the Mission Bay team could apply what they learned.

Each day had a theme and goals. The first day focused on understanding VDC best practices in each of the participating companies, establishing the vision for VDC use on the project, and planning the use of VDC to address a specific challenge anticipated on the project. Representatives of each organization described what they believed was their best application to date following a template they had been given before the workshop. UCSF Medical Center leaders shared their vision for the Mission Bay Hospitals project along with what they were hoping for from the deployment of VDC. Becky Wheeler of the Jet Propulsion Laboratory (JPL) explained how JPL developed and used an approach they labeled "Extreme Collaboration" to dramatically reduce the time to design space missions for the National Atmospheric and Space Administration (NASA). CIFE researchers studying the JPL process named it integrated concurrent engineering (ICE). Inspired by JPL's accomplishments, the Mission Bay Hospitals project team discussed how they could apply the ICE method of bringing together specialists from multiple disciplines and functional groups to collaborate intensely, leveraging BIM and other information to solve the tough problems they faced.

The second day focused on VDC and metrics. The goals were to establish the importance of and culture for metrics; to introduce participants to key metrics for latency, work backlog, and costing; and to introduce key VDC methods to establish these metrics and track them. Participants learned the importance of metrics for understanding how well work processes, including BIM and schedule simulations and information flow, were working, and to forecast cost and schedule outcomes. Breakout groups were formed so that participants could decide on what factors were within their control and what metrics would be useful. By the end of day 2, most participants were struggling to make sense of all the new terms and ways of working they had learned about. The CIFE faculty stayed with George Pfeffer, DPR's project executive, after everyone else had left, to revise presentations and breakout group topics to make them tangible so that participants could see themselves using VDC as the vehicle for integrating their knowledge and effort.

The going continued to be rough until the afternoon of day 3, when people began to see how VDC could help them. The theme of the day was "production planning and control with VDC." The goals were to understand how VDC methods supported production planning and control and to develop an approach for the Mission Bay Hospitals project. Roberto Arbulu of CIFE member Strategic Project Solutions (SPS) shared his and SPS's experience applying lean production methods in design and construction. Roberto described new ways of planning production and using VDC to insure that materials and assemblies would fit and perform as intended. The breakout groups thought about how they could measure the effectiveness of their planning efforts. By the time participants gathered for wine and cheese at the end of the day, the conversations were about how they could put these new ideas into practice in a way that could really help the project.

Day 4 focused on "putting it all together and implementing VDC." Again, participants broke into small groups to draft a BIM implementation plan and decide on metrics for target costing, clash resolution, response latency, and planning reliability. The team working on the hospitals building drafted their own charter. Team leaders decided how to group people in the Big Room. Later in the day, they mapped the flow of information between functional groups working on the three buildings that made up the Mission Bay Hospitals project. The workshop ended with participants presenting their plans to the project sponsors. Everyone was upbeat and optimistic as they finished their session at CIFE. Nine months after the workshop, team members were surveyed to determine the success of the workshop, and people who had been skeptics said the workshop was one of the most helpful early events in the project.

8.6.2   Operating as an Integrated Organization

The Mission Bay Hospitals team applied what they learned at the CIFE workshop on their project and made significant innovations building upon what they learned. Team leaders worked together to put into place structures to allow decision-making at the lowest possible level, rather than forcing team members to go through multiple time-consuming approval levels. In order to develop detailed designs for production, the team co-located in a Big Room on the job site, and became collaborators, as Stuart Eckblad was counting on them to do.

Clusters were determined for systems and functions: site; structural; exterior; interiors; mechanical, electrical, and plumbing; and equipment. Building teams were organized first for virtual building and later for construction of the hospitals, outpatient clinic, and energy center.

They were composed of project managers, designers, building information modelers, MEP engineers, general foremen, and representatives of the general contractor and facilities operators. Team members pull-planned work from target dates for delivering design packages while meeting cost-reduction goals. A special group was established to review modular and prefabrication opportunities. The BIM team completed the BIM Execution Plan and followed it as they modeled and coordinated thousands of square feet each month. Figure 8.19 shows the flow of information and decisions within the ICDC organization.

The image shows the overall organization of the UCSF Mission Bay Hospitals project in the ICDC. Team leaders organized the project into production teams that worked on producing and coordinating the BIM, design teams, and project control groups responsible for cost, schedule, and quality. These teams elevated big issues to three "captains," who could refer really big questions to three top

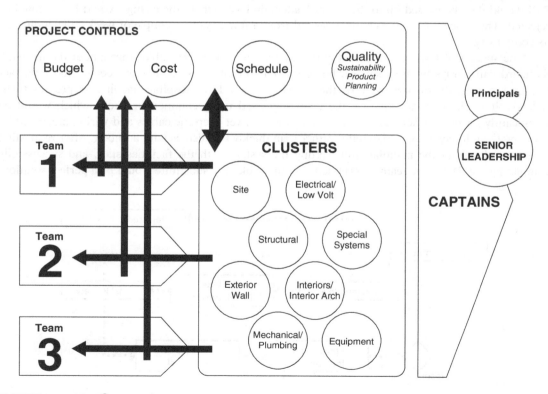

FIGURE 8.19 ICDC© process flow diagram showing the integrated organization for the project. Licensed by The Regents of the University of California on behalf of its UCSF Medical Center; courtesy of Stantec Architecture Inc; illustration recreated by DPR Construction, Lyzz Schwegler.

senior leaders. Together, the captains and senior leadership teams made decisions to keep the project moving forward.

Project team leaders at all levels focused on reducing latency. That's why they allowed the building team BIM specialists to make decisions without supervision from the cluster or senior leadership level if they did not impact cost, schedule, or sustainability goals. Because the modelers were working together in the same physical space, most issues could be resolved together by multiple specialists during one session, rather than being passed from person to person over a longer period of time. Overall, 95 percent of decisions were made within 30 minutes, and the rest within a day. DPR project executive Ray Trebino felt that many of the issues that the integrated project team solved within one day would have taken several weeks in a traditionally organized team.

Leaders developed and drove the adoption of the process for documenting, analyzing, and deciding on value engineering proposals. Larger decisions, which impacted cost, schedule, or sustainability goals, were passed up to the cluster and senior leadership levels for approval using a process and template developed by UCSF Mission Bay Hospitals project leaders called Project Modifications and Innovations© (PMI),[3] which is shown in Figure 8.20. First, the building team would discuss and vet the idea, gathering as much input as possible from all clusters that could be impacted, along with the facility operators, architects, and so forth. If a building or cluster lead signed on as a "Sponsor," the PMI would then be passed up to the senior leadership level, where the change would be accepted or rejected. The process for large changes ranged from a few days to a couple of weeks, depending on its complexity.

One successful PMI centered on the tray system, which carries cables throughout the hospital. Standard cable trays were expensive, but one common alternative, J-hooks, seemed too unattractive and potentially disorganized to the UCSF Mission Bay Hospitals facilities representatives. A low-voltage designer working on the team suggested using common brackets, which would be significantly lower cost both to purchase and to install, yet keep the cables and wiring organized by allowing them to lay more horizontally, rather than thickly bundled as with a J-hook. After requesting some samples from the manufacturer, vetting the system with the facilities staff, and successfully completing the PMI©, the team determined that the cable tray substitute would be a perfect solution.

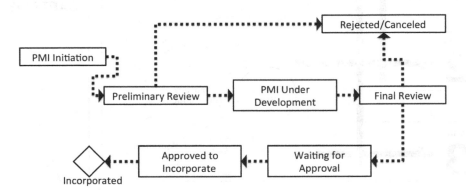

FIGURE 8.20   Project Modifications and Innovations Process©. Licensed by The Regents of the University of California on behalf of its UCSF Medical Center; courtesy of Stantec Architecture Inc. and Cambridge CM; illustration adapted by the authors.

The cable tray solution was implemented using an ICE session. ICE sessions were used on several other items to bring rapid closure to issues and opportunities.

ICE sessions were held to deal with inadequate ceiling space above Caesarean section operating rooms and intensive care units on the third floor of the hospital building. The cause was the very heavy structural steel required to support the recessed roof gardens directly above the area. The good news was that the issue had been discovered during their pre-BIM rough coordination phase; the bad news was that it had to be resolved quickly to maintain the permit submittal schedule. Worse still, the stakes were very high, including potentially compromising the clinical program by changing the use of the space to allow for the ceiling height to be lowered by at least one foot.

Three 4-hour sessions were held with leaders from every entity involved with the design and construction of the hospitals building: UCSF Medical Center staff, architecture, engineering, GC, and trade contractors. As this group walked out of the large conference room after the first session, they believed they had viable paths to solution. The roof garden would stay and the ceiling height would be maintained by resizing steel beams, rerouting above-ceiling HVAC and power, and shifting the location of the operating rooms (ORs) to a different area without sacrificing the clinical program. When the same group met three days later, they were ready to develop the new design to the point that it could be described to UCSF Medical Center stakeholders. The group then worked toward confirming rough order of magnitude (ROM) cost and schedule impact and develop the system designs in more detail in the third ICE session approximately two weeks later. Every participant realized that they had saved months and significant dollars through their intense collaboration within the span of three weeks. They knew from their collective hundreds of years of experience in designing and building that a fragmented delivery team with no awareness of ICE could not have handled the problem the way they did.

Team members reported each week on how they were performing relative to their latency target: 80 percent of issues resolved in 30 minutes or less within teams, and 80 percent of all others resolved in four hours or less in the Big Room. Design cluster teams consistently tracked and reported progress towards reducing costs. These teams also tracked numbers of requests for information (RFIs), material submittals, and change requests and compared these against key performance indicators (KPIs) they had set. Performance relative to metrics was reported each week in the project executive meeting and charts posted on the Big Room wall for everyone to see, including the many visitors to the project.

The team achieved the cost reductions UCSF Medical Center had set as goals during the design development and construction documents phases. In fact, two progressively lower cost targets were met. This success allowed the UCSF Mission Bay Hospitals project to proceed with building its full program rather than having to reduce it because of the $200 million budget overrun forecast by their third-party estimator before constructors were brought on board. Fully modeled and coordinated design packages were submitted and permitted on schedule. The team met all but one of its performance targets for incentive compensation.

8.6.3   Redesigning the Integrated Organization

Project team leaders redesigned their organization to meet the challenges of planning and managing construction, including the many design changes they expected from stakeholders intent on maintaining UCSF Medical Center's reputation as one of the top medical research and teaching institutions in the world. Their goal was the same: make the best possible decisions quickly and provide clear direction to the build teams. The leadership's guiding principle also remained the same: authority to decide should reside with people having the visibility, information, and experience to have the best understanding. For Stuart Eckblad, success was directly proportional to how well team members related and shared their knowledge; better relationships would lead to greater innovation and solutions within the budget and the time he had to deliver the project.

There were thousands of issues and questions on a project this large. Most of those were resolved by the parties working collaboratively in the Big Room and documented in a confirming RFI. However, if an issue had a schedule, cost, or design impact, or required additional outside consulting, the team would bring the issue into the Project Solutions Group© (PSG),[4] an organizational mechanism developed by Stuart Eckblad together with UCSF Mission Bay Hospitals team leaders. Project leaders for the UCSF Medical Center, the construction manager, and the design and construction firms met daily except for Fridays in the PSG to address issues as quickly as possible. The goal was always to solve the problem. During the course of construction, the PSG helped the Mission Bay Hospitals project team, among other things, reconfigure 100,000 square feet in the outpatient building, upgrade interiors, and redesign the city street in front of the Medical Center as a plaza—and all without extending the schedule.

Leaders wore several hats throughout the project. GC and design team project managers led clusters and the three buildings' teams. Some of these people participated with others in the project captains and senior leadership team. Each building team leader stayed on during construction to manage that particular subproject and surface issues for the PSG. Construction teams, organized by floor, took over from the area BIM teams for each building. On-site construction administration (CA) teams replaced the cluster teams to be the "first responders" for design questions.

During construction, the ICDC open office space doubled in size to accommodate the general contractor's staff. Inspectors and a larger construction manager staff took the place of all but the CA architects in the original space. More than a dozen office trailers were located along a boardwalk to house trade partner staff and provide additional meeting rooms. Slides showing quantities of work installed were added to the weekly progress update to the project owner, architect, contractor (OAC) meeting, printed as posters and displayed along a wall in the new area of the ICDC.

Stuart Eckblad, Cambridge CM, and DPR project leaders worked with the UCSF Medical Center counsel to redraft the UC standard contract, including general and special conditions and the scheduling specification, to incorporate Lean construction planning and scheduling practices. In particular, the new specification allowed for a high-level contract schedule supported by the Last Planner® System methodology of phase schedules, short interval work plans, pull scheduling, and weekly commitments updated daily. This extended the integrated organization to work crews through daily huddles to coordinate work. The huddles were the forum for field issues to surface quickly and for solutions to be worked out on the spot, if at all possible, with all trades present.

8.6.4 Performance

The Mission Bay Hospitals project team produced extraordinary results, especially compared to previous projects on the UCSF Mission Bay and other UC campuses, including:

- Reduction of the original cost estimate by $200 million;
- Construction of all of the original scope;

- Addition of $55 million of scope changes for the medical program, without impact to budget or schedule;
- Significant redesign of the outpatient building (OPB) with no budget impact; and
- Completion eight days ahead of schedule.

8.6.5 Keys to Success

The success of the UCSF Mission Bay Hospitals project can be attributed to commitment, preparation, and a culture which emphasized shared goals and innovation. It began, as is often the case, with the owner; in this case, a strong and experienced advocate for collaboration and integrated delivery in the person of Stuart Eckblad.

While the processes were mapped by team members and given names in the ICDC, Stuart continued working with his construction manager, Cambridge CM, and design team on collaborative methods for keeping the project within budget and schedule, even before he was able to bring the constructors into the team. The culture that had been formed became stronger during the VDC workshop as team members saw how they could leverage modeling and simulation together. All of this took place on the foundation of collaboration that Stuart Eckblad helped lay many years earlier.

Stuart and project leaders are working to share what they learned with the industry, particularly the culture and tools for collaboration, and the impact of the Big Room. Stuart explains the learning curve this way: "When we started not everyone wanted to come to the Big Room, but when we ended, no one wanted to leave" (Stuart Eckblad, personal communication, December 8, 2014, April 26, 2015, and April 15, 2016).

8.7 INTERCONNECTIONS

Integrating the organization is the critical link in the chain of integrating project delivery. The result of not getting this right will be that the project falls short of expectations. We've seen an unfortunate drama repeat itself. The team assembles with one after another person joining in the first weeks. Although things seem disorganized, people are optimistic, expecting things to fall into place. The disorganization persists because people are working in a brand-new network dealing with tough problems, often without any idea of how they should organize themselves. Without exception, this apparent chaos and the angst that goes with it results from leaders who don't have a map through the wilderness. Often, project leaders seem frozen as they stare at what Digby Christian depicted as the "Dragon of Uncertainty." What we've seen many times is project managers, whether for the owner, architect, or general contractor, nodding their heads in seeming agreement in the kickoff meeting, and quickly reverting to what they know about managing their own team/silo, the way they've always done it. They are eager to put everyone to work and don't want to waste valuable time in daily coordination meetings because they are convinced that they already know what has to be done. The idea of periodic reflections seems like another waste of valuable work time.

The point we are making is that people have learned how to create integrated organizations, and it can be messy and off-putting to owners and project managers who are not prepared for it. And we know for certain that rubber banding to the business-as-usual management that Peter Senge describes will not bring success. Disciplined practice, which takes patience, and sometimes outside coaching does work.

The three projects we've discussed are examples of the possibilities that open for leaders and team members who have integrated their project organization. All three teams accomplished great things. Hence, it is worth the investment to integrate the project organization. People thinking and acting in the way we've described in this chapter can leverage all of the other elements of the Simple Framework. In fact, these other elements come naturally. It's worth remembering the words we quoted from Peter Senge at the beginning of this chapter. It's not the labels people choose that are important; it's their actions.

8.8   REFLECTIONS

Owners will increasingly demand more from their project delivery teams. Owner and construction management organizations built to ensure compliance to requirements will continue down that path. Profit margins for the designers and architects who work in this model will likely decrease because the focus is on reducing price of these services. An increasing number of owners, perhaps those less invested in the compliance model or dissatisfied with its results, will encourage their teams to organize and work in an integrated way. Overall, these efforts will produce better results even though they may not come close to their potential because of how little experience practitioners will bring to these efforts.

A certain number of design and contracting companies will gain confidence in their abilities to deliver greater value while also reducing their risk because they have learned how to contribute as members of integrated organizations. Inspired by the possibility of greater success, these companies will seek projects using IPD agreements. Design-build companies will begin to integrate their project teams. Designers and general contractors may find ways to integrate project teams operating under CM@R agreements.

People who have learned to lead and work in integrated organizations will gain confidence and improve their practices and behaviors. This will create a virtuous cycle leading to better outcomes for owners. Insurers will design new instruments and offer those to integrated project teams. Technology companies will develop new applications to support BIM integration, integrated budgeting, quality by design, hazard awareness, scheduling, and production management.

Expectations will rise for integrated organizations. Meeting them will require builders to collaborate with their subcontractors and suppliers. These organizations, many of which are already ahead of general contractors, will form virtual, integrated organizations with their suppliers. Standard operating procedure will be for organizations to agree on methods, processes and protocols, and data standards. BIM component libraries will be shared across teams from day one of new projects. Eventually, big material suppliers will realize the purchasing power of these supply networks and be forced to respond. Much of what integrated organizations must do after they form will be done prior to projects. Integrated supply chain organizations will pursue and win projects and proceed to deliver exceptional outcomes routinely.

8.9 SUMMARY

Requesting that managers, leaders, including owner representatives and key stakeholders, read and discuss the Five Big ideas is a good starting point. The Five Big Ideas points are powerful and clear alternatives to the old way. Just getting down to business, which will most likely be as usual, is absolutely not the right thing to do.

We recommend a series of discussions based on the four main points of this chapter because we have not seen these ideas presented together. These dialogues, with everyone sharing their understanding and questions cannot be brief check-ins and should take place in person. This kind of work requires time and effort, which will be repaid many times over through significantly better team performance and results. In summary, we recommend the following:

- Start with the Five Big Ideas.[5]
  - If that is as far as people will go, this can make a big difference, especially if combined with knowledgeable coaching. These conversations must extend to everyone on the team, eventually including the trades-people who will be creating the value.
- Connect people's actions, information, and decisions.
- This is seeing the project as a network, not as a collection of companies.
  - Distinguish the type of interdependencies within subteams/groups within the network.
  - Recognize the "hidden effort" required for coordination; never accounted for in staffing plans drawn up by company executives who have never worked in an IPD team. This hidden effort is a schedule killer if team members refuse to get out of their silos.
  - Design, estimate cost, and build in multidisciplinary/cross-functional teams.
  - Ensure that team members and leaders meet to describe actions, roadblocks, questions, and decisions in tiered daily scrum meetings.
  - Honor and strengthen the network of commitments.
  - Use language effectively to make offers, request and reliable promises, assessments, and assertions.
  - Attend to individual and collective moods.
  - Listen for possibilities. These will emerge from dialogue.
- Build a learning organization. Celebrate questioning, problem discovery and solving; teach people the best method for this.
  - Encourage people to develop their personal vision, and create a shared vision for the project.
  - Practice dialogue, not debate or discussion. Expose and question individual and mental models. We suggest starting with four, as follows: (1) full utilization of each team member results in shortest schedule and lowest cost; (2) reducing cost for every budget item results in lowest total project cost; (3) the more work put in place early, the better; and (4) managers at the center must drive the project forward by planning for others.

- Think in systems; never blame the individual. Look for causality in the system, that is, the way things work. Do not be satisfied with simple cause and effect analysis. The first answer to why something happened is rarely correct. Ask why five times with the group affected to reach the root cause.
- Connect work through the project organization to customer value.
  - Use the POP mental model to connect work through the organization to what is being built.
  - Map the flow of value. Not investing in value-stream mapping will put the project at high risk of schedule and cost overruns, and quality problems.
  - Always work within the Plan-Do-Study-Act learning cycle. The Shewhart Cycle can be used for strategic planning, both in design and construction.
  - Structure work to produce greater value. This is looking a second time at the product in terms of how it can be best fabricated and assembled. Asking these questions repeatedly at lower and lower levels of detail guarantees better solutions.
  - Use integrated concurrent engineering to make big decisions as a team together with customer stakeholders and decision makers.

We realize this list is long and daunting. Of course, it's far better than showing up on game day having never practiced together. We believe it's important to realize that all of these elements work together and reinforce each other. They are synergistic and create a whole greater than the sum of the parts. One discipline or practice will lead to another; improvement in one area will help others. They contribute to a virtuous circle.

We also have two pieces of advice:

1. Pair younger people who seem to "get it" with older people with experience. Ask the younger ones to assume responsibility for doing as much of what we've described as possible.
2. Work hard on changing the experienced people's minds and practices. If they can't make the shift, change them out with people who are willing to learn. Although this seems harsh and difficult, we've rarely seen a project where this wasn't necessary. Unfortunately, in most of those cases, the second change out was not made soon enough or ever and the rest of the team and outcomes suffered.

NOTES

1. The collaborative planning approach that defines the production approach for achieving flow for a phase of work (Hal Macomber).
2. Integrated Center for Design and Construction© (ICDC) is a collaboration space and infrastructure developed by the UCSF Medical Center.
3. Project Modifications and Innovations© (PMI) is a method developed by the Mission Bay Medical Center team for the UCSF Medical Center.
4. Project Solutions Group© (PSG) is a method developed by the Mission Bay Medical Center team for the UCSF Medical Center.
5. Available on the Lean Construction Institute web site, www.leanconstruction.org.

REFERENCES

Ballard, G., Koskela, L., Howell, G., & Zabelle, T. (2001). "Production system design: Work structuring revisited." White paper 11.

Chachere, J., Kunz, J., & Levitt, R. (2003). *Can you accelerate your project using extreme collaboration? A model based analysis.* Center for Integrated Facility Engineering (CIFE), Stanford University, CIFE Technical Report 154.

Davis, C. (1998). *Listening, language and action.* Retrieved from http://stratam.com/assets/articles/Listening_Language_Action.pdf, March 6, 2005.

Flores, F. (2013). *Conversations for action and collected essays: Instilling a culture of commitment in working relationships.* CreateSpace, https://www.createspace.com/3952130.

Latzko, William J., and David M. Saunders. "Four days with Dr. Deming: A strategy for modern methods of management." Long Range Planning 29.4 (1996): 594–595.

Lencioni, P. (2012). *The advantage: Why organizational health trumps everything else in business.* Hoboken, NJ: John Wiley & Sons.

Levitt, R. E., Kunz, J. C., Luiten, G. T., Fischer, M. A., & Jin, Y. (1995). *CE4: Concurrent engineering of product, process, facility and organization.* No. 104. Technical report.

Levitt, R., & Kunz, J. (2002, September). Design your project organization as engineers design bridges. CIFE technical paper, Stanford University.

Macomber, H. (2004a, March 23–24). Five Big Ideas that are reshaping the design and delivery of capital projects. Document prepared for the Sutter Health Lean Project Summit, Concord, CA.

Macomber, H. (2004b). *Securing reliable promises on projects: A guide to developing a new practice.* Retrieved from http://www.reformingprojectmanagement.com.

Macomber, H., & Howell, G. (2003). Linguistic action: Contributing to the theory of lean construction. *Proceedings of the 11th Annual Meeting of the International Group for Lean Construction,* pp. 1–10.

Macomber, H., Howell, G. A., & Reed, D. (2005). Managing promises with the last planner system: Closing in on uninterrupted flow. *13th International Group for Lean Construction Conference: Proceedings,* p. 13. International Group on Lean Construction.

Mann, D. (2014). *Creating a Lean culture: Tools to sustain lean conversions.* Florence, KY: CRC Press.

Rother, M., & Shook, J. (1999). *Learning to see.* Cambridge, MA: Lean Enterprise Institute.

Senge, P. M. (2006). *The fifth discipline: The art and practice of the learning organization.* New York, NY: Broadway Business.

Sutherland, J., & Sutherland, J. J. (2014). Scrum: The art of doing twice the work in half the time. New York: Crown Business.

Thompson, J. D. (1967). Organizations in action: Social science bases of administration. New York, NY: McGraw-Hill.

CHAPTER 9

Leading Integrated Project Teams

"Individual commitment to a group effort—that is what makes a team work, a company work, a society work, a civilization work."

—Vince Lombardi

9.1 INTRODUCTION

Integrated project delivery (IPD) is built around teams. Early involvement of key participants is a core IPD concept, and these early participants, like virtually all IPD participants, are organized in teams. Teams provide the right knowledge at the right time, stimulate creativity, and lower the barriers among the many project participants. Multidisciplinary teams create alignment, commitment, and engagement. Moreover, team processes improve decision making and increase support for the strategy chosen. And IPD is not only executed by teams; it is led and managed by teams as well. Without teams, IPD does not function.

But the mere existence of teams does not guarantee success. For some tasks, individual or group action is more efficient than multidisciplinary teams. In fact, an individual's performance may actually decline in group settings due to social loafing,[1] the bystander effect,[2] groupshift,[3] and groupthink.[4] In addition to these dangers, teams require additional training and management. Clearly, teams must be organized, managed, and motivated properly to gain the benefits of teams while avoiding these performance pitfalls.

This chapter combines research on team performance with observations of teams on IPD projects and suggests practices from other industries that might be useful for IPD teams. It also examines the challenges and opportunities of team-based decision making, drawing on current research and practices. This chapter does not address, except in passing, several related concepts such as target value design, simulation, visualization, collaboration, and co-location, as these are given fuller treatment in other chapters.

The research demonstrates that many differences in cost, schedule, and quality outcomes are attributable to the strength of interorganizational relationships. Projects with a greater depth of team integration, observable by their participation in high-quality interactions, generally saw reduced schedule growth and increased intensity. Design and construction teams that were highly cohesive reported lower cost growth, with a better turnover experience for the owner and higher perceived system quality. These findings strongly suggest that team integration and group cohesiveness are desirable attributes in effective project teams (Molenar, Messner, Liecht, Franz, & Esmaeili, 2014).

9.2   WHAT ARE IPD TEAMS?

If individuals are the atoms of IPD, teams are the molecules. They combine the strengths of the individual team members to create outcomes that they could not achieve individually.

Not every group of individuals is a team. If the need for interaction and creativity is low, work groups can effectively process parallel work. But synergistic teams are the engine of IPD. These teams are fundamentally different from work groups because they require both individual and mutual accountability. IPD teams are committed to a common purpose, process, and outcome for which they hold themselves mutually accountable (Katzenbach & Smith, 2005). Stephen Robbins distinguishes work groups and work teams, stating:

> Work groups have no need or opportunity to engage in collective work that requires joint effort. So their performance is merely the summation of each group member's individual contribution. There is no positive synergy that would create an overall level of performance greater than the sum of the inputs.
>
> A work team, on the other hand, generates positive synergy through coordinated effort. The individual efforts result in a level of performance greater than the sum of those individual inputs (Robbins, 2011).

IPD teams are diverse. They are cross-functional and multidisciplinary. They are largely self-managing and self-coordinating (although they need leadership and mentoring) and draw on the talents of their members. IPD teams vary from smaller groups focused on solving specific problems to larger groups working on brainstorming, coordination or scheduling. They are active—and sometimes raucous—but their passion is directed for the project rather than at each other.

IPD teams are creative, efficient, and they get the job done.

9.3   WHAT DOES SUCCESS LOOK LIKE?

You walk into the Big Room. In a corner, a group is gathered around a wall talking constantly as they are adjusting dates and deliverables in a pull schedule. Another group is huddled around computer monitors evaluating different options for a mechanical routing. You realize as you look at them that you don't know who they work for. As you move closer, you overhear enough to know that someone is an engineer, another is a facility manager, and several trades are involved in the decision. A logo on one of their shirts betrays their firm—without that you wouldn't know.

In another group, everyone is looking toward a woman. She must be the leader. But as you draw close to that group, you realize that she is asking questions, not giving orders. And she is pulling information from each team member and carefully listening to their response. Moreover, she is asking others to add or comment on the responses. Slowly, she is helping the team evaluate all of the considerations, hear the minority and contrasting views, evaluate them, and draw to a consensus. Looking over at a wall, you see a list of the key project values and you realize that the criteria the team just used to make their decision aligned with the project values. You were right—she is a leader, a team leader.

There is a buzz in the room and no librarian to keep everyone quiet. But although you hear excited voices, they are passionate, not angry. If you stopped to ask anyone how it was going, they would explain the problems and the challenges, and then tell you that this is why they got into design and construction. To solve problems and to do something significant.

9.4   HOW CAN THIS BE DONE?

9.4.1   Leading IPD Teams

IPD projects are led and managed by committees with members representing the key participants. These management structures are usually contractually defined and, while there are differences in terminology and the details of decision processes, all of these management groups have similar responsibilities with regards to IPD teams. We find six responsibilities particularly critical.

The first responsibility of IPD leaders is to develop a clear and common understanding of the project values and goals. Most commonly, this is done in workshops that include all key stakeholders and participants. The values are developed, refined, and valued. Goals are established consistent with the values, and where possible, metrics developed to assess progress toward the goals.

The importance of this first step cannot be overestimated. As the IPD teams engage on the project they will need to make an almost infinite number of decisions, many small, some quite large. Choosing among alternatives will require clear and consistent criteria. The high-performance building won't be achieved if the lighting and electrical group has different criteria than the mechanical or the building envelope groups. All must be working toward the same values and goals.

The second responsibility of IPD leadership team is to clearly communicate the values and goals to all project participants and then continually reinforce the goals and values through repetition and recognition. On-boarding, discussed in greater detail below, is a process for reinforcing and communicating the goals and values as new members are brought into the teams.

The third responsibility is to create a functional physical and virtual space for co-location. This includes joint or interconnected digital networks, software, and collaboration systems. Because of its importance, co-location is discussed in a separate chapter.

The fourth responsibility is the definition or composition of the necessary teams for the project. This will vary during the life of the project and some tasks will require discrete specialist teams, and others, such as scheduling, will require participation from all parties. But whatever is required, the project management group will create and adjust the teams throughout the project.

The fifth responsibility is to provide training and mentoring for the project teams. In some instances, teams may need an outside facilitator to help them achieve consensus. In this context, "outside" may be an independent consultant or a skilled facilitator within the project, but not part of the specific team. Team members may also need assistances in using specific tools, such as Choosing by Advantages or pull planning. IPD supervisors should heed W. Edwards Deming's admonition that "the job of management is not supervision, but leadership" (Deming, 1982). And, as observed by Atushi Niimi of Toyota, the difficulty in teaching foreign managers "is that they want to manage, not teach" (Larman, 2008).

The final responsibility is to monitor and adjust team dynamics. Some people are just not team players and others can quench creativity due to their dominance. However skilled the person may be, if he or she is reducing team effectiveness, the project management must intervene and either adjust the person's behavior or replace them.

The responsibility to choose, mentor, and monitor teams is one of the most important tools of the IPD project management. In *Good to Great*, Jim Collins observed that great leaders initiate change by assembling the right team even before they decide where their organizations are going. Collins counsels that you need to get the right people on the bus, the wrong people off the bus, and the right people in the right seats (Collins, 2001). And Amabile notes that matching personnel and assignments is one of the most significant decisions a team leader can make (Amabile, 1998). In the end, the leadership and management of teams is about people.

9.4.2   Team Composition

Teams require a variety of skills, backgrounds, and personalities. A well balanced team needs members with technical expertise, problem solving and decision making capability, and interpersonal skills such as the ability to listen effectively, provide feedback, and resolve conflict (Katzenbach & Smith, 1992; Robbins, 2011). Because few individuals have all of these capabilities, the team members should be chosen to assure that these capabilities are represented within the team (Hackman, 2011; Robbins, 2011). A good strategy is to choose two or three members that excel technically, assess their leadership and interpersonal skills, and then add members to balance the team. Conscientiousness is another key attribute and teams with more conscientious members perform better (Robbins, 2011). Because the team's work may change as the project matures, team leadership must be sensitive to the changing requirements and readjust team composition accordingly.

The technical nature of the project affects team composition. Hospitals have complicated mechanical, electrical, and plumbing (MEP) systems, and hospital teams consequently emphasize this requirement. Theaters and university classrooms will have critical acoustic, display, and information technology requirements. Projects with challenging environmental requirements may need to include specific energy efficiency or sustainability expertise. These projects will appropriately build teams around these requirements.

The type of work being done also effects team composition. If a team is facing novel, complex problems, the team members need to be chosen for intelligence and creativity. These high-ability teams are more adaptable to changing situations and can effectively apply existing knowledge to new problems (Robbins, 2011). But this same team will be less successful on routine tasks, whereas teams with more modest capabilities will remain focused and productive. High-ability teams also tend to work best when led by equally high-ability managers, or self-organized and self-managed (Robbins, 2011). High-ability teams are preferred if project success requires innovation and creativity.

Teams should also be composed of individuals with differing backgrounds, viewpoints, and experience (Hackman, 2011). The most creative teams are not homogenous (Amabile, 1998). This is one advantage of involving trade contractors, end users, and maintenance personnel early in design. Each brings a different perspective to the design problems being addressed. Daniel Pink suggests: "Set up work groups so that people will stimulate each other and learn from each other, so that they're not homogenous in terms of backgrounds and training. You want people who can readily cross-fertilize each other's ideas" (Pink, 2009). Not only does this diversity provide more information to inform the design, the tension between perspectives stimulates greater creativity within all the individuals.

Personality should also be considered when choosing team members. Some percentage of people are not inherently collaborative (Benkler, 2011), and others do not like working in teams and will opt out if they have the chance (Robbins, 2011). Employees trained in command-and-control structures can have difficulty transitioning to team structures and either want to be told what to do or want to tell others what to do. They may not have the patience for the more deliberative processes used in IPD teams. And as noted previously, corporate culture is pervasive and the effect of the team member's home culture can undermine the member's effectiveness in the collaborative team.

Many firms have used personality data, such as the Meyers-Briggs Type Indicator (MBTI), and while testing may be useful for some purpose, there is little data supporting the use of Meyers-Briggs for choosing team members (Hackman, 2011; Robbins, 2011) or assessing team performance. There is evidence that psychological attributes are correlated with individual success and the interaction of team members' personal characteristics is undoubtedly significant, but balancing a team profile appears to be currently more art than science. Nonetheless, this is an approach worth considering if the personality traits measured are correlated to job performance and team interaction.

It is also important to weed out members who are undermining team performance. Research shows that a bad actor can damage the team, and unless the behavior is corrected, the team member needs to be removed (Robbins, 2007). As Glenn Ballard (cofounder of the Lean Construction Institute) has often stated: "You have to either change the people or change the people."

9.4.3 Team Structure

As project size increases, the basic approach (although not the basic theory) must change. One flexible team simply isn't large enough to do the work. Work should be structured to fit the team rather than increase the team to fit the work. This means that the core leadership team must create a structure that keeps working teams reasonably compact, does not have responsibility gaps between teams, allows contemporaneous coordination, and provides alignment to overall goals.

Several structures are commonly used to divide work scope.

If the project is susceptible to geographic division, then the basic team approach can be used—but each team is responsible for a section of the project. For example, work can be divided by wings, floors, phases, or structures. Within their geographic division, each team is responsible for all functions and disciplines.

Area responsibility teams will need to be provided an overall systems approach and will need to coordinate with other area teams. The overall systems approach is developed by a separate team that evaluates options and provides diagrammatic direction to the area teams.

Unless a project is quite small, no team can do everything. Thus, a key element of team organization is the structure of teams within teams. In most projects, specific teams are created that handle the design and eventually the construction of specific elements, systems, or physical areas of the project. A team working on the mechanical systems for a floor of a large hospital, for example, would generally report to a team responsible for MEP systems. Similar approaches could be used for foundations and structures, framing and exterior skin, or other systems. One strategy for team boundaries is to assess areas of historical failure (intersections between work such as slab edges) and assure that the team contains personnel with responsibility for both sides of a problem interface.

Some tasks, particularly scheduling and coordination, may require a "committee of the whole" to ensure that all necessary inputs are received and considered and the appropriate commitments made. There can also be issues that stretch across many boundaries that require swarming or "super-team" efforts. But these gatherings are generally ad hoc and short lived, or in the case of scheduling sessions, are regular, but of limited duration.

Coordination between area teams can be handled in two ways. First, the overall systems team can have coordination responsibility. This is not preferred as it releases the area teams from coordination responsibility and violates the rule of designing coordination into the process rather than testing for coordination after work has been performed.

A better approach is to place coordination responsibility on the teams and use team member overlap or regular Big Room coordination (or a combination of these techniques) to assure that the design being developed is coordinated. The Big Room coordination meetings should not only focus on coordinating the work that has been done, but should also engage in discussion of what design work will be done by each team in the interval before they meet again. This should be sufficiently detailed to allow the teams to uncover and solve potential coordination issues before detailed design is performed.

A second approach to larger scale projects is to divide the work on a systems basis, such as dry or wet mechanical systems. This has the advantage of providing an overall view of the specific system and allows all members to have high levels of subject matter expertise. It reduces diversity, however, and creates additional coordination issues between functional systems and between physical elements (penetrations and physical conflicts).

Systems-based teams can coordinate through the Big Room process, with joint meetings between interrelated teams, such as dry mechanical and framing. In the Big Room meetings, the teams analyze problem issues, clash detect their work, explain what work they plan to accomplish before the next joint meeting, and create the list of decisions and deliverables each requires of the other.

Team coordination can be enhanced by regularly posting design information in visible locations, such as corridors and walls. Although this information may be available digitally on servers, having the information present where it is regularly seen by other teams is a more effective tool. It may be worthwhile to create a prominent physical area where the teams post their current work such that other team members can see at a glance where each team is going.

Regular pull scheduling is also a useful coordination exercise because it focuses on the decision interchanges between teams. In order to pull schedule to a milestone, the teams must request and promise information and deliverables from each other, which exposes coordination issues.

On larger projects, information management can become a significant obstacle, particularly in sharing building information modeling (BIM) data between different software tools. Moreover, even if systems are adequately interoperable, the information needs to be appropriately categorized, labeled, tracked, and archived. This requires project standards and procedures. In addition, if information will be repurposed, then the parties that will create and use the information must agree on how design and construction elements will be represented in the building information models. On larger projects, the information requirements are sufficiently challenging to require a separate team focused on the project's information requirements.

9.4.4 Team Size

Team size should match purpose. Larger groups, 12 or more, are better at developing alternative project solutions, but are less effective in getting things done (Robbins, 2011). Smaller groups have limited skill sets, and their lack of diversity limits knowledge and creativity (Robbins, 2007).

A good rule of thumb is to keep working teams between five and nine members (Robbins, 2011). Several experts also recommend that the team be no larger than necessary to accomplish its task (Hackman, 2011; Larman, 2008). If the task is too large for an efficient team, the task should be broken into subtasks. Keeping the team small reduces the information loss among members and creates greater individual accountability. Because the members of smaller teams know what each member is doing, it is hard for any team member to slacken his or her efforts without other team members noticing the imbalance.

When teams have excess members, cohesiveness and mutual accountability decline, social loafing increases, and more people communicate less. Members of large teams have trouble coordinating with one another, especially under time pressure (Robbins, 2011).

9.4.5 Team Diversity

IPD teams should be multidisciplinary, particularly during the design and preconstruction period and should generally be cross-functional. Multidisciplinary teams are composed of members with differing training and experience. Cross-functional teams are composed of members with differing responsibilities. For example, a design phase team composed of architectural, mechanical engineering, mechanical contracting, and general contracting expertise is multidisciplinary, but they may all be focused on design during that phase. A cross-functional team would include representatives from procurement, cost management/estimating, and operations, as well as those responsible for design and construction. Their functions vary as well as their backgrounds. For example, a cross-functional IPD team should jointly design a portion of the project and should also be responsible for managing the cost, meeting the schedule, constructing, and commissioning that work. Scope, schedule, and budget should be tightly bound within the team, and not delegated to separate departments.

Cross-functional teams have been highly successful in manufacturing and software design. Boeing, Toyota, IBM, and others have used teams with members from the different internal groups with entire responsibility for a product, or a portion of a product, from conception through creation and including sales and marketing. Cross-functional teams composed of people from design, engineering, production, and sales is a key component in W. Edwards Deming's recommendations for Western management. (Deming, 1982). As noted by Stephen Robbins:

> Cross-functional teams are an effective means of allowing people from diverse areas within or even between organizations to exchange information, develop new ideas, solve problems, and coordinate complex projects (Robbins, 2011).

Craig Larman cautions that cross-functional integration of management is not sufficient.

True cross-functional integration in large product development is rare. Instead, we have frequently encountered cross-functional project management groups with management representatives of the different functional areas. They do not work. True cross-functional integration occurs at the working level (Larman, 2008).

Whenever possible, the team should have responsibility for all components necessary to achieve the project goals and should be responsible for coordinating with other teams. Responsibility for a discrete, complete portion of the project reduces errors at the interfaces between disciplines and promotes ownership and pride in the whole (Deming, 1982). As noted by J. Richard Hackman:

The better a team's work is designed (that is, the extent to which members have collective responsibility for carrying out a whole, meaningful piece of work for which they receive direct feedback), the higher a team's collective internal motivation (Hackman, 2011).

Current IPD teams have generally assembled around related systems, such as MEPF or foundations and structural systems. These have then provided their recommendations or work to a higher level team with broader responsibilities. The higher level team operates as an information accumulator and a group that passes work down to the functional teams. There is evidence from software development that suggests that teams should take direct responsibility for coordination rather than relying on a coordination layer (Larman, 2008), although self-coordination may be difficult in larger projects.

9.4.6   Team Stability

Personnel turnover can increase waste and limit team effectiveness. Construction teams are often short lived, with members moving in and out of the team as work increases or slacks. This practice is contrary to research on project teams.

Research findings overwhelmingly support the proposition that teams with stable membership have healthier dynamics and perform better than those that constantly have to deal with arrival of new members and the departure of veterans (Hackman, 2011). Other researchers, notably Ralph Katz (1982), have concluded that research and development (R&D) teams increase their effectiveness for a period of three to four years.

Many construction projects are completed in less than the time required to develop optimal team dynamics, and even in these projects, handoffs from design to preconstruction and on to construction increase project turnover and shorten interaction time. However, there are several strategies to counteract this effect and improve knowledge transfer and performance. First, the proposing team should be the executing team, with little change to the core personnel. Not only is less information lost on transition, the stable team does not need to rebuild relationships and trust. Second, if an owner has multiple projects, it should consider engaging a specific team to do sequential projects provided that the team shows continuous improvement and utilizes substantially the same personnel. Third, firms should identify IPD experts who work with each IPD project to carry lessons learned among projects and to guide less experienced teams. If the firms do not yet have skilled "gurus," they should consider augmenting their teams with consultants that have that experience. Finally, firms should actively incorporate the learning from ongoing projects into their training programs. The knowledge of how to do IPD should be institutionalized. If you are lucky enough to have teams that are stable for long periods, then it may be wise to begin mixing in new members or have the team engage in a "creative disruption," to counteract the performance slump associated with long-term projects.

The importance of team stability has led a few serial builders to consider using the same team on multiple projects on a "yours to lose" basis. Essentially, as long as the team can demonstrate continual improvement compared to their prior project and the market, then they will be awarded the next job. If a team member cannot improve, or appears to be taking advantage of its preferred position, it can be removed by the owner or the team.

Keeping the Team Together

After Lawrence & Memorial had nearly finished its first IPD project, the very successful L&M/Dana-Farber Cancer Center, it decided to use the exact same team (with the exception of one person) to execute its next project. Although board members questioned why the new project wasn't advertised for proposals, L&M's facility team believed that the savings in having a functioning team far outweighed any potential advantage of going to market—and when a differing site condition problem occurred—having a team that already knew how to jointly resolve problems, saved a significant percentage of the project budget.

9.4.7 Training (the Team)

Training should occur continuously throughout the project and should complement the team's self-evaluation and improvement processes. Coaching should have a dual focus: helping individual members learn ways they can strengthen their personal contributions and, at the same time, exploring ways the team as a whole can best use its resources (Hackman, 2011). Whenever possible, the project should commence with a boot camp that begins the processes of developing trust among teams and management and trust between team members themselves. This is also the opportunity to assess the strengths and weaknesses of team members, improve interpersonal skills, address training opportunities, clarify goals and expectations, and enhance the team's ability to use the tools and techniques required for the project.

Research on team training shows that the greatest effect is obtained when coaching interventions address three task performance issues: (1) the level and coordination of member effort, (2) the appropriateness to the task and situation of the performance strategies the team is using, and (3) the degree to which the team is using the full complement of its members' knowledge and skills (Hackman, 2011).

For example, many team members have limited or incorrect understanding of Lean principles. Rather than lecture on Lean techniques, the team members should be actively engaged in tasks and training relevant to the project such as pull scheduling to a project milestone, mapping the value stream of a problematic process, reaching a joint decision using a structured decision method, or documenting a root cause analysis using a Plan-Do-Study-Act (PDSA) A3 format. Working jointly with the assistance of a skilled teacher increases the team's baseline capability while building the relationships necessary to work together—and unlike "ropes course" bonding, the team actually creates something relevant to them and the project.

Team effectiveness requires clear communication and the ability to dissent without damaging relationships. As discussed in the Team Motivation and Creativity section later in this chapter, modest levels of conflict regarding task and process heighten creativity. But the ability to freely and energetically differ without creating damaging personal conflict requires strong communication skills. Communication should be a focus of the initial boot camp with training focusing on listening, clarity of communication, and dispute resolution.

First, team members need to understand what others are saying. Covey counsels that to communicate clearly, listen first (Covey, 2006). Many people are not good listeners and need to be reminded how to listen. Speaking clearly is also a rare skill, especially when the information may not be well received. Yet the very basis of reliable commitment is a clear understanding of the conditions of satisfaction and the license to disagree. Unless there is honesty, improvement will not occur.

Honesty can be brutal, but it needn't be. The second skill is the ability to dissent or critique without personalizing the dispute. When issues are important or emotional, the slightest personalization triggers defensive reactions that trigger counterreactions that reinforce the personal dispute and overpower the substantive discussion.

The third skill is dispute resolution. When a disagreement has clearly been stated and understood, the team members need to know how to resolve the disagreement, or decide among options, while preserving the dignity of the team member whose suggestion will not be followed. The dissenter should disagree with and support the decision.

Healthy communication skills should not be presumed and must always be monitored. Communication skills and dispute resolution should be addressed in the boot camp and reinforced throughout the project. Team leaders should model listening skills, defuse damaging conflict, and counsel team members whose communication techniques are ineffective or destructive.

9.4.8   Setting Goals

Goals are critically important to team success. When they are challenging, they inspire creativity. When they are overwhelming they create defensiveness. They need to be specific, appropriately challenging, related to team performance, and measurable (Katzenbach & Smith, 2005; Robbins, 2011). General goals, such as "be the best" are ineffective because they do not guide behavior, defy measurement, and do not promote accountability. In contrast, a specific goal such as reducing heating, ventilating, and air-conditioning (HVAC) cost by, for example, reducing the number of smoke dampers from x to y gives the team a specific goal on which to work. Research also shows that challenging goals result in higher levels of performance (Amabile, 1998; Robbins, 2011). One study noted that the goals were most effective when the team thought it had a 50-50 chance of success (Hackman, 2011). Amabile notes that the perfect match should stretch the employee's abilities. The employees should not feel bored, nor should they feel overwhelmed (Amabile, 1998). When you start a project, the IPD team should be breathing heavily, but not hyperventilating.

9.4.9   Purpose

Teams are more motivated if their work product fulfills a valuable purpose. Team leaders can strengthen commitment by focusing on the value of the work to others. The value can be inherent in the project, such as a center to care for veterans, or can reside in the appreciation of end users or project participants who rely on the team's work. In health care projects, for example, frontline users, such as nurses and doctors, can explain to the team how they will use the facility to treat patients. Similar approaches can be taken with schools and many other types of projects. The project's positive purpose can be reinforced by photographs or other information that shows how the facility will affect others. Even if the project does not fulfill a grand purpose, the leader can explain how the team's work benefits the project as a whole and the customers of the team's work to regularly show their appreciation.

9.4.10 Supporting Team Members

To assess why an employee is not performing to her or his best level, look at the work environment to see whether it's supportive. Does the employee have adequate tools, equipment, materials, and supplies? Does the employee have favorable working conditions, helpful co-workers, supportive work rules and procedures, sufficient information to make job-related decisions, adequate time to do a good job, and the like? If not, performance will suffer (Robbins, 2011).

The necessity of adequate resources may seem obvious, but in many instances teams are hampered by lack of adequate hardware or software, insufficient administrative or technical support, or lack of time or other restrictions that distract team members from their primary purpose. Not only are the restrictions inefficient, they are frustrating, disheartening, and imply management disinterest in the team's tasks. Team leaders should be sensitive to how team members' time is being spent and should use Lean tools and processes to reduce nonvalue activities.

9.4.11 Team Motivation and Creativity

All projects require motivated teams, and many require innovation, too. Fortunately, the principles that improve motivation also improve creativity. Team leaders who want both should start by creating a work environment that improves motivation and engagement and then layer additional factors that stimulate innovation.

It may seem soft and fuzzy to suggest that job satisfaction and job performance are related. However, there are at least 300 studies that suggest the correlation is quite strong and long-term studies show a high correlation between employee engagement and productivity (Robbins, 2011).

The most important thing leaders can do to raise employee satisfaction is to focus on the intrinsic parts of the job, such as making the work challenging and interesting. Although paying employees poorly will likely not attract high-quality employees to the organization, or keep high performers, managers should realize that high pay alone is unlikely to create a satisfying work environment. Interesting jobs that provide training, variety, independence, and control satisfy most employees. There is also a strong correspondence between how much people enjoy the social context of their workplace and how satisfied they are overall. Interdependence, feedback, social support, and interaction with coworkers outside the workplace are strongly related to job satisfaction, even after accounting for characteristics of the work itself (Robbins, 2011).

Work satisfaction is perhaps the most significant motivating factor. As one of the authors overheard an engineering supervisor say: "Never tell engineers to work late. Just make the job so interesting they never want to leave."

Generally, people are motivated by what interests them and what they believe is important. Having the authority to use their skills as they think best, is also important.

Teams work best when employees have freedom and autonomy, the opportunity to utilize different skills and talents, the ability to complete a whole and identifiable task or product, and a task or project that has a substantial impact on others. The evidence indicates that these characteristics increase members' sense of responsibility and ownership over the work and make the work more interesting to perform (Robbins, 2007).

Recognition also motivates team members. For professionals, the recognition may count more than a tangible token (Hackman, 2011). The most important recognition is from the team members themselves and is more powerful than externally generated feedback (Robbins, 2011). Recognition from management can be helpful provided that it reinforces team activity. Recognition of a "star," especially if tied to money or perquisites, can lead to corrosive competition within the team.

Much has been written about the ineffectiveness of external rewards, such as cash. Commentators as early as Deming have stated that monetary rewards for outstanding achievement may be counterproductive (Deming, 1982). Daniel Pink's (2009) bestseller, *Drive,* was based on the concept that intrinsic motivation was much more powerful than external rewards and that external rewards can be counterproductive. In fact, several researchers have found that pay for performance actually decreased engagement and creativity (Deci, 1971, 1972; Robbins, 2011; Pink, 2009; Amabile, 1998). If external rewards are used, they are most effective if unexpected—a treat rather than a payment (Pink, 2009). And if rewards are used, they should be based on team performance, not individual success (Robbins, 2007).

Creativity is often associated with dramatic achievements in art or science, with breakthroughs and stunning structures. For IPD teams, creativity is developing efficient and elegant solutions at every level of execution and encompassing revolution and evolution. Properly managed teams are an essential component to increasing project creativity.

Current theory posits three major elements to creativity: expertise, creative thinking skills, and intrinsic motivation (Amabile, 1998; Robbins, 2011). In most projects, there is not enough time to significantly improve the critical thinking skills of team members. As a practical matter, team leaders must try to select team members who already have these attributes.

Expertise, however, can be affected by management practices. First, the creation of cross-functional teams increases the expertise of the team as a whole. By providing personnel with a broad range of training and experience, the team has access to more information and more options. This is one of the reasons that creativity experts suggest diverse teams (Amabile, 1998; Larman, 2008; Pink, 2009; Robbins, 2007).

If you want to build teams that come up with creative ideas, you must pay careful attention to the design of such teams. That is, you must create mutually supportive groups with a diversity of perspectives and backgrounds. Why? Because when teams comprise people with various intellectual foundations and approaches to work—that is, different expertise and creative thinking styles—ideas often combine and combust in exciting and useful ways (Amabile, 1998).

But diversity can also lead to conflict because of different experiences and work styles. Kept within limits, conflict can actually boost creativity by stimulating higher levels of innovation and exchange.

Conflict on a team isn't necessarily bad. Teams completely devoid of conflict are likely to become apathetic and stagnant. Thus, conflict—but not all types—can actually improve team effectiveness. Relationship conflicts—those based on interpersonal incompatibilities, tension, and animosity toward others—are almost always dysfunctional. However, on teams performing nonroutine activities, disagreements among members about task content (called task conflicts) stimulate discussion, promote critical assessment of problems and options, and can lead to better team decisions. How conflicts are resolved can also make the difference between effective and ineffective teams. A study of ongoing comments made by 37 autonomous work groups showed that effective teams resolved conflicts by explicitly discussing the issues, whereas ineffective teams had conflicts focused more on personalities and the way things were said (Robbins, 2011).

Thus, the conflict that may be created by diverse teams is an asset that should be carefully managed. Teams should be taught how to interact forcefully without engaging personally and team leaders should keep conflict within bounds and focused on task or process. In addition, team members must recognize and respect the knowledge and contributions of each team member.

Team leaders can also increase knowledge through continuous education and information exchange. This can occur formally, as in regular teaching sessions, and can occur informally through periodic information exchanges or even information updates on walls or banners. The goal is to disperse information broadly throughout the team.

The final element, intrinsic motivation, should flow from the nature of the work itself and should be a result of proper team and work organization. "When people are intrinsically motivated, they engage in their work for the challenge and enjoyment of it" (Amabile, 1998, p. 80). And intrinsic motivation is a crucial driver for creativity.

Although creative motivation is intrinsic, it can be affected by management practices. According to Amabile's research, challenge, freedom, resources, work group features, supervisory encouragement, and organizational support all affect intrinsic motivation. The challenge is matching the proper employees to tasks that are challenging, but within their reach. Freedom is allowing the team to determine how to achieve the assigned or agreed goal. Resources include providing sufficient material support, but also the right amount of time to address the problem. Real and realistic (not arbitrary) time pressure spurs creativity, but truly impossible deadlines are seen by the team as not worth attempting. Work group features include having diversity of knowledge, but also diversity of skills, such as problem solving and interpersonal skills. Supervisory encouragement acknowledges the importance of the team's work and keeps the team engaged even if results are not immediately evident. Moreover, the leader must not allow criticism to kill good ideas—or prevent their expression—before they can be fairly developed and analyzed. Finally, organizational support mandates information sharing and collaboration and sweeps aside political issues that may undermine the creative efforts (Amabile, 1996, 1998).

Although creativity cannot be reduced to a formula, there are management practices that increase a team's ability to develop new and innovative solutions to existing problems. Luckily, most design and construction participants are intrinsically motivated and want to be proud of their projects. The challenge for team leaders is to remove impediments that prevent them from doing so.

9.4.12 Decision Making in Teams

Teams offer the opportunity to have a collective intelligence that exceeds the intelligence of any team member. But without careful leadership and management, this opportunity will not be achieved.

Over the past several decades, researchers have examined the biases and mental shortcuts ("heuristics") that undermine accurate decision making. Essentially, behavioral economists and psychologists propose that humans use two separate decision making methods. As described by Nobel Prize winner Daniel Kahneman (2011) in *Thinking Fast and Slow,* System 1 operates rapidly and almost without conscious thought. System 2 is engaged when we stop and carefully consider an issue, logically weighing the advantages and disadvantages. Needless to say, using System 2 for every decision we make would be exhausting, and we default to System 1, which often works acceptably.

System 1 is easy to use because it simplifies problems and applies "rules of thumb" based on our instinctive biases and because of this, System 1 is easily fooled. Data anchors, priming, regression to the mean, question substitution, availability, framing, the law of small numbers, and many other influences and heuristics disturb accurate decision making and can make decisions seem easier (engaging System 1), but less accurate (Kahneman, 2011). Moreover, Kahneman and his colleague Amos Tversky demonstrated that humans, even when using System 2, did not decide issues rationally. Humans tend to be risk-adverse (almost twice as concerned with losses than gains) except at the fringes of probability. At the fringes, humans will accept a highly unlikely bet if not doing so insures a significant loss. This is "double or nothing" gambling logic. When a potential loss is large, humans don't like to take risks, even with very favorable odds. Moreover, whether an outcome is a gain or loss, is a psychological, not an economic, calculation. Figure 9.1 is based on Kahneman (2011).

Interestingly, when negotiating IPD agreements, we have found that teams, especially those first undertaking IPD, fixate on the potential loss of profit rather than the more probable gain. This is entirely consistent with the differing slope for losses, and the upshot for management is that understanding and mediating the sources of bias, while consciously engaging System 2, results in better decisions.

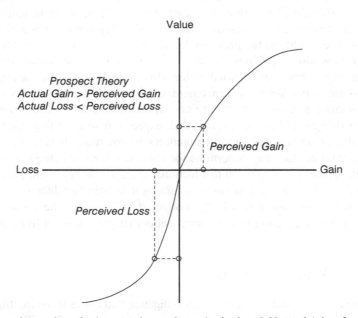

FIGURE 9.1   Prospect theory view of gain versus loss and perceived value. © Howard Ashcraft.

Test It Yourself

Give the following problem to a few people and ask for the answer:
A bat and ball cost $1.10.
The bat costs one dollar more than the ball.
How much does the ball cost?
Ask a few others to solve the following problem.

$$X + Y = \$1.10$$

$$X = Y + \$1.00$$

$$Y = ?$$

The first formulation invites the use of System 1. The second invites System 2, even though they are the same problem.

But what about teams? It turns out that teams can actually amplify errors in decision making, but they can also be better. Based on our experience and current research on teams, team decision making can achieve this goal by using the following techniques.

Techniques for optimizing team decision making.

- Engage System 2 to reduce the effect of "shortcut" thinking. The Plan-Do-Study-Act cycle, often summarized in an A3 report, is an excellent way to engage System 2. Choosing by Advantages (Suhr, 1999) also engages System 2 thinking.
- Confront biases. Understand how a decision might be swayed by built-in biases.
- Aggregate information. One of the major benefits of teams is the availability of more information.
- Have inquisitive and self-silencing leaders. The stronger the leader, the less time he or she should spend talking and the more time he or she should spend listening or asking real questions.
- Reward group success. Emphasize that individual success is not a metric. The team goes forward only if all succeed. This is one of the bases of IPD.
- Design a team decision methodology that reduces the effects of social pressure, ensures expression of minority opinions, and adequately incorporates the wisdom of teams.
- Assign roles (devil's advocate, expert, etc.) to team members and have them defend their position.
- Change perspective. Don't ask how a design/construction team should address a problem; ask how someone from a completely different perspective would assess the problem.
- Consider voting and averaging (or the Delphi method) when trying to determine an estimate.
- Finally, use real data whenever it is available.

9.5   INTERCONNECTIONS

Teams are the primary operational element of the integrated organization. Drawn from the various component organizations within the IPD project, they stretch across corporate boundaries and bring reality to integration. Because work in IPD is performed by teams, they provide the muscle for the integrated organization. The more general concept of enhancing collaboration and the specific tool of co-location are closely related to teams.

9.6   REFLECTIONS

Unused human potential is often described as the "eighth waste." Leaders focusing on this eighth waste will necessarily focus on improving team performance, and this will lead to examining the evidence about team dynamics and decision making. The future leaders will understand that their role is to enable teams to perform better. As observed by Craig Larman (2008) in a software context, team managers should build teams to build projects.

But to achieve this end, individual firms will need to reward employees for their ability to enhance team performance, rather than their personal competence or performance. This will be resisted because many of the managers of firms achieved their positions on individual performance and will feel undermined and threatened by a new set of metrics that may not favor their managerial style. But as collaborative projects continue to succeed, it will be increasingly difficult to ignore the importance of teams and the knowledge and skills necessary to improve team performance within the integrated organization.

9.7   SUMMARY

Teams are the fundamental building block of the integrated organization. Managed well, they can achieve results impossible for the individuals alone. But leading and managing teams will require understanding the strengths and weaknesses of teams and requires understanding how they should be created, supported, and nurtured. Moreover, it requires a new type of leader who understands that project success is achieved by improving team performance.

NOTES

1. Social loafing occurs when the team member is not internally motivated and personal accountability is diluted by focus on team performance. If social loafing occurs, the average performance of the team is less than the individual performance of its members. This was first documented by Max Ringelmann in the 1920s and confirmed by later studies (Karau & Williams, 1993).
2. The bystander effect occurs when an individual does not take action because he or she believes that someone else will. It is also called the Genovese syndrome because of alleged inaction by multiple witnesses to a brutal and protracted murder.

3. Groupshift can distort decision making by exaggerating individual positions through reinforcement and lack of individual accountability. This results in group views that are more extreme than the average of individual opinions (Robbins, 2011).

4. Groupthink occurs when individuals will not challenge beliefs or assumptions that are perceived as inherent in group self-identification. Group pressure results in conformance to group norms (Janis, 1972).

REFERENCES

Amabile, T. M. (1996). *Managing for creativity*. Boston, MA: Harvard Business School Press.

Amabile, T. M. (1998, September–October). How to kill creativity. *Harvard Business Review*, 76(5), 76–87.

Benkler, Y. (2011, July–August). The unselfish gene. *Harvard Business Review*, 89(7–8), 76–85.

Collins, J. (2001). *Good to great: Why some companies make the leap … and others don't*. New York, NY: HarperBusiness.

Covey, S. M. R. (2006). *The speed of trust*. New York, NY: Free Press.

Deci, E. L. (1971). Effects of externally mediated rewards on intrinsic motivation. *Journal of Personality and Social Psychology, 18*(1), 105–115.

Deci, E. L. (1972). Intrinsic motivation, extrinsic reinforcement and inequity. *Journal of Personality and Social Psychology, 22*(1), 113–120.

Deming, W. E. (1982). *Out of the crisis*. Cambridge, MA: MIT Press.

Hackman, J. R. (2011). *Collaborative intelligence: Using teams to solve hard problems*. Oakland, CA: Berrett-Koehler.

Janis, I. L. (1972). *Victims of groupthink*. Boston, MA: Houghton Mifflin.

Kahneman, D. (2011). *Thinking, fast and slow*. New York, NY: Farrar, Straus and Giroux.

Karau, S. J., & Williams, K. D. (1993, October). Social loafing: A meta-analytic review and theoretical integration. *Journal of Personality and Social Psychology, 65*(4), 681–706.

Katz, R. (1982). The effects of group longevity on project communication and performance. *Administrative Science Quarterly, 27*(1), 81–104.

Katzenbach, J. R., & Smith, D. K. (1992). *The wisdom of teams*. Boston, MA: Harvard Business Press.

Katzenbach, J. R., & Smith, D. K., (2005, July–August). The discipline of teams. *Harvard Business Review, 71*(2), 111–120.

Larman, C. (2008). *Scaling Lean and agile development: Thinking and organizational tools for large-scale scrum*. Boston, MA: Addison-Wesley Professional.

Molenar, K., Messner, J., Leicht, R., Franz, B., & Esmaeili, B. (2014). *Examining the role of integration in the success of building construction projects*. Austin, TX: Charles Pankow Foundation/Construction Industry Institute.

Pink, D. H. (2009). *Drive: The surprising truth about what motivates us*. New York, NY: Riverhead.

Robbins, S. P. (2007). *The truth about managing people*. Indianapolis, IN: Pearson FT Press.

Robbins, S. P., & Judge, T. A. (2011). *Essentials of organizational behavior* (11th ed.). Upper Saddle River, NJ: Prentice Hall.

Suhr, Jim, (1999), *The Choosing By Advantages Decision Making System*, Westport, CT: Greenwood Publishing Group.

Sunstein, C., & Hastie, R. (2015). Wiser: Getting beyond groupthink to make groups smarter. Boston, MA: Harvard Business Review Press.

CHAPTER 10

Integrating Project Information

"Truth is a fundamental principle that is needed for any form of building or construction. So also is honesty."

—Sunday Adelaja

10.1 WHY BOTHER?

Decisions in a complex design and construction project are made constantly. As the project progresses and conditions (such as the design and schedule) change, the owner and the project team must respond swiftly. And the decisions need to be well informed. Team leadership must have ready access to the latest relevant information including cost, scope, schedule, and quality. If the information is scattered throughout the project, in different formats, and located on different systems, project leadership is flying blind.

Integrated information coordinates information from all disciplines to provide an accurate representation of project reality. It allows project leaders to understand the current conditions and to allocate resources at their disposal in order to achieve specific project outcomes. It provides the tools for anticipating the consequences of their decisions.

Integrated information also provides all project participants with the necessary information to perform their responsibilities. In Lean production, a worker's tools and resources are organized and readily at hand. Similarly, integrated information organizes project information and makes it readily available to all.

Integration requires exchanging information among disciplines. Without integrated information, critical information can become siloed within a discipline and not understandable to others. Integrated information allows information to freely flow among disciplines creating the possibility of integrating processes and organizations.

Integrated information is the neural system of integrated project delivery (IPD).

10.2   WHAT IS INTEGRATED INFORMATION?

Integrated information has five characteristics:

1. It uses a common language for sharing the information so that it can be understood by all parties. This requires protocols, naming, and interoperability standards.
2. It is readily accessible by all who require the information. Ideally, it is stored in an organized data library so that the information resides in one space (although that space could be virtual).
3. It is unique and reusable. Data reflect the needs of all users and are structured to contain the information required by different parties. For example, there should be a single source of information about a wall, and it should contain the information required for the architect, estimator, framer, and others.
4. It has a source of truth, to allow the user to determine its reliability.
5. It is aggregated from cross-functional sources to provide a current and accurate representation of the project.

The modern automobile is a good example of tightly integrated information. Although a car is sold by a manufacturer, it is actually composed (much like a construction project) of parts and systems that are created by a wide variety of manufacturers. Many of these components are "intelligent" in that they create, monitor, or respond to information. They can only speak with each other because they share a common set of data protocols—the common language of our definition.

The information is stored in a single location that is readily accessible to all components. The modern car contains a microprocessor that serves as the information hub. Using Siemens terminology, this is the engine control unit (ECU). The information reported to the ECU is available to all who need it. The stability control unit, for example, can receive information regarding tire slip angle and acceleration to allow it to assert control. The engine receives information about operating conditions to allow it to tune itself to changed circumstances. The service manager can extract data regarding service intervals. The mechanic can review real-time operational data and error flags to assist maintenance and troubleshooting. And the driver receives information, discussed further below, that contains information from multiple systems to allow her to make informed decisions.

The information is also unique and reusable. In the preceding examples, there were not separate data stores for the mechanic, the driver, or the engine. Rather, the same information was accessed for different purposes using different views. The mechanic was reading the information through a diagnostic tool, and the driver was viewing information on the dashboard.

The information has a source of truth, or at least a system to flag reliability problems. The Siemens ECU, for example, compares data from multiple locations for reasonableness. If measurements are conflicting or out of range, it posts a flag, "Improbable Data," to caution that the data are unreliable.

Finally, it aggregates information from multiple sources to create a view with the information the driver requires to make informed decisions. This is the "dashboard" that every car, and every project, must have. The dashboard provides real-time information to the driver on what the speed is, how much fuel is left in the car, and how far it can travel. The Global Positioning System (GPS) system in the car provides real time feedback on the time it will take the driver to reach his destination and where he is currently at. This allows the driver to make adjustments during the journey. For example, if he is running low on fuel he can stop to fill the tank, if he is traveling over the speed limit he can slow down, he can even take a different road if the road he is traveling on is congested. The dashboard:

1. Provides a clear picture of current reality and aggregates the historical information for the stakeholder. The driver knows the speed, outside temperature, fuel left, distance traveled, and so on.
2. Exposes performance across multiple factors or disciplines. For example, it shows speed, fuel left in the car and the distance left to travel.
3. Reveals interdependencies that are difficult to see. For example, the combination of fuel left, miles per gallon and the GPS unit showing the distance yet to travel allows the driver to see interdependencies and understand whether he can make it to his destination given the distance left to go and the fuel left in the car.
4. Updates continuously. The latency of the information availability is practically zero. This means that while the driver is driving, the information is relevant to the current conditions.
5. Adjusts continually in response to the changing environmental conditions.
6. Predicts possible outcomes. For example, the GPS unit will inform the driver that he still has so many more miles to go to reach his destination.
7. Aids the driver in making decisions based on the current reality and predicted outcomes. For example, if the dashboard shows the driver is going above the current speed limit, then he knows that he is breaking the rules and needs to slow down.

In a project context, the dashboard provides an accurate snapshot of current project conditions and makes predictions, based on project budgets, schedules, productivity data and similar information, concerning the project's direction and outcome. A complex project has too much information for managers to assess without a simplifying, but accurate tool. Because the dashboard is a data view that allows managers to make decisions, it should provide information relevant to their three opportunities for change ("levers")—Product, Organization, and Process—described in Chapter 5.

In addition to the important characteristics above, Integrated Information in the architecture, engineering, and construction (AEC) industry should be designed to:

1. Allow exploration of multiple alternatives, and reveal the consequences of selecting an alternative across many factors.
2. Reveal dependencies and constraints from multiple points of view, and therefore predict the true outcome of the project. Then the owner and team leaders can be confident in making decisions and accepting (or not) various alternatives.
3. Support decision making at the team as well as the manager level.

Garcia used the DEEPAND framework to describe the work teams do in meetings to reach decisions (Garcia, Bicharra, Kunz, & Fischer, Garcia, Bicharra, Kunz and Fischer, 2003). The DEEPAND tasks are Describe, Explain, Evaluate, Predict, Alternative formulation, Negotiate, and Decide. Although the meeting value resides in the later tasks, the first two tasks absorb most of the meeting time because they are spent determining what information is relevant and whether it is current and accurate, rather than doing value adding work, that is, the performance prediction of multiple disciplines, formulation of alternatives, negotiation of trade-offs between the multiple alternatives, and decision making. One of the goals of integrated information is to remove the waste often associated with the first two DEEPAND tasks.

10.3   WHAT DOES SUCCESS LOOK LIKE?

The goal of an integrated information system is to accelerate and increase everyone's understanding of the current status, so project team members can frame problems better, see possible solutions quicker, make more accurate predictions and ultimately make better decisions.

All project teams use a variety of methods, technologies and various applications to accomplish the tasks of producing design, estimating costs, generating schedules, predicting energy performance and generating various design options. The assumption is that the use of these methods, technologies, and applications means project teams have a great integrated information system. But many times this assumption is not true and the use of disparate methods, technologies, and applications ends up creating silos of information, which leads to poor decision making by the project team. On the contrary, a successful integrated information system is one that supports the work of an integrated team and organization by revealing the dependencies between the team members, thus deepening team members' understanding of the work being done by others.

Building information modeling (BIM) representing all the disciplines in a single model is a good example of integrated information. For example, the UCSF Mission Bay Hospitals project team developed an integrated BIM that combined the models generated by various disciplines into a single model which was used to understand the coordination between the various systems and resolve coordination issues before construction. This integrated BIM allowed team members from the various disciplines to understand the dependencies of their work on the other disciplines. Figure 10.1 shows the integrated BIM for the patient rooms in the UCSF Mission Bay Hospitals project. It includes walls, metal studs, ceilings, mechanical, electrical, plumbing, fire protection, medical gases, and seismic bracing, and allowed project teams to understand dependencies of their work on others in real time.

An integrated information system also provides real-time status for design completion, cost, quality, and other factors important for the owner across the project disciplines and member organizations.

An example of this is the cost and budget tracking dashboard that the UCSF Mission Bay Hospitals team developed to keep track of the progress toward achieving the owner's cost targets. The dashboard was updated with information from the various "cluster" groups or disciplines every week and the information was used to guide the project design process with almost real-time feedback to the project team. The dashboard wall, shown in Figure 10.2, was a very prominent feature in the Big Room during the project design phase. The dashboard was updated every week and provided access to near real-time information on costs for the project team to make informed decisions toward reaching the target cost.

An integrated information system also supports decision making with a holistic view across all relevant factors and the right information available in making these decisions.

The Temecula Valley Hospital project team created an integrated cost workbook to enable target value design and delivery. The owner established an aggressive and specific cost target prior to design, and all decisions had to be consistent with this objective. During design development, members of the

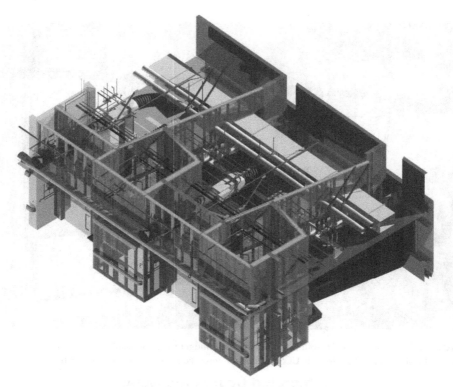

FIGURE 10.1 Integrated BIM at the UCSF Mission Bay Hospitals patient rooms. Licensed by The Regents of the University of California on behalf of its UCSF Medical Center; courtesy of DPR Construction.

integrated team started evaluating the various design options and preparing cost studies. They quickly realized that they needed an integrated view of this cost information to decide among alternatives but didn't know how to go beyond maintaining a project budget. The problem was the constant ebb and flow of that budget as many alternatives were considered, and ultimately rejected or accepted. Delivery team managers found that they could not answer the question of whether their company would earn the profit to which they were entitled through the IPD contract. As the team mobilized on site and began working together daily in their Big Room, they adopted a tool developed by Ken Lindsey, the project executive for Southland Industries, the HVAC and plumbing trade partner. Ken's Excel workbook became a place where he and his counterparts captured both potential savings or cost increases to their scopes of work and the project as a whole. This allowed the owner and delivery team to see the current and anticipated final cost. It also allowed the team to understand what needed to be done and what additional cost savings were required to meet the target cost and profitability goals. The workbook created a single source of truth for everyone.

Figure 10.3 shows the information that the IPD team partners, including the owner, shared and discussed each month. They identified risks and opportunities in one of the three main worksheets along with initiatives they could take to eliminate waste and save money. They tracked budget transfers, increases and savings openly with information from their company job costing systems. What made the

FIGURE 10.2   UCSF Mission Bay Hospitals project cost and budget tracking dashboard. Licensed by The Regents of the University of California on behalf of its UCSF Medical Center; courtesy of Howard Ashcraft.

Integrated Cost Forecasting
Monthly Budget & Billing Workbook

Budget Management	Production Management	Cost Accounting
• Risks of Increased Cost • Opportunities to Reduce Cost	• Anticipated Savings • Realized Savings	• Transfers • Increases • Savings
Risks & Opportunities	**Path Back to Budget**	**Budget (Partners / Scopes)**

Production Management
Factors & Metrics

Safety, Quality & Cost Factors	Time Factors	Metrics
• Work Processes / Methods • Supervisor & Technical Skills • Built-in Quality Methods • Safety Training / Awareness	• Planning & Coordination • Constraints Removal • Network of Commitments	• Operations time (Video Studies • Last Planner™ System (TA, TMR, PPC) • Crew Productivity • Rework % • Inspections Passed % • Total Recordable Injury Rate

FIGURE 10.3   Integrating cost forecasting and production management. Adapted from Ken Lindsey, Southland Industries.

system work was the safety, quality, and production information the IPD partners were continually gathering and seeing in the *gemba* (Japanese term for workplace) walks they took together as part of their weekly team meetings. Without knowing how safe and productively their crews were putting work in place correctly the first time, they could not have forecast the effort and cost to complete. Their cost projections wouldn't have been any better than on most projects.

This offers an important lesson on what is important about integrated information. Each of the individual entities had their own system for cost tracking. But the multiple, separate systems did not give the owner and team an accurate projection of overall final cost. Once the team agreed on a common cost format and adopted the "Joint Budget & Billing Workbook" to represent all cost impacts, team managers had the information they needed to make good decisions.

10.4 HOW CAN THIS BE DONE?

Integrated information is not about technology per se. It is about accessing the same information quickly, with less effort. In this way, it supports the goal of Lean, which is to reduce waste to increase customer value. The prerequisite for integrating information is agreement on organization, standards, and protocols for creating and sharing it. Integrating information depends on people trusting others to do the right thing with the information. Integrating information and sharing it can accelerate the building of trust.

10.4.1 Start Early

Information integration must be supported, and preferably promoted and used, by top management. Because integrating information requires a long term investment across multiple business functions, individuals can do very little without management support.

An example of an integrated information system designed as an experiment, which can be used on multiple projects rather than for a particular project, is the Integrated Cost Accounting and Production Planning system (iCAPP) developed by Ram Ganapathy of DPR's Construction Technologies team in collaboration with Lean project teams in Phoenix. The system is based on the use of detailed BIM, parametric estimating assemblies from the estimating database and Lean production planning techniques used to track work. Ram and the project teams developed a process to break up the BIM into the various work locations so that the quantities in the work locations can be identified from the BIM. The assemblies (installation work packages) from the cost database were then used to calculate production rates for work that foremen were being asked to complete within project locations, and then the same information was used to track production by location and procurement packages. This system provides project teams productivity rates and quantities for planning remaining work based on actual quantities of work put in place according to their plans up to the current date.

Figure 10.4 shows the iCAPP system that uses quantities from BIM and estimating assemblies to generate cost and anticipated production rates by location and then uses the same information to track production work in the field.

Integrating project information is best done prior to starting design, not when a team is under time and money pressure, trying to solve the technical problems of design and construction. At the very least, integrating information should be tackled immediately, along with designing the organization and

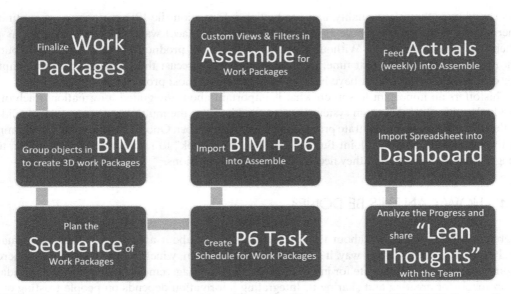

FIGURE 10.4 DPR's Integrated Cost Accounting and Production Planning system (iCAPP). © DPR Construction, by Ram Ganapathy.

processes for the project. The value of designing an integrated information system at the beginning of the project is threefold:

1. It not only creates a better information environment that creates the transparency needed for the team to do its work; it also
2. Starts building the trust and collaborative culture necessary to integrate the organization and process to create a highly valuable building; and
3. Avoids miscommunication and waste caused by inconsistent and incompatible systems. Once teams get committed to their individual ways of working, it will be difficult to integrate.

10.4.2 Develop Common Standards

Aggregating information is much simpler if it is recorded consistently. The Temecula Valley Hospital example discussed previously is an example of how common standards for cost tracking allowed aggregation of data and a deeper understanding of the challenges and opportunities to achieve target cost. A similar approach should be taken with other information to enable information to be aggregated or transferred among information systems. Depending on the size and complexity of the project, this may require standardization of how data are captured, what data are captured, how data are structured, how data are named, who can access the data, and with what usage rights.

Standards also need to be applied to software used on a project. The team should decide on a common platform or platforms, or should at least define standards for software interoperability.

This includes versions of software, as well as the product. In some instances, a software upgrade may change data structures that render data unreadable with earlier versions. In one project, a team member unilaterally decided to upgrade all of its BIM software and caused a several week delay due to resulting data incompatibility.

A simple, but common, problem on projects is tracking communications. Information can be exchanged by paper copy, e-mail, phone, wikis, blogs, messaging, posting, and even physical mail. Ironically, the existence of a rich set of messaging tools can lead to total confusion. If some parties e-mail, others text, and some post their information to the project portal, no one will ever be sure that the information they are viewing is complete and current. Moreover, some of the systems, such as attachments to e-mail, bury information so deeply that it might as well be lost.

The upshot is that to integrate information effectively, you need to develop and enforce standards about the information itself and about how the team interacts with that data.

10.4.3 Create Clarity

On any project, there will be accidental overlaps, rework, and mistakes. An example of traditional rework is the overlap in work between mechanical engineering design and fabrication drawings (or models). A well-constructed BIM Execution Plan is an example of a tool that reduces waste by bringing clarity to roles and deliverables. It usually describes:

1. The goals of the use of BIM on a project, which might include coordination, cost estimating, layout, generation of shop drawings, facility management, and other uses.
2. The roles and responsibilities of the players who are involved in the BIM process and detailed description of who will do what.
3. The level of development (LOD) that the models will be developed to and the sequence in which this LOD will be developed and the responsibility for developing the LOD.
4. The naming conventions and interoperability of the various systems that will be used as well as the source of truth for the information for the various systems incorporated into the BIM.
5. The protocols to share the models and also processes to use the models for their stated purpose.

This way the whole team develops a common understanding of how BIM will be used on the project. It is also a good example of what an integrated information system ought to do.

10.4.4 Enforce Standards

Integrated information systems offer tremendous opportunities for enabling better and more effective communication. But they have a major weakness. No system is fully integrated if a key participant doesn't use it. And there is an almost overwhelming preference to use methods and systems you are familiar with. Under pressure, team members will resort to using their familiar tools in a familiar way—to get their work done even if it makes work more difficult for others. For project management, this implies three rules:

1. Choose systems that are the same as those used by the most important team members (if possible) to avoid backsliding;

2. Don't obsess about having the latest or greatest technology. Use the most effective technology that you can get everyone to use—even if that means it isn't cutting edge; and

3. Once you decide on a standard, enforce it. Unless everyone is on board, you can't achieve reliability and, in turn, integrated information.

10.4.5   Combine Methods and Tools to See from Different Perspectives

One of the important characteristics of an integrated information system is its ability to accurately forecast outcomes and therefore inform the project team to make better decisions. The next level, requiring much better information, is to predict outcomes. Prediction depends on an integrated team's ability to build virtual models and simulations that are good abstractions of reality, and then use these models continuously throughout the project to make the necessary changes and adjustments.

The capabilities and limitations of the prediction methods must be considered relative to the issues that need to be explored on a project. For example, a Critical Path Method (CPM) schedule is a model that is commonly used in the AEC industry to predict and forecast project completion. Although it is a widely used, it is a model that only partially considers the constraints project teams face in a design and construction project and turns many assumptions into deterministic values. For example, it fails to consider the impact of coordination and reciprocal work in the design phase and it inadequately considers time-space conflicts during construction and the impact of those conflicts on production activities. A four-dimensional (4-D) model is a much better abstraction when it comes to understanding time-space conflicts during construction. A Monte Carlo simulation that considers resource capacities, resource skills, and utilization is a much better model of the coordination activities that project teams have to perform and how those activities affect the potential project completion date. A line of balance method provides a much better visual guideline to the project team to understand the differences in production rates between multiple trades and the options the team has to impact the flow of work between various work areas of the project and between the various trades. An integrated information scheduling system encompasses all these methods and allows the project team the ability to make changes when they are needed.

Many times we see that project teams blindly adopt complicated scheduling specifications that require a very detailed CPM schedule from the start. The CPM is not created by the people responsible for doing the work and makes assumptions about future conditions that cannot be reliably predicted. And, as discussed earlier, it is only an incomplete version of reality that disregards the benefits other methods bring. Recognition of the limits of CPM is leading to a more evolved approach to integrating schedule information so that the right people have the right information at the right time using the most appropriate method. Similar concerns apply to prediction methods for other performance objectives like cost and CO_2 emissions.

10.5   EXAMPLES AND BENEFITS OF INTEGRATED INFORMATION SYSTEMS

10.5.1   Integrating Information for Target Value Design

One of the keys to target value design (TVD) is integrating function, use, and cost of a building from an owner's perspective. A big part of this is making cost an input to design. It turns out to be harder than most people expect. Haahtela, operating in Finland and Sweden, has developed a different approach than anyone else we know. It's based on years of experience as cost consultants and disciplined work to integrate the right information.

Haahtela, led by founder Yrjänä Haahtela and partner, Ari Pennanen, are cost experts able to leverage their own large database, rivaling that of any other major construction cost consultant. They differ from others in using cost and information about client building requirements to steer design rather than estimate its price tag.

Over the years, Haahtela has come to believe that value for their customer does not come principally through design. They believe that value is the ability to fulfill the strategic goals of the organization. It follows that understanding these goals is the Haahtela point of departure. Haahtela's Ari Pennanen helps clients realize that the transformations the client makes will increase value. Students gain knowledge, hospitals heal, lawyers settle disputes, restaurants are meeting places, and so on. Ari and Haahtela team members help clients consider the value-adding activities their spaces should support to become enablers of transformation.

Haahtela works hard to understand the owner's space program in the context of the organization and activities of the users of the spaces. They are Lean thinkers and believe they must "go see." They observe what people currently do in their spaces and learn what they want to change. Haahtela staff often map their client's activities and work processes.

Once Haahtela understands function and actual and desired use of space, they create a "BIM Prior to Design." Their BIM objects, including the quality of systems and components, are linked to their cost database. Space is mapped to function, which reflects the activities that increase transformation and value their clients provide to their customers. No architectural design has occurred at this point, but Haahtela can inform their client how much it will cost to deliver the space program. In nearly every case, the cost exceeds the available funds, what is called "allowable cost" by TVD practitioners. Haahtela then proposes changes to the space program, often advocating for multiuse to increase utilization. Almost always, owners change the space program and sometimes are able to allocate or raise more money. With Haahtela's help, having integrated information about function, use, and cost, the client is able to begin their project with a high-value space program and an achievable target cost.

Cost is a fixed parameter when the design team is engaged, before that work begins. Quality is steered by Haahtela working with the delivery team. The Haahtela team remains engaged throughout the project to provide rapid feedback on the cost of proposed solutions so that quality is aligned with target cost. This kind of steering and rapid feedback would be extremely difficult, if not impossible, to achieve without integrated information.

10.5.2   Integrating Information to Automate Design Tasks

In the previous section, we described the need to build simulation models that are as close to reality as possible and then run them in order to generate predictions about various outcomes for project teams to consider in their decision making process. As noted but as rarely found in practice today, these simulations must share a common information basis across the many disciplines involved in the design and construction of the facility. With the current advancement in cloud computing, where massive banks of computers are available to be rented inexpensively for purpose-built processing, the project teams can now automate the process of design based on multiple design inputs and models.

For example, Aditazz, a start-up in Silicon Valley, is utilizing methods that have been honed in the electronic design automation (EDA) space and applying those methods to generate the most optimal functional requirements that can drive a conceptual layout of a medical facility based on thousands of simulations that consider inputs such as location, demography, and workflow of the various functions performed in the hospital. What used to take architects and their subconsultants weeks or even months to produce can be completed at a much higher quality using the Aditazz Realization Platform in a matter of days or hours. The Platform, illustrated in Figure 10.5, integrates client requirements with targeted building performance data to rapidly iterate and automatically generate layout options for a healthcare facility. Subsequent changes and any impact of changing inputs can then be simulated in a matter of minutes or seconds. This capability will improve exponentially in the years to come where many more inputs across diverse verticals such as finance, engineering, materials science, environment, schedule, quality, safety, and so forth, can be taken into account simultaneously to produce the most optimal design suited to the most efficient and safe construction methods while maximizing sustainability. It is impossible to accomplish all of the above with today's traditional methods because just compiling the necessary information for these design tasks from fragmented sets of information would probably take longer than an entire design cycle carried out with the Aditazz Realization Platform.

10.5.3   Integrated Information for Rapid Prototyping

Integrated information is useful only if it gives perspective on cost, quality, scope, and other aspects. So far, this chapter has focused on digital information. Physical models, mock-ups, or prototypes are another example of the result of integrating information. Prototypes should allow a project team to look at trade-offs in a more systematic manner. Prototypes address whether or not a given design is buildable, operable, usable, and sustainable, and provides the ability to test whether to not it fits those qualifications.

Prototypes must be quick and easy to assemble. If it takes too long to see the ramifications of various options, they are not useful. The whole purpose of the prototype is to inform the final design, so it must be created and reviewed in time for resulting decisions to be incorporated.

Prototypes can be virtual or physical. The process of developing them is typically just as important as the actual prototype itself. Through the act of creating a prototype, the team not only learns to work together but also receives input from many stakeholders, including end users and operators. For example, the team building the Seattle Children's Hospital built physical mock-ups for doctors and nurses to reenact surgical and other procedures, revealing patient and staff flows. And Cook Children's Hospital in Fort Worth engaged nurses, doctors, and staff in a rapid prototyping of clinic spaces, facilitated by the project architects, using cardboard and tape, shown in Figure 10.6.

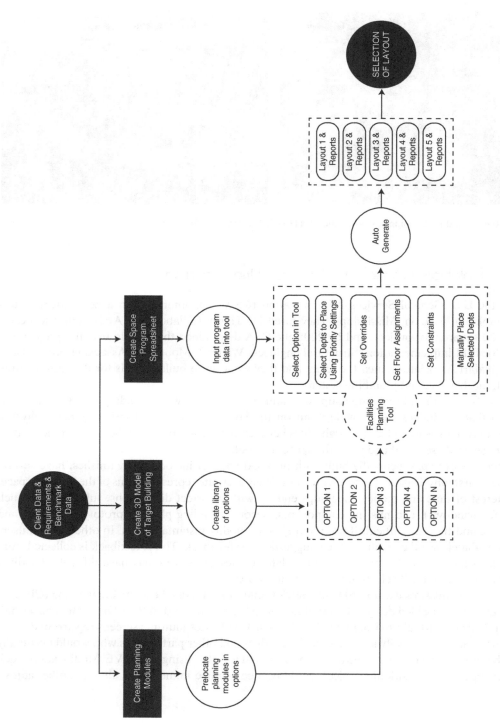

FIGURE 10.5 Aditazz Realization Platform from client data and requirements. © Aditazz; illustration recreated by DPR Construction, Lyzz Schwegler.

221

FIGURE 10.6   Rapid prototyping at Cook Children's Hospital. © Howard Ashcraft.

10.5.4 Prototyping Digitally for Common Understanding

In 2002, the U.S. General Services Administration (GSA) was interested in a new federal courthouse complex in Jackson, Mississippi. They hired H3 Hardy Collaboration Architecture to design the 410,000 square foot project, and by 2005 the GSA determined they would use that project as a test case for BIM implementation. Structural engineers Walter P. Moore and Associates and consulting firm Ghafari Associates joined the team, and they went on to build models for the architectural, structural, mechanical, and plumbing systems.

Teams often build full-scale mock-ups of planned courtrooms when building new federal courthouses, so that the judges can provide input on the layout and design. These very basic physical mock-ups allow judges to understand sight lines between the judges' bench to the jury and the witness stand, and generally see firsthand what the space will look like.[1]

But those prototypes, generally built with plywood and not including any finishes, have several drawbacks.[2] First, physical mock-ups are not completely accurate representations of the finished space. In the interest of time and money, mock-ups end up with incorrect dimensions, substitutions (such as paper for windows), and missing components. Second, building physical prototypes is slow and cumbersome and they are difficult if not impossible to modify instantaneously. In other words, users cannot immediately see the results of their suggestions or feedback. Third, feedback is collected from users in small groups and has to be integrated later by designers who may have difficulty handling potentially conflicting feedback from the different groups.

The GSA commissioned Stanford University's Center for Integrated Facility Engineering (CIFE) to build a 3-D computer-aided design (CAD) model, which could be used in Walt Disney Imagineering's Computer Assisted Virtual Environment (CAVE). The CAVE uses multiple screens to surround viewers and immerse them virtually in the model. The judges and other participants who would eventually work in the courtroom space gave feedback on the design together using the CAVE. Modifications such as color changes and repositioning objects could be incorporated immediately, although other adjustments such as dimensions had to be done later. Note that on this project the two sets of information were integrated:

1. The BIM that combined all necessary disciplines and became the common basis for the review by the future users; the key part of this review was the prediction in each user's mind of how the space will work for them.
2. The feedback collected from the users, which was assembled in an integrated way because all users looked at the same scenarios in context of the same BIM at the same time.

CIFE assessed the effectiveness of the CAVE sessions using the DEEPAND framework it had developed and determined that the sessions were more productive than traditional design review methods (Majumdar, Fischer, and Schwegler, 2006). The group spent more time on activities such as making predictions and suggesting alternatives, rather than being told how to interpret the design.

10.5.5   Integrating Design Development and BIM

At the UCSF Mission Bay Hospitals project, project director Stuart Eckblad strongly supported integrating design documentation and BIM, rather than allowing the BIM effort to follow and result in a separate and possibly conflicting information set. Team leaders agreed, especially DPR, as the GC, along with the mechanical, electrical, and plumbing (MEP) subcontractors. Stantec (then Anshen + Allen / AA), the project architect, and Arup, the MEP engineer, accepted the challenge.

All parties also endorsed the principle that the BIM for the project should be co-created with participation from the trade contractors that were going to build the facility. This, in turn, required very close coordination between the design and build teams to ensure that the contractor's detailers preserved the design intent and provided the most constructible solutions.

The designers and builders had to solve the problem of interoperability between the many different software applications being used, each with their own object libraries, often customized by their BIM specialists. Their solution was to use Autodesk's NavisWorks for coordination and live with the inevitable loss of information. This would allow the modelers to build virtually on a very aggressive schedule to meet the permit submittal schedule that the UCSF Medical Center had negotiated with California's Office of Statewide Health Planning and Development (OSHPD). In this way, OSHPD could allow construction to start before they had reviewed every design document. Submitting the design packages on schedule was the early critical path for completing the entire project on schedule.

The architects and engineers had to work closely with the BIM detailers while other team members were proposing value engineering ideas to reduce an estimated $200 million budget overrun. Many of the decisions were significant and needed to be reviewed by UCSF Medical Center administrative, clinical, and facilities managers and subject matter experts, which required time. At the same time, the design team was incorporating changes while completing design development across a large, world-class, teaching and research facility.

The UCSF Mission Bay Hospitals team had to overcome what seemed like an impossible situation: interoperability between BIM software; finalizing the design of a complex set of buildings while value engineering that design; and building virtually to submit constructible design documents for review on a tight schedule.

Fortunately, the team was organized to allow decisions to be made at the lowest responsible level within defined limits. This allowed designers and modelers to resolve most issues by talking directly within the project Big Room. If they needed to involve others, or others needed their expertise, they could use the organizational structure created to support knowledge sharing, described previously in the UCSF Mission Bay Hospitals case study in Chapter 8. Project leaders interfaced with "cluster teams" designing and managing scopes of work across the acute care building housing the three special-purpose hospitals, the outpatient building, and adjacent energy center. Four BIM production teams, one for each building and the site worked with the cluster teams. This allowed anyone to escalate an issue that they couldn't resolve at their level. Figure 10.7 shows these interconnections, which were critical to the team's success in meeting the aggressive permit submittal schedule.

The next step was to design a work process to support the completion of Design Development while incorporating value engineering changes and hand-offs to the BIM production teams modeling by floor areas divided according to Life Safety smoke compartments. Figure 10.8 shows that design/BIM work process.

Once the process was established, designers and BIM production team members decided on who would do what with which tools. Table 10.1 shows team member responsibilities and the software applications they used.

Along with BIM tools and commitment to follow an agreed process, the team needed a way to exchange the large amounts of data and information they would create hourly. Design and BIM managers adopted the "Federated Model Management" approach, shown in Figure 10.9, because this would allow team members, some working remotely, to develop models and submit them to a repository where they could be accessed by anyone who needed to see that work product and use it for coordination with their work.

The next step for coordinating design and BIM was to select a software platform that supported Federated Model Management. The managers decided to license Bentley Systems' ProjectWise application because it would allow deployment of servers at various locations so designers and trade detailers could develop and save the models locally to their hard drive, without having to go over the Internet. Team members could work productively rather than constantly wait for uploads and downloads.

Once team members had agreed on their strategy, process, roles and responsibilities, and tools, they were able to develop a template schedule they could repeat through all of the levels of the buildings to hand off design information for modeling and coordination, and subsequent production of the 2-D drawings required for OSHPD review. Figure 10.10 shows that template, which the team used successfully to deliver quality documents on schedule. OSHPD reviewers were able to approve the design submittals with only a single back-check in most cases, allowing construction to move forward as planned. The exception was when the UCSF Medical Center changed the program for two floors in the outpatient building, which required BIM to be done again, and another back-check. The changes were made efficiently and because everyone, including the OSHPD inspectors, was on site, the team was able to document everything quickly. The Big Room and the open relationship with the inspectors had a lot to do with this success. In this way the UCSF Mission Bay Hospitals project was built on a platform of integrated information.

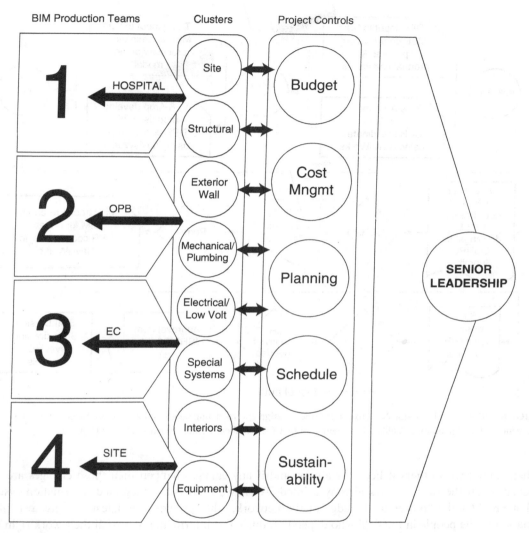

FIGURE 10.7 UCSF Mission Bay Hospitals project team organization supporting the BIM production teams. Licensed by The Regents of the University of California on behalf of its UCSF Medical Center; courtesy of Stantec Architecture Inc and Cambridge CM; illustration recreated for publication by DPR Construction, Lyzz Schwegler.

10.5.6 Integrating Information to Support Production

A recurring problem on construction projects is that not everyone who needs the latest information has easy access to that information. Project teams can only be successful if each and every one of the people working on the project has access to accurate, current, and complete information. We see project teams spending enormous amounts of time copying and recopying information on paper and then sending it out to project team members in the field. The moment something is printed and

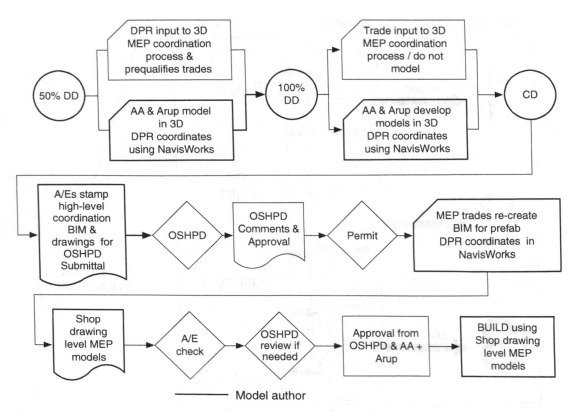

FIGURE 10.8   UCSF Mission Bay Hospitals project design and modeling coordination process. Licensed by The Regents of the University of California on behalf of its UCSF Medical Center; courtesy of DPR Construction.

then shared with others it becomes a copy and if any changes happen then those changes are not reflected on the paper copies already printed. What we need is an integrated information system that provides the "power to the edge of the network," that provides the latest and greatest information to the people in the field who depend on this latest information to install their work right the first time.

Superintendents and foremen were equipped with iPads on the UCSF Mission Bay Hospitals project. This allowed them to bring the most current information out to where systems and components were being installed and value was being created. They and their crews had access to all 2-D drawings remotely as PDF files and could see all requests for information (RFIs). All the PDF drawings were synchronized across the iPads with the central document repository. The drawings were marked up with the most recent information; for example, the latest RFI was annotated on top of the drawing The drawings also showed information that was relevant to the field personnel that was develope from the BIM. The drawings showed exact location of metal studs, with dimensions from grid lin and openings in the wall framing so the field personnel knew exactly what they were looking at a what needed to be done. No one needed to waste time searching and comparing documents to find t most up-to-date version, and they also did not need to search drawings from multiple disciplines

TABLE 10.1   UCSF Mission Bay Hospitals Project Modeling Responsibilities and Tools

Team	Models They Will Create/Responsibility	BIM Tools Used
Architect	Design model of the building	Revit Architecture 2010
	Producing 2-D submittals for OSHPD and other cgency review	AutoCAD 2010
Structural engineer	Design model of the structure	Revit Structure 2010
	Producing 2-D submittals for OSHPD and other agency review	Tekla Structures
MEP designers	Design models and criteria for MEP systems	Revit MEP 2010
	Producing 2-D submittals for OSHPD and other agency review	AutoCAD MEP 2010
MEP subcontractors	Fabrication level models of the MEP systems	CAD Duct
	Help producing the 2-D documents for A/E teams for agency review	CAD Mech
		CAD Pipe
		QuickPen 3-D
Fire protection subcontractor	Fire protection models	SprinkCAD
	Help producing the 2-D documents for A/E teams for agency review	
General contractor	Model of miscellaneous steel details	Timberline
	Model of self-perform work such as drywall, concrete, rebar	Navisworks
	Coordination of MEP systems	Solibri
		Tekla
		Strucsoft
		Revit Architecture 2010
		Revit Structure 2010
		Innovaya Visual Estimating

Licensed by The Regents of the University of California on behalf of its UCSF Medical Center; courtesy of DPR Construction.

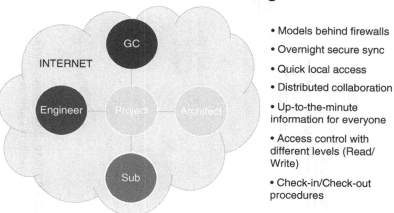

Federated Model Management

- Models behind firewalls
- Overnight secure sync
- Quick local access
- Distributed collaboration
- Up-to-the-minute information for everyone
- Access control with different levels (Read/ Write)
- Check-in/Check-out procedures

FIGURE 10.9   The Federated Model Management architecture for BIM collaboration. © DPR Construction.

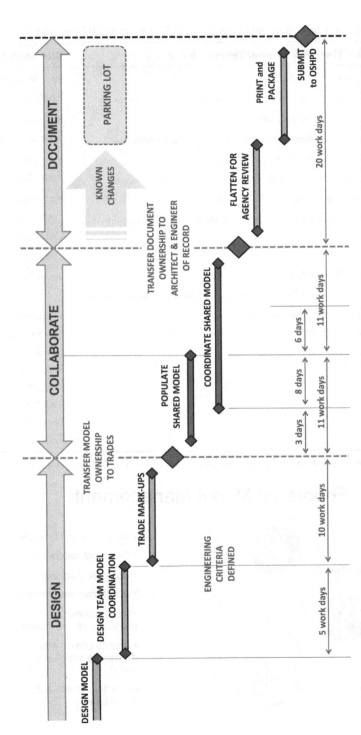

FIGURE 10.10 UCSF Mission Bay Hospitals project template for design, model coordination, and preparation of permit submittals. Licensed by The Regents of the University of California on behalf of its UCSF Medical Center; courtesy of DPR Construction.

FIGURE 10.11   Field foreman looking at BIM for framing and in-wall utilities details. Licensed by The Regents of the University of California on behalf of its UCSF Medical Center; courtesy of DPR Construction; courtesy of Mirror Mirror Digital Media, Inc.

search for information on utilities in the wall or openings. All of that information had been coordinated in the BIM and was contained in the 2-D drawings on the foreman's iPad. The key superintendents were also trained to use the 3-D model, and when a problem arose, they were able to look at it directly with project engineers available to troubleshoot.

In Figure 10.11 a field foreman uses an iPad to look at the detailed model of the wall framing and utilities that will be installed in the wall after they have completed their work. This gave field personnel information where and when it was needed.

Figure 10.12 shows the layout information the field drywall foremen were seeing on their iPads. They had shop-drawing-level BIM based on approved details at their fingertips. The difference between this and the 2-D drawings was that all the dimensions the foremen needed were shown in the way they wanted them.

10.5.7   Integrating Project and Production Schedules

The UCSF Mission Bay Medical Center integrated project and production scheduling. If you visit a DPR Construction project, you might hear someone say, "We need the right people planning in the right level of detail at the right time." They call this the "3 R's." This was on the minds of the DPR leaders after being selected as the General Contractor and became a primary goal as they moved into the Integrated Center for Design and Construction© (ICDC), the very large project co-location site for

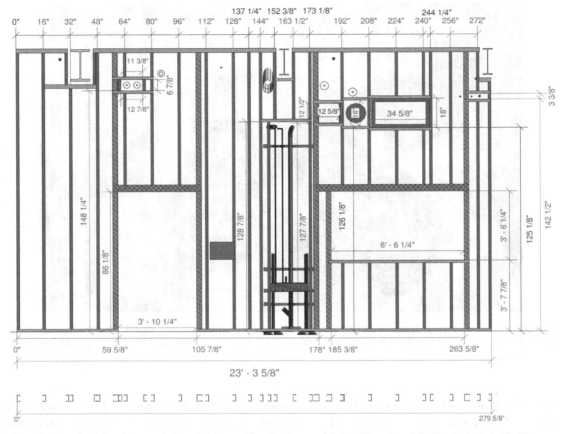

FIGURE 10.12   A drywall "spool sheet." Licensed by The Regents of the University of California on behalf of its UCSF Medical Center; courtesy of DPR Construction.

design and construction of the UCSF Mission Bay Hospitals project team. DPR was concerned that the University of California's standard requirement calling for a very detailed CPM schedule that was to be produced at the outset of the project would undercut the team's ability to engage the trade contractors (right people) in planning work closer to the time frame that the work was actually to be executed (right detail, right time). The team was also concerned that the traditional CPM management techniques did not align with the desire to incorporate lean management processes to maximize efficiency in both planning and execution of the work. Stuart Eckblad, project director for the Mission Bay Hospitals, had challenged the entire project team to "think differently" about execution to prevent incurring a long delay in delivering the project, as had been experienced on similar large-scale medical projects such as the Ronald Reagan UCLA Medical Center.

After a series of meetings in which Mike Gundrum, DPR's project risk manager, explained the advantages of combining CPM with the Last Planner® System, university officials agreed to replace

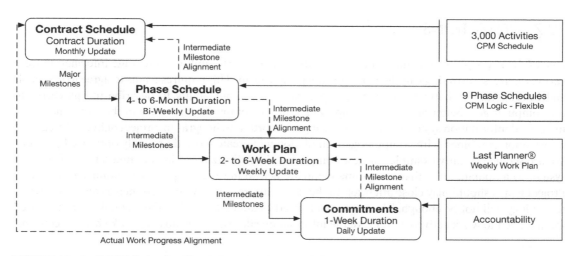

FIGURE 10.13   UCSF Mission Bay Hospitals project framework for connecting project and production schedules. Licensed by The Regents of the University of California on behalf of its UCSF Medical Center; courtesy of DPR Construction.

their standard scheduling requirements with one that Gundrum proposed: the application of phase schedules developed within two months of the beginning of each phase of work supported by six-week look-ahead schedules and weekly work plans to show the work trade foremen believed was ready for their crews to start and complete without interruption. Figure 10.13 shows the new process, which the team successfully implemented. Although the project was large and complex, scheduling included GC and trade superintendents and foremen, which is often not the case. This resulted in better control at all management levels and enabled the project team to complete the project early.

10.6   INTERCONNECTIONS

Integrated information is an absolute must to support the work of an integrated organization. It is the foundation on which the project teams can integrate their processes and allows for the project teams to understand the interdependencies amongst the team members and make informed decisions.

Integrating information is key to integrating processes among the various team members, and enhances project collaboration. The examples of an integrated BIM for coordination or the integrated cost workbook discussed in this chapter are examples of integrating outputs from the processes performed by various team members to provide a unified view to the project participants. An integrated information system is a key to support the work of a co-located team, and simulation and visualization rely on the use of an integrated information system. In short, integrated information is the backbone for the work of an integrated project organization to achieve highly valuable buildings.

10.7   REFLECTIONS

One of the key aspects of an integrated information system is the ability to integrate information from various sources into a single unified view for the project team. This requires teams to agree to protocols and interoperability between the various systems used. As our ability to describe the processes to the computer gets better and as the cheap availability of computing resources in the cloud increases, integrated information systems will be increasingly important to integrating project delivery. Integrated information combined with computerized simulation will create more thoroughly optimized projects. We are seeing the early examples of how companies like Aditazz are using an integrated information system to automate tasks that were done manually before in order to produce optimized outcomes. Project teams should play close attention to the use of information systems on their project. Just using technology will not be enough. The teams really need to understand what the technology will be used for and need to work actively to remove the information silos so that teams can make better decisions.

10.8   SUMMARY

Integrated information accomplishes four tasks:

1. It provides a common language, through defined data structures and protocols, that enable the fluid exchange of information and ideas.
2. It provides each participant with current and accurate information necessary needed by each participant.
3. It makes data visible.
4. It aggregates into dashboards information from multiple disciplines into dashboards, thus allowing project management to effectively assert operational control.

To effectively accomplish these tasks, integrated information must:

1. Use a common language for sharing the information so that it can be understood by all parties. This requires protocols, naming, and interoperability standards.
2. Be readily accessible by all who require the information. Ideally, it is stored in an organized data library so that the information resides in one space (that space can, of course, be virtual).
3. Be unique and reusable. Data reflects the needs of all users and is structured to contain the information required by different parties. For example, there should be a single source of information about a wall, and it should contain the information required for the architect, estimator, framer, and others.
4. Have a source of truth to allow the user to determine the reliability of the information.
5. Be aggregated from cross-functional sources to provide a current and accurate representation of the project.

Effective information integration is a critical component in allowing project participants to transform into a virtual organization with a common purpose.

NOTES

1. Judging BIM, *Civil Engineering*, March 2011.
2. Conceptual design review with a virtual reality mock-up model, Joint International Conference on Computing and Decision Making in Civil and Building Engineering, 2006.

REFERENCES

Garcia, A., Bicharra, C., Kunz, J., & Fischer, M. (2003). Meeting details: Methods to instrument meetings and use agenda voting to make them more effective. In *Meeting of the Center for Integrated Facility Engineering*, Stanford (no. TR147).

Majumdar, Tulika; Fischer, Martin; and Schwegler, Benedict, R. (2006). "Virtual Reality Mock-up Model." *Building on IT, Joint International Conference on Computing and Decision Making in Civil and Building Engineering*, Hugues Rivard, Edmond Miresco, and Hani Mehlhem (eds), June 14-16, 2006, Montreal, Canada, 2902–2911.

CHAPTER 11

Managing with Metrics

"Information only has value if it affects behavior."

—Steven Spear

11.1 WHAT ARE MEASURABLE VALUE AND CONTROL? HOW DO THEY RELATE?

As we have seen so far in this book, a team can and must decide how to shape and integrate its work processes, how to collaborate, and how to leverage technology. These decisions determine the allocation of resources—most importantly, what the team pays attention to—and ultimately shape the selected design solution and its performance. Hence, the effectiveness of these decisions and the efficiency with which they are reached must be measured. Through measurement, a team can gain control over the objectives of a project and how to achieve them.

Measurement and control are directly related. In this context, we use the term *control* as defined by Peter Drucker (Drucker, 2008). *Control* is a verb. To control is to direct the project to desired outcomes. Control is forward thinking, focused on achieving what should be. Control should not be confused with project controls. Project controls are the measurement and analytical tools we use to determine state, that is, what was and what is. We use project controls (measurement and analysis) to allow us to apply the correct control (direction) to manage the project.[1] The challenge is to define metrics that relate to the project's goals and that provide information to project managers that enables them to assert control over the factors that affect project outcome.

The relationship between project controls and control is at the heart of Deming's "Plan-Do-Study-Act" cycle for continuous improvement. *Planning* involves assessing the current state (project controls). Unless we measure we do not have a basis for improvement. *Doing* involves applying control to achieve a new outcome. *Studying* is using project controls to determine if the control changed the outcome. Are we achieving, or have we come closer, to our goals? *Acting* is applying successful control to the project generally. And then the cycle repeats. It is a continuous interplay between measurement/analysis and informed action.

All projects have outcomes, planned or not. If we avoid setting targets based on objectives, don't make as good a plan as possible, don't measure, don't compare results to predictions, and don't adjust our plans based on our measurements, then we leave project outcome to chance. But if we want to control outcomes and reliably deliver good projects, we must use relevant metrics to inform decisions.

11.2   WHAT DOES SUCCESS LOOK LIKE?

As discussed in Chapter 5, the project team assembles the right people to reveal and clarify the client's goals, which are then translated into specific objectives and metrics. As shown in Figure 11.1, the client goals will generally result in business, use, and operations metrics that relate to the project as it is used by the client. The team will also develop project goals, objectives and metrics that address buildability, which includes cost and schedule constraints, as well as objectives and metrics related to process and production management. It will focus on managing the project through controllable factors that are combinations of actions and metrics used to affect outcomes.

Team leaders will invest time and resources on the methods and activities to achieve the agreed objectives. For each set of methods/activities for controlling outcomes, those doing the work will decide on the metrics they will use to evaluate whether the controls are effective and then will adjust the controls and test again. The project team will use this "Predict-Test-Adjust" cycle, which, according to MIT Professor Steven Spear (2009), explains instances of superior performance in the industries he has studied. The project team members understand that to create a valuable building, the collection of organizations must:

- Agree on how to measure the value they are creating for the owner;
- Align how they'll produce the value; and
- Measure production and progress along the way.

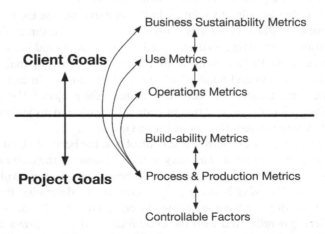

FIGURE 11.1   Translating client goals into project objectives, outcome metrics, and controllable factors.

Owner representatives will work alongside team members to develop and agree on how to work efficiently to produce the desired outcomes. Everyone understands the time they will save by first mapping work processes to see when and how individual team members will execute and hand off their work. They learn how to work as an integrated organization rather than operating as a collection of companies each doing things the best way for themselves.

Team members will pay particular attention to the decisions that owner stakeholders need to make. While it seems like these could be mapped in a single workshop, experience has taught that decision mapping is ongoing because of the deeper understanding of needs and opportunities people gain during the design process. Identifying decisions, coupled with deciding on a method for making them, allows the team to pull the work of exploring many more alternatives. Agreeing on decision criteria allows team members to focus on what counts, rather than wasting time on things that don't matter to stakeholders.

Teams that know how to make predictions based on facts quickly see where they are falling short and adjust their processes to provide the value their owners seek to consistently deliver high-performing buildings. They are the "learning organizations" Peter Senge (2006) described in his foundational text, *The Fifth Discipline*. Team members are encouraged to raise concerns and doubts and expose problems, which are simply the difference between desired outcomes and actual achievement. They are trained in effective communication and problem solving as they join the team so they can discover root causes. Team leaders know they are responsible for creating and strengthening a culture to support Spear's "Predict-Test-Adjust" methodology for everyone who joins the project along the way, right up the very last trades on the job.

11.3   HOW DOES A PROJECT TEAM MEASURE AND CONTROL THE DELIVERY OF VALUE?

11.3.1   Deciding What to Measure

As explained in Chapter 7, knowledge from all life cycle processes must be integrated to design and construct a building that is buildable, usable, operable, and sustainable. To recap from Chapter 5, a sustainable building (the result of the design and construction activities of a project team) and sustainable building (the process of designing—or virtually building—and constructing a building) must balance and maximize the economic, environmental, and social equity performance goals, often referred to as EEE performance, that exist for every facility. Economic goals include metrics of first cost, life cycle cost, energy costs to operate the facility, and income generated by the facility. Environmental goals include metrics like habitat availability for certain species, storm water retention capacity, carbon dioxide (CO_2) emissions over the life cycle or in a particular project phase, and similar considerations. Social goals can be quite broad, but might include safety during construction, development of human capabilities, inclusion of certain groups in the facility life cycle, and community interaction. Some goals are expressed in monetary terms and others in other units or assessments. Table 11.1 provides examples of metrics for first cost, life cycle cost, and income earned.

Some metrics focus on the cost of facilities and other metrics measure the income of the facility. A highly valuable building design considers both cost and income of a facility when making decisions that affect design, construction, and operation and use. Again, this sounds obvious but is rarely done.

TABLE 11.1 Sample Metrics related to First and Life Cycle Cost and Income Earned

	First Cost (FC) Related To Initial Investment, Design and Construction	Life Cycle Cost (LCC)/Period—20, 30 or *n* years	Income Earned (e.g., per person-occupancy-hour $, students educated, patients treated)
Economic ($, hours)	First cost/sf	LCC/sf	Revenue generated, LCC/income metric
Energy (kBtu)	kWh construction/sf	kWh/sf	kWh/income metric, kWh produced/week
Quality	% conformance to explicitly stated design intent, normalized by relative weight of each quality item	Productivity cost of workers who must compensate for quality deficits	Productivity cost of workers who must compensate for quality deficit/income metric
Safety (Incidents or lost-work hours)	Incidents/Msf built; project cost of workers' compensation insurance	Incidents/yr/Msf operated; operational cost of workers' compensation insurance; /hr. of operation; /work hours of operation	Increase in facility value due to safety features
Schedule Duration	Cost of design and construction services plus interest on any loans	Speed of achieving as-designed facility performance	Additional income (loss) because of schedule gain (delay)
Schedule Conformance (%)	Cost of contingency to account for schedule variability	Effort for preventive maintenance	Productivity gain of building users from preventive maintenance

Provided by Martin A. Fischer, Stanford University.

Consider this example from a major global owner of facilities that, among many other facilities, designs, builds, and operates many retail facilities (Figure 11.2). This owner wanted to incentivize the operators of these stores for energy-efficient operation, that is, operation with low energy intensity. They measured the stores' yearly energy consumption in kWh/m^2 and found that store B used 1 (normalized) units of energy measured with this metric while store A with the same merchandise and comparable climate used 1.5 (left side of Figure 11.2). They urged store A to learn from store B on how to be more energy efficient. Then they compared the two stores with an income metric like kWh/m^2 per transaction and found that store A used 1 (normalized) and store B used 1.8; that is, based on this metric, store A was more energy efficient and store B had to learn from store A, or store A would get the bonus and not store B. This example illustrates that focusing on cost-based metrics—such as kWh/m^2—alone (as is often done because costs can typically be counted more easily) may not reward the behavior that is really wanted; after all, a retail facility is built to generate income through enabling transactions.

Choosing and making measurements is a first step toward control. But to be useful, the information must be aggregated and summarized to allow project management to see a complete picture of project performance. From this holistic perspective, project management can make decisions that

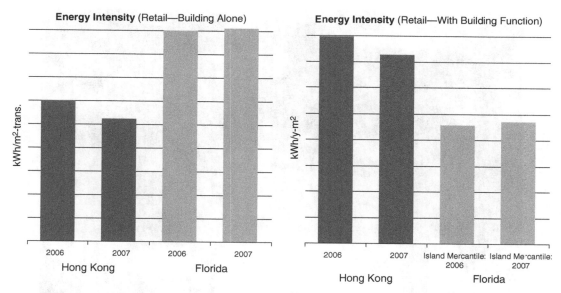

FIGURE 11.2 Two ways to look at performance: cost or income. © Walt Disney Imagineering R&D.

are consistent with project goals and that optimize competing values. A project dashboard reduces information clutter, creates synergies among information sources, and informs decisions.

11.3.2 Prioritizing Stakeholder Goals and Connecting Them to Design Solutions

Discovering and Understanding Stakeholder Values and Value

Professor Simon Austin at Loughborough University, United Kingdom, and his research colleagues developed an approach to developing stakeholder values, "Value in Design" (VALiD; Thomson & Austin, 2006) that was subsequently developed by ADePT Management Ltd., a Design Management consultancy. In three stages, the VALiD methodology assists project stakeholders in defining individual and organizational values, establishes project-specific value criteria, and provides an assessment mechanism for stakeholders to measure emerging design solutions against the previously defined value criteria. Figure 11.3 shows a group talking about their values relative to a plot showing universal values (Schwartz, 1992).

During the second day of a two-day workshop, stakeholders develop a set of value criteria and agree on targets in scorecards showing a range from minimum to maximum for each one, forming a dashboard. The act of deciding on ranges is intentional and follows from VALiD's definition of value as Benefits minus Sacrifices in relation to available resources, shown in Figure 11.4.

In Stage 3, stakeholders judge the value proposition offered by the emerging design solution at key points in the process using their own judgment and summarize it in their collective dashboard, shown in Figure 11.5.

The VALiD methodology incorporates the values and judgments of project stakeholders and allows this information to be used to make informed decisions regarding design alternatives.

FIGURE 11.3   VALiD Stage 1: Understanding values for the J. F. Tuttle Middle School, Crawfordsville, Indiana. Courtesy of Jamie Hammond, Adept Management Ltd.

Seeing Alternatives from Stakeholder Perspectives

John Haymaker's research at Stanford University's Center for Integrated Facility Engineering (CIFE) was strongly influenced by two observations. First, he recognized that design teams are always involved in decisions and that these decisions are directly related to the questions they asked. William McDonough pointed out that sustainable design was about the questions you ask, and the decisions you choose to make. Second, the actual processes for bringing people together and formulating, making, and documenting decisions are ad hoc, inefficient, and largely ineffective. John also realized that significant time and value are lost by groups that are just trying to choose the right questions to ask, assemble the information responsive to the questions, and then quickly and confidently use the information to make decisions.

Haymaker collaborated with John Chachere, a doctoral candidate in the Department of Management Science and Engineering (MSE), the home of Dr. Ron Howard and the Decision Analysis movement. Haymaker and Chachere worked with Stanford Capital Planning and Management and several local design and construction teams to prototype a design decision-making methodology called Multi-Attribute Collaborative Design, Analysis, and Decision Integration (MACDADI). One version of the MACDADI process is diagrammed in Figure 11.6, and described as follows.

No:1

Functionality > Use

Meets space requirements of users

Teachers (Stacey Guard)

Stakeholders: School Management, School Operations, Teachers, DB Team A, DB Team B, DB Team C

Worst

The size, shape and layout of internal spaces are below users' required standard

Best

The internal spaces exceed users' required standard

	1	2	3	4	5	6	7	8	9	10
CE										
T										
CZ										
V1										
V2										
V3										
V4										

Does the building allows easy supervision?
Is each space sufficient and does it provide suitable flexibility? (e.g., compare the width of proposed space with those in an existing school, is there space to prepare healthy meals and snacks, is there counceling and nursing space)
Are the hallways and stairwells open and well lit with good sized lockers?
Does the design provide all of the spaces and sizes contained in the owners description of Educational Performance Requirements?

FIGURE 11.4　Stage 2: Stakeholder value scorecard. Courtesy of Jamie Hammond, Adept Management Ltd.

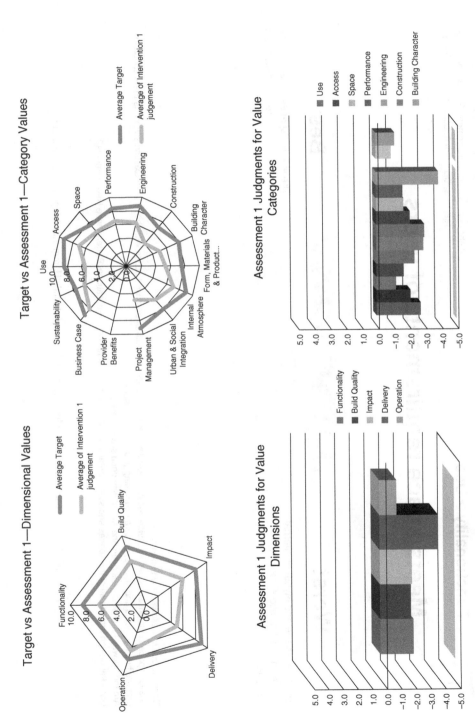

FIGURE 11.5 Stage 3: Assessment dashboard. Courtesy of Jamie Hammond, Adept Management Ltd.

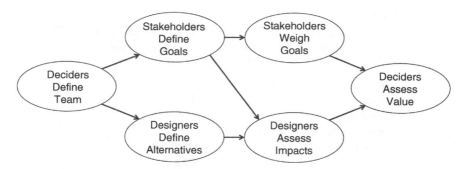

FIGURE 11.6   An overview of the MACDADI process. Courtesy of John Haymaker.

A difficult aspect of executing a collaborative decision model is gathering and organizing all decision rationale from so many participants. Haymaker created an online tool called Wecision that enables groups of people to more easily construct and share different types of collaborative decision models, such as MACDADI, Choosing by Advantages (CBA), and others (Design Process Innovation, 2015). Following is a short example of one decision in Wecision constructed with a large owner and construction firm as they decided which aspects of a large overseas construction project lent themselves to prefabrication.

After deciding what decisions to model, the first step in Wecision involves identifying and weighing the importance of project stakeholders. Figure 11.7 shows that the stakeholders of concern in this decision are the designers, owner project manager, owner construction manager, and operations and maintenance staff. Project teams often remark that this process of explicitly identifying and weighing the importance of the many stakeholders who are impacted by the decisions is a valuable step, but not often taken in practice.

When a stakeholder model has been completed, stakeholders are asked to develop and define a set of goals. When the list of goals has been developed and accepted by the group, each stakeholder is asked to define their priorities over the goals. Figure 11.8 shows a weighted goal model created by the stakeholders. The model demonstrates which goals are most important to which stakeholders, helping to identify potential conflicts between stakeholder groups, and becoming a useful starting point for the design team to begin generating designs.

Next, designers generate and analyze the multidisciplinary impact of alternatives. Alternatives are gathered and their performance on each goal is catalogued into a matrix using a normalized score to

FIGURE 11.7   Example of a stakeholder model for deciding which elements to prefabricate on a construction project. Courtesy of John Haymaker.

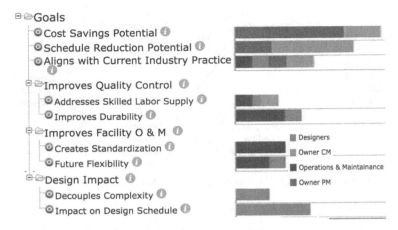

FIGURE 11.8 Weighted goal model reveals stakeholder priorities. Courtesy of John Haymaker.

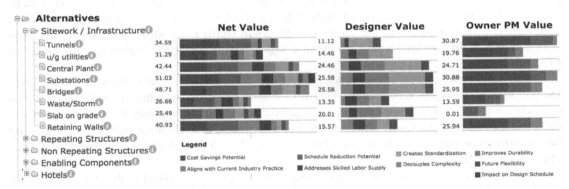

FIGURE 11.9 Value for all stakeholders, and individual stakeholders, in a decision. Courtesy of John Haymaker.

allow comparison between goals of different metrics and scales. Final value is then calculated by multiplying the impact of an alternative on the goals with the weights of stakeholders and their goals. In this way, alternatives that perform well on highly prioritized goals generate more value, while alternatives that perform well on goals that are not as important, generate less value. Figure 11.9 shows a graph that summarizes the value for each individual stakeholder and for all of the stakeholders combined.

The act of clarifying rationale affects different project participants in different ways. Stakeholders broadly support the process of defining and clarifying goals. A stakeholder on one project remarked, "Every project should start this way." Designers often express skepticism of the ability of a rationale capture system to keep pace with the fast moving and qualitative aspects of design processes, making the future development of tools like Wecision a priority. Decision makers can find the flattening of power and broad communication of their rationale difficult to accept, but must weigh this against the potential benefits in decision quality and consensus building.

From Haymaker's perspective, some decision challenges are so full of uncertainties that a coin flip is the best you can hope for. Others are so well defined and understood that they can be fully automated;

for example, to determine exactly how much steel is needed, where to save the most material, or how long the overhangs for the windows should be to save the most energy. But there are many other types of decisions that require different levels of support. Some need to engage large and diverse sets of stakeholder groups. Others need to look through a large and complex set of alternatives. Some require fast analyses that are qualitative, or only correct within an order of magnitude; others require very precise analyses and quantification of uncertainties. A new design age is dawning in which design and construction teams are learning to work together and use decision methods more efficiently and effectively (John Haymaker, personal communication, March 8, 2013). Moreover, with a set of decision tools available, the project team can choose an approach that best suits the circumstances, team, and decision to be made.

As early as possible in a project, the methods to evaluate the design of the facility and its life cycle organization and process against each performance goal or metric also need to be established and agreed upon by all the key stakeholders. Furthermore, the framework and process to use these performance predictions needs to be set up. It is unlikely that the team will identify all the right performance goals and ways of using particular performance predictions right away. However, it will be far easier to adjust them during the project if the team had the experience of setting them up in the first place. Also, after the experience on several projects, the starting point will become much stronger.

11.3.3 Setting Performance Targets and Designing to Them

Cost Performance/Target Value Design

As the design develops, two key methods for accomplishing the continued integration of the evolving design with user values are target value design (TVD) and continuous commissioning. TVD depends on users establishing specific values for performance targets for the economic, environmental, and social goals of a project (Ballard & Zabelle, 2000). Although this sounds obvious and is what we have been advocating in this book, it is infrequently practiced. The key point of TVD is that it reverses the relationship between facility design and value (including cost). In typical practice, costs (and sometimes value) are estimated given a particular design. In TVD practice, cost is an input to design rather than an output. For example, building systems design is based on the desired value for the whole facility and the cost allocated to particular systems before design commences.

A key aspect of facility performance requires special mention here. The purpose of a facility is, of course, to meet or exceed the users' performance expectations and support the activities of the facility users in the context of their activities (e.g., learning in an educational facility). This link between a facility and the activities of its users is often lost in the typical space program (or brief) created in the early project phases. Formal workplace planning methods can connect the strategy of the facility client with the activities of the building users and the space program and facility design solutions and can better guide the trade-off between strategic goals, (business) activities, and spaces that arises on virtually all projects (Pennanen, 2004).

In the early 2000s, leaders of the Finnish project management company Haahtela realized that they had to help their customers define spaces in relation of business strategy or vice versa. They developed an approach, Strategic Workplace Planning, and parametric analysis software to represent a customer's business plan as activities that require spaces with certain characteristics. Essentially, they found a way to link project definition directly to a building's future users' business plan and activities.

The big idea is to use the client's metrics on revenue generated by activity, space required by those activities, and expected space utilization to establish value targets. Haahtela was then able to develop a building's space program based on the highest value activities, including sharing spaces in new ways. Working with what their client could afford to spend, that is, the "allowable cost," Haahtela could estimate market cost ("expected cost") and establish a target cost before starting design. Their success in helping deliver value using Strategic Workplace Planning has been impressive. For example, Strategic Workplace Planning analysis called for designing multiuse spaces and increased utilization by renting out the therapy pool for the Arcada Polytechnic in Helsinki, shown in Figure 11.10. The board of directors adopted this program, which allowed the project to move forward into design and successful construction (Ari Pennanen and Yrjänä Haahtela, personal communication, November 13, 2013).

Operations and Life Cycle Performance

As a facility design evolves, "continuous commissioning" repeatedly validates the design options against all user values (Laine, Hänninen, & Karola, 2007). If they differ, the design is adjusted. Figure 11.11 illustrates continuous commissioning from the perspective of energy performance throughout design and construction, into facility operation. Note that the BIM is updated as the design gets detailed and built; actually, the detailing should occur with BIM tools. The BIM is

FIGURE 11.10   Arcada Polytechnic: 7 million euros over budget because spaces were underutilized. Courtesy of Haahtela.

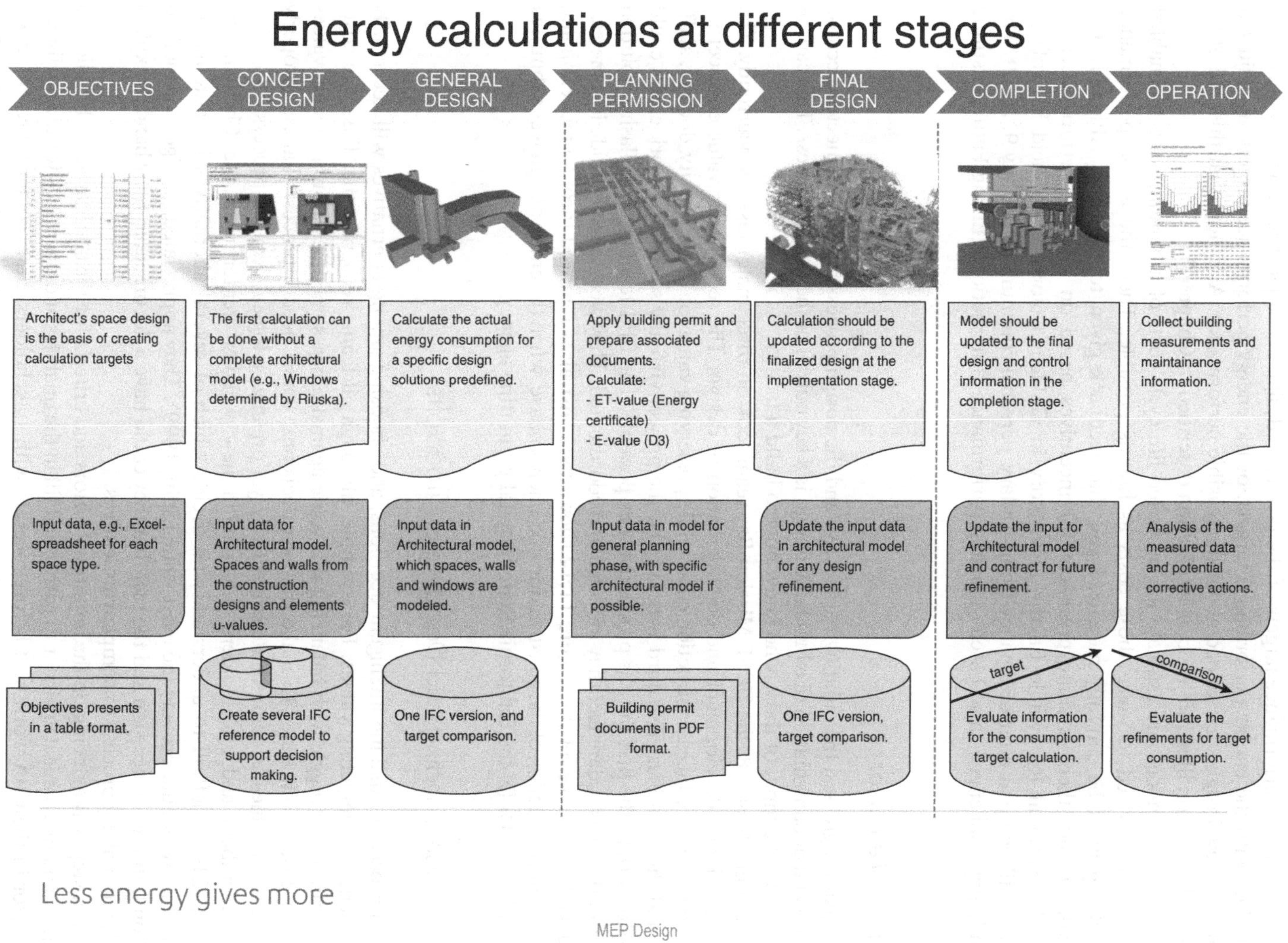

Energy calculations at different stages

OBJECTIVES	CONCEPT DESIGN	GENERAL DESIGN	PLANNING PERMISSION	FINAL DESIGN	COMPLETION	OPERATION

Architect's space design is the basis of creating calculation targets

The first calculation can be done without a complete architectural model (e.g., Windows determined by Riuska).

Calculate the actual energy consumption for a specific design solutions predefined.

Apply building permit and prepare associated documents. Calculate: - ET-value (Energy certificate) - E-value (D3)

Calculation should be updated according to the finalized design at the implementation stage.

Model should be updated to the final design and control information in the completion stage.

Collect building measurements and maintainance information.

Input data, e.g., Excel-spreadsheet for each space type.

Input data for Architectural model. Spaces and walls from the construction designs and elements u-values.

Input data in Architectural model, which spaces, walls and windows are modeled.

Input data in model for general planning phase, with specific architectural model if possible.

Update the input data in architectural model for any design refinement.

Update the input for Architectural model and contract for future refinement.

Analysis of the measured data and potential corrective actions.

Objectives presents in a table format.

Create several IFC reference model to support decision making.

One IFC version, and target comparison.

Building permit documents in PDF format.

One IFC version, target comparison.

Evaluate information for the consumption target calculation.

Evaluate the refinements for target consumption.

target

Comparison

MEP Design

Less energy gives more

FIGURE 11.11 Continuous commissioning for the energy performance of a facility. Courtesy of Granlund.

evaluated against the performance targets of the facility as design develops. The design is adjusted to bring it in line with the energy performance targets, or the energy performance targets are adjusted up or down as new insights are gained about the feasible performance. Also note that this and other examples in this section illustrate process integration described in Chapter 7.

The team must model and estimate the projected life cycle costs of a building, and be enabled and, ideally, incentivized to make those costs as low as possible. Then, the actual performance of the building must be carefully measured and that feedback given to the team. Although it is not yet widespread, there is some precedent for incentives based on building performance: for example, the National Renewable Energy Laboratory in Golden, Colorado, withheld half of the project fee until one year after occupancy (John Andary, personal communication, July 9, 2012). The Living Building Challenge is also based on actual performance (International Living Future Institute [ILFI], 2015).

Coordination Performance

As the project progresses, the building perspective and the design process must be connected because the creation and documentation of a buildable design is a key output of the design process. This cannot happen without the input of those who know how to build a facility.

As noted in Chapter 8, the UCSF Mission Bay team took on three major tasks concurrently. The first was completing development and documentation of design. The second was value engineering (VE) to significantly reduce construction costs while preserving value for the university, doctors, staff, patients, and their families. The third was to build the project virtually so that the work crews could install systems right the first time as productively as possible. The BIM team tracked clash resolution for the 935,000 gross square feet they modeled and coordinated. They were entirely guided by metrics and produced impressive results. Figure 11.12 shows the constraints removed through modeling and coordination at completion of the construction documents phase, when the entire design was submitted for approval by OSHPD, the California State hospital permitting agency.

11.3.4 Understanding and Deciding on Trade-offs

Design is a process of making intelligent decisions, and choosing among alternatives will necessarily involve balancing competing values. For example, although all teams recognize that life cycle costs are important, they struggle with balancing this value against other, sometimes more pressing, values such as first cost or schedule. To make the best decisions, teams need a process that can properly incorporate team values, accommodate subjectivity, yet organize a logical decision process.

Team members and stakeholders designing a new wellness center used the Choosing by Advantages (CBA) method to understand and decide which of several alternatives for automating operable windows would provide the most value (Suhr, 1999). They appreciated the rigor the method imposed on framing the decision and the fact that they could have a document in their hand to explain their decision-making process to company executives.

The first step was agreeing on the important factors and criteria for the alternatives. This allowed the architect and mechanical engineer to gather and fill in the qualities and quantities for the attributes of each factor in the CBA worksheet before meeting with stakeholders. Stakeholder representatives, in-house engineers, facilities operations and maintenance managers, and the delivery team members

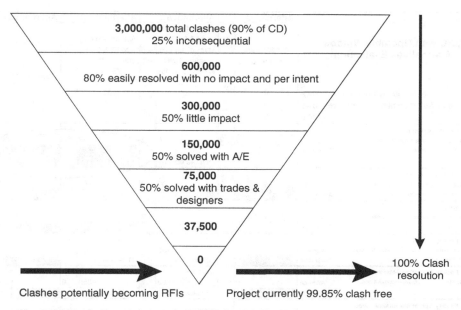

FIGURE 11.12 Constraints removed through modeling and coordination. Licensed by The Regents of the University of California on behalf of its UCSF Medical Center; courtesy of DPR Construction.

were able to flag the attributes with the least important advantages and the ones with the greatest advantages. Then they decided on the one attribute for a particular factor they thought was the "paramount advantage" to be the top of their rating scale. The decision makers followed up by assigning a numeric value to every other attribute for all the factors relative to the paramount advantage. They totaled the scores for each alternative and then talked about whether they had made the right choice. This process actually extended over four meetings and was the basis for their go-forward recommendation for operable windows with automation on levels 2 and 3, and the stairwell (Alternative 3 in Figure 11.3).

The CBA method allowed the team to assure that it had considered all factors relevant to them and to assess the relative importance of the advantage of each alternative relative to those factors. The relative advantages were jointly agreed, leading to a consensus regarding the correct decision. Moreover, the logical analysis did not mask the subjective nature of some assessments, but allowed them to be considered with other objective data. CBA can be viewed as an optimization routine for decisions with many (potentially competing) factors. Figure 11.13 was created using CBA to make a trade-off decision between different approaches (design sets) of using operable windows and mechanical cooling systems together. The alternatives are headers in the top row. Decision "factors" and criteria are listed in the left-most column. The data (called "attributes" in CBA) are entered for each alternative and factor in the cells. The "paramount advantage" is circled and given the highest score. All other advantages are scored relative to that attribute. Total scores are listed at the bottom. In CBA, the advantages are always identified before cost, whether first/capital or life cycle costs are considered. In this case, the team calculated the capital cost premium for each alternative after determining the relative advantages.

HVAC with Operable Windows Alternatives Evaluation		Alternative 1 Windows are not operable.		Alternative 2 Automation only on L2 & L3 + stairwell for Life /Safety & NO Windows on 4th Floor		Alternative 3 Automation only on L2 & L3 + stairwell for Life /Safety		Alternative 4 100% of windows are operable through automation.	
# of Manually operable windows		0		30		61		0	
# of Automatically operated windows		0		48		48		109	
Total		0		78		109		109	
Factor: User Experience Criteria: Maximize individual thermal comfort and environmental control.	Attribute Advantage	0	0	78 78 more options	5	109	10	109	10
Factor: User Experience Criteria: Minimize physical effort while maximizing equitable ease of use	Attribute Advantage	0	3	−30 30 more levers	2	−61 61 more levers	1	0	3
Factor: Wellness Criteria: Promote overall health, wellness and healthy living	Attribute Advantage	0	0	78 78 more	7		9	109	9
Factor: Facility Operation Criteria: Minimize the need for staff to close windows manually.	Attribute Advantage	0	6	−30 30 less task	2	−61 61 less tasks	4	0	6
Factor: Facility Operation—Maintenance Criteria: Minimize parts replacement costs.	Attribute Advantage	0	9	windows = −388 553 less parts	7	windows = −388 947 less parts	6	windows = 156 −656	0
Factor: Facility Operation—Maintenance Criteria: Minimize system control/BMS complexity (when it's working)	Attribute Advantage	0 109 less points	4	−48 61 less points	2	−48 61 less points	2	−109 109 more points	0
Factor: Total Cost of Ownership—Energy Criteria: Minimize energy consumption.	Attribute Advantage	0 Greatest energy	0	4	5	4	5	4	5
Total Importance			22		30		30		33
Capital Cost Premium				$ 421,560		$ 499,000		$ 793,000	

FIGURE 11.13 Using CBA to evaluate alternatives and choose the best for operable windows with different HVAC configurations. Courtesy of DPR Construction, by Jason Brenner; adapted by the authors.

Elucidating and prioritizing stakeholder goals and objectives, evaluating design alternatives using these stakeholder objectives, and setting up transparent decision-making processes are critical to the success of an IPD project. Although the actual performance for the project objectives is only finally determined when the project is placed in use, project teams need production metrics and controllable factors to shape and guide a team's work every day.

11.3.5 Using Metrics to Control Outcomes: Controllable Factors and Production Metrics

"Controllable factors" are the actions a team commits to manage and control to achieve the outcomes they have promised to deliver. Controllable factors are the key to producing value because they convert the possibilities of creative vision and strategy into action. People decide what they should do, do it, then measure what they accomplished so they can learn and improve.

A controllable factor inextricably links an action the team chooses to do with the measurement of the performance of the action. In other words, a controllable factor is like a coin with an action on one side and a metric on the other. One side is work activity, applying a method and following a process.

The other is prediction and measurement. Intentionally choosing to focus attention on a particular action, predicting outcomes, doing the work, and measuring performance, which are the steps for every controllable factor, makes it possible to learn and improve. These are the steps W. Edwards Deming taught Japanese industrial leaders in the 1950s.

The main controllable factors are the actions a team can take, in particular, how it organizes (integrated organization and collaboration), how it manages processes and production (integrated processes and production management), and how it represents design and construction information and uses technology (integrated information and visualization/simulation).

Deciding: Benefits versus Time/Cost

Every project team needs to decide on what it can and must control on a day-to-day or week-to-week basis and come to an agreement on the controllable factors of the project. These controllable factors must be measured and the measurements shared with the project team so that it can reflect on the performance to date and improve the process if necessary. Examples of controllable factors include:

- Meeting participation (%).
- Project scope coordinated with a 3-D model (%).
- Level of development (LOD) of the BIM per discipline (% / LOD).
- Quantity takeoff per scope from a 3-D model (%).
- Using 3-D and 4-D models for constraint identification (%).

A project team can decide to do or not to do the action or work entailed with a controllable factor. For example, it can decide to coordinate the structural, mechanical, electrical, and plumbing (MEP) scopes with a 3-D model at a particular level of development. In deciding to do so the team believes that the benefit of carrying out this action exceeds the costs incurred for, in this case, the 3-D modeling and coordination activities. The result of a 3-D model–based coordination might be a faster resolution of spatial clashes between the structural and MEP scopes, which contributes to the on-time completion of the design and construction phases. Metrics like the speed of resolution of spatial clashes are known as production metrics because they measure how well the application of the controllable factors improves the production of the building, which, in turn, contributes, in this case, to the on-time completion of the project or, more generally, a project objective.

Note the chain of events: (1) the team decides on an aspect of the project it wants to control—in this case how it coordinates the geometry of the structural and MEP scopes with 3-D models; (2) the team defines one or more metrics to account for the value they expect to achieve—in this case faster resolution of spatial clashes; (3) the team tracks and reports the number of clashes detected and time to resolution each week. This allows team members to improve their processes and performance as they go along, as opposed to simply reporting the number of change orders resulting from system clashes at the end of the MEP rough-ins. Regardless of whether those numbers are low or high, by the time these results are known, it is too late to do anything about it.

Tracking Use

To reliably meet a project objective a team wants to control, in this case how it coordinates the geometry in some way, it needs to measure results of key processes during the project, such as the rate at which spatial clashes are resolved. To guide the allocation of resources on the current and future projects, the team also needs to track whether a controllable factor was indeed deployed and, if so, at what cost; that is, did the team coordinate the particular scopes in 3-D and with what effort and effect?

These metrics can be measured as a project progresses, for example, each day or week or month depending on the speed of management required; that is, how quickly and frequently should project managers learn about performance and be able to intervene, and the speed of reporting that's enabled by the project's information environment; that is, how quickly can data be aggregated into the desired measurements for reporting on performance.

Targeting Improvement

A project needs to set targets for required or expected performance and for acceptable performance for the objectives it chooses. For example, the project team might have a target of 80 percent off-site fabrication for MEP with 70 percent being the lowest acceptable percentage. Or it might have a target of zero change orders due to conflicts during construction. Note that for some of these metrics, the "game is over"; that is, the target can no longer be met once it has been exceeded. For example, if there is a change order during construction due to conflicts the target of zero can no longer be met. On the other hand, if the MEP off-site fabrication rate is at 65 percent, the team can possibly still make design changes to increase the number of subsystems that are prefabricated.

Figure 11.14 shows the connections between the factors of production that are under the control of the project team, including the BIM, how to manage processes and production (Project Production Management [PPM]), and the nature and frequency of collaborations (integrated concurrent engineering [ICE]) and project performance objectives (i.e., buildability) and client and user objectives (i.e., usability, operability, and sustainability).

Figure 11.14 can be read from right to left as the arrows indicate, but also from left to right. Following the arrows, the integrated project team translates the client goals into project objectives (as described in Chapter 5). To make it more likely that the team meets the project objectives at the end of the project it needs metrics that guide it along the way—production metrics. To meet the production targets, the team needs to decide what actions to take, i.e., wisely select the most impactful controllable factors. Read from left to right, the figure shows that the selection of a set of controllable factors should improve performance of the project during its virtual and physical production, which, in turn, should improve the project objectives and ultimately the client objectives.

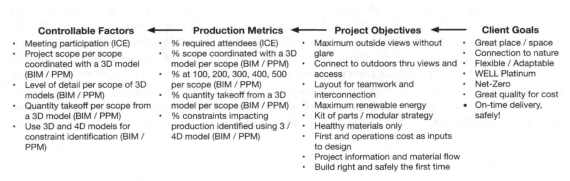

FIGURE 11.14   Metrics connecting production to client goals.

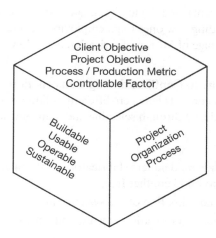

FIGURE 11.15 Ways to categorize performance metrics.

The performance metrics can be grouped in three ways: (1) by criterion they address, that is, buildable, usable, operable, and sustainable; (2) by degree of influence by the project team, that is, controllable factor, process and production metrics, project objectives, and client objectives; and (3) according to the categories in the Product-Organization-Process (POP) matrix, that is, metrics can be related to the product, organization, or process, as shown in Figure 11.15. Plan Percent Complete (PPC) is the percentage of tasks actually completed within a given period by work crews compared to what their foremen/women had promised. PPC, for example, is a buildability metric, a process/production metric; and a process function and behavior in the POP matrix. Embodied energy is a sustainability metric, a project metric, and a product function and process. Response time to occupant calls is a usability and operability metric, a client metric, and an organization function and behavior.

Focusing Attention and Effort on the Right Things to Control

So far we have introduced the ways to bring formal objectives and metrics into the day-to-day work and progress on projects. We now give specific examples of objectives that are typically important on projects.

Percentage of the MEP Systems Fabricated Off Site While a team can decide on the percentage of MEP prefabrication and, in this way, use MEP prefabrication as a controllable factor, it is more useful as a production objective with a specific target percentage informed by the time and space available for on-site installation and off-site fabrication capabilities and capacities. Controllable factors like use of 3-D/4-D models for spatial and temporal conflict identification and resolution, percentage of a discipline's scope modeled in 3-D, participants in and frequency of ICE sessions to resolve spatial conflicts, engagement of a fabricator capable of model-based prefabrication during the preconstruction phase, and so on, can be managed by project teams directly and help make a prefabrication target reality. The percentage of MEP prefabrication can be measured in percent of the weight or cost of prefabricated assemblies vs. components assembled at the work face, or in on-site vs. off-site work hours, or another measure the team finds useful. It can be assessed daily or weekly (or as often as the team wants to see this measurement), and the team can take corrective action by paying more attention to the controllable factors or adding new ones (e.g., employing auto-routing methods for piping) if the measured and predicted percentage of MEP prefabrication is too low.

Number of Change Orders Due to Conflicts during Construction   Many teams aspire to keep the number of change orders due to conflicts between different building systems to a minimum or even set a target of zero. This can be achieved through selecting and managing appropriate controllable factors that include:

- Participation of the engineers, detailers, fabricators, and builders involved in all the building systems the team wants to build conflict-free.
- Timely and detailed 3-D coordination of the systems; and
- Use of the 3-D information in fabrication, layout, and construction planning.

Note that all these factors are controllable; that is, the team can decide to do these actions or not. In addition to measuring how well these actions are done, for example, how timely the coordination really was, the team can also measure the production metric "change orders due to conflicts during construction" as construction progresses. Note how zero "change orders due to conflicts during construction" cannot be controlled directly but is an expected result of carrying out the controllable factors.

Hours Lost Due to Safety Incidents   There is, of course, no more important goal than to have zero safety incidents. Just like the objective of "zero change orders due to conflicts during construction," the objective of "zero lost hours due to safety incidents" won't just happen without conscious effort. The project team has to decide what to do to achieve this objective. For example, it could only hire subcontractors with a stellar safety record, carry out safety analyses during design and construction planning, engage crews in 4-D model-based safety analyses, and have meaningful daily safety meetings. The team can then measure daily whether work hours were lost due to safety incidents and find out the root causes for the incidents.

Percent of Budget Items within a Certain Range of Budgeted Cost   Budget reliability and conformance is an important objective on virtually all projects because a project with high budget reliability for all the important items has allocated the budget appropriately. Of course, budget overruns are bad because they mean invariably that more money than expected needs to be made available for the project or an aspect of the project that still lies ahead must be curtailed. But budget underruns are also not always desirable because the project team could, most likely, have allocated the funds saved in other areas had it known sooner. For example, on a large corporate campus, the owner's design-construction management team was so worried that the new campus construction would not cost more than estimated that the team was overly conservative in their estimates and the project came in 10 percent under budget. While the team first looked like heroes, that assessment quickly vanished as the building users moved in and found that many items on their wish list had not been included, reducing the usability and operability of the new campus.

Like the other metrics in this section, this metric can be measured as regularly as the team would like as the project progresses but is also the result of controllable factors the team decided to implement. Examples of potentially useful factors for achieving high reliability, say plus or minus 2 percent for the 100 most expensive or riskiest budget items, include:

- Percentage of quantities derived from a coordinated 3-D model;
- Alignment of the budget with procurement and construction activities to enable a rapid comparison as actual numbers come in; and
- Explicit discussion of critical uncertainties during design meetings, and so on.

In general, achieving budget reliability requires a three-step estimating and budget validation process to get the budget for the current project as accurate as possible and create the foundation for higher budget reliability on the next project. First, the right budget items have to be included. Then, the right quantity of materials and work for each budget item has to be predicted, and finally the production rates and unit costs for each budget item have to be correct. Without separating learning about the accuracy of the budget items, their quantities, and the unit rates, it is impossible to untangle the reasons for discrepancies between the budget and the actual cost and the estimating basis, which means that the budgeting process cannot be improved in fundamental ways.

Requests for Information (RFIs) Due to Conflicts During Construction The number of requests for information caused by conflicts between building systems during construction is a good measure of the quality and clarity of the coordination and documentation of the design. Many teams strive to have zero RFIs and many have achieved this performance or have come close. They have typically decided to invest in controllable factors similar to those to achieve change order and off-site fabrication production targets.

Minutes per Day Superintendents Spent Resolving Issues between MEP Trades Many superintendents pride themselves on resolving issues in the field quickly and decisively as soon as they arise. This is, of course, an important quality in superintendents, but an even better approach to construction than relying on the superintendent's firefighting capabilities is to have as few issues between MEP trades as possible. A recently completed project found that the detailed and timely coordination of the MEP systems with all the parties involved in its design, detailing, fabrication, and installation reduced the time superintendents had to spend resolving issues in the field from 180 minutes per day on comparable projects to 20 to 30 minutes.

The main issue with resolving issues in the field when they arise is that this time is rarely budgeted and something else often falls between the cracks, leading, most likely, to other issues that require immediate resolution. Furthermore, imagine how much better a project could run with 150 minutes of additional attention by the superintendent each day—projects employing many of the strategies and methods we are outlining in this book have seen such dramatic improvements. In addition to the MEP systems having been coordinated in 3-D, achieving a low number of minutes superintendents have to spend resolving issues in the field, most likely, also requires the coordination of the fabrication and installation process in 4-D to recognize and resolve as many issues as early as possible.

Rework Hours Compared to Total Hours Few things kill the morale of crews and the ability of a project team to complete a project safely and on budget and schedule faster than lots of rework because rework typically has to be done in a hurry and it costs extra to do things twice. Moreover, craftsmen take pride in their work and having their work replaced and redone undermines that sense of pride. For these reasons, tracking rework and the reasons for rework is an important production metric to achieve a project's buildability objectives. Like the other production metrics discussed among these examples, achieving no or very little rework will hinge on the project team's selection of the right controllable factors, such as hiring crews with demonstrated competency in the required work and a complete design before materials are ordered.

Using All of the Last Planner® System Metrics PPC is a process and production metric that is used on many projects. Each week the Temecula Valley Hospital (TVH) team coordinating work in the field looked at plan percent complete (PPC), the percentage of tasks the trade work crews had completed compared to what they had promised. For example, eight tasks were committed for the current week and only six were completed on time, which means that the PPC = 75 percent. The superintendents and general foremen for the trade partners also considered "Tasks Anticipated" (TA) and "Tasks Made Ready" (TMR). They calculated the percentage of tasks that were planned to be done the next week that were on the look-ahead plan from the previous week. For example, if only 6 of the 10 tasks to be done in the following week were on the work plan the week before; TA = 60 percent. The most important metric for indicating how effective the project team was in removing constraints affecting production was Tasks Made Ready, the percentage of tasks in the construction look-ahead plan the week before that were ready so trade foremen could commit their crews to work on them in the following week. For example, only 7 of the 10 tasks planned the week before were made ready in time so the foreman could commit to getting them done in the following week; TMR = 70 percent. Figure 11.16 gives an idea of the questions that tracking these metrics provoked.

Using Key Performance Indicators Team leaders decided in their CIFE VDC Workshop, described in Mission Bay Hospitals case study in Chapter 8, to track response latency, along with how quickly

FIGURE 11.16 Tasks anticipated, tasks made ready, and plan percent complete. Courtesy of CDReed.

TABLE 11.2   UCSF Mission Bay Hospitals Project Key Performance Indicators

Metric	Target
Response Latency	Teams shall deal with 80% of the "issues" within 30 minutes.
	80% of the remainder shall be resolved within 4 hours in the Big Room.
	The other 20% of the remainder shall be resolved within 2 days.
PCI	Potential change impacts (PCIs) shall be approved within 60 calendar days from start of pricing.
RFIs	98% of RFIs shall be answered without the need for revision or resubmittal.
Submittals	95% of Submittals shall be reviewed without the need for resubmittal.
PPC	80% of tasks shall be completed by the date planned.

Licensed by The Regents of the University of California on behalf of its UCSF Medical Center; table provided by the authors.

the team would approve material submittals and deal with design questions and potential changes. They also asked the team to report on safety and how well team members and especially trade foremen were able to keep their commitments to deliver what they had promised. Table 11.2 shows the metrics the UCSF Mission Bay Hospitals team used on the project, called key performance indicators (KPIs), to evaluate the performance of the project's organization and processes.

Focusing On and Measuring Quality to Drive Improvement   The Sutter Health Eden Medical Center team wanted to raise the bar significantly for quality by establishing project objectives and corresponding metrics. For quality, they measured and reported the number of inspections passed the first time. Negotiating the integrated form of agreement required delivery team members to justify direct and indirect costs. Expectations for rework and productivity had to be taken into account as the team calculated how close it was to achieving target cost. The integrated form of agreement and target value design inspired the second quality metric: actual rework versus anticipated. This required the team to track rework, making it visible.

At the 100 percent design and 78 percent construction complete mark, 97 percent of inspections had passed the first time and rework was down 80+ percent from anticipated. Rework had dropped from an expected 7 percent to 0.5 percent for mechanical, from 10 percent to 2 percent for plumbing, from 10 percent to 8.5 percent for electrical, and from 5 percent to 0.5 percent for framing and drywall. Repairing damage of finished walls and other installations was the only thing that did not decline appreciably.

Digby Christian, the project manager for Sutter Health, and Ralph Eslick, the construction manager for DPR Construction, attribute this success to several factors, all of which required commitment, continuous problem solving, and diligence by the delivery team. First and foremost, Digby convinced everyone involved in designing the new facility to see their goal as delivering problem-free instructions for fabrication and installation, as opposed to permittable drawings. Eliminating rework was critical for achieving certainty of scope, budget, and schedule for Digby.

Eliminating rework required the team to model and coordinate even more scopes of work than they had originally planned. DPR project engineers found themselves modeling miscellaneous metals, four different exterior skin systems to assembly level detail, roof deck insulation, and edge of deck anchors; most of which were produced from 2-D vendor drawings. Modeling the tapered roof insulation resulted in discovery of some major waterproofing issues and a complete redesign of the slope of the podium roof. DPR project engineers generated layout drawings from 3-D models for concrete shear walls and embeds. They also put coring for skin system components and electrical and plumbing penetrations onto a single drawing to help field crews. DPR engineers became experts in producing installation drawings to meet OHSPD requirements for in-wall backing, supports for medical equipment, and patient lifts in 66 specific locations.

Measuring Value-Add and Redesigning Work Processes The Temecula Valley Hospital team filmed work crews as they made the first of a number of repetitive installations and analyzed how much time the workers spent actually installing work as opposed to processing or moving materials long distances, waiting, or fixing mistakes. Work crews were shown the video and a pie chart created from the metrics and asked how they thought they could be more efficient. The crews always saw opportunities and improved by the time a "Second Run Study" was done after they redesigned their work process. Productivity increased and labor cost, by far the largest in construction, went down.

DPR Construction's Southern California Self-Perform Drywall Group, has continued to do this on other projects. Jason Herrera and Anthony Munoz reported results at the Lean Construction Institute 2015 Congress (Herrera & Munoz, 2015). They analyzed the time spent by framing and plaster crews placing larger, energy-efficient exterior windows as part of a hospital renovation. Herrera and Munoz counted the time spent framing and installing the windows, replacement membranes, backing and plaster coats as value, meaning something that directly contributes to meeting the customer's conditions of satisfaction (CoS). They measured and classified time spent planning, preparing, handling material, measuring, and cutting as necessary non-value-adding tasks, meaning that these tasks did not directly contribute to meeting CoS, but could not be avoided in the situation. Herrera and Munoz also tracked time spent searching for tools, waiting, rework, uncoordinated transportation, unnecessary movement and producing more than what was needed. They categorized this as waste because it contributed nothing that the client would pay to have done.

Figure 11.17 shows the method the DPR Drywall Group is using. Managers and crew members watch the video and simply count the seconds for work that adds no value and is waste, tasks that are unavoidable but do not add value directly, and those which contribute value directly. This focus on value helps foremen and their crews think about working differently so they spend less time doing things that add no value, and find ways to do the necessary support work better. This results in more time spent adding value and almost always leads to reducing the time to produce a piece of work from start to finish, called "cycle time."

11.3.6 Communicating Performance to the Project Team

Figure 11.18 shows "visual control" in action, inspired by Toyota. The UCSF Mission Bay Hospitals charts depict the progress being made each week to reduce cost, find innovative solutions, decrease wait time for answers, and increase the number of tasks completed as planned (PPC). Each cluster team reported every other week staggered so that half had new information each week. The budget trend (1) and progress toward reaching the target cost (2) charts were updated each week for leadership meetings and prominently displayed so that every team member and visitor could see them. A big contributor to finding savings were Project Modifications and Innovations© (PMI) proposals (3) from

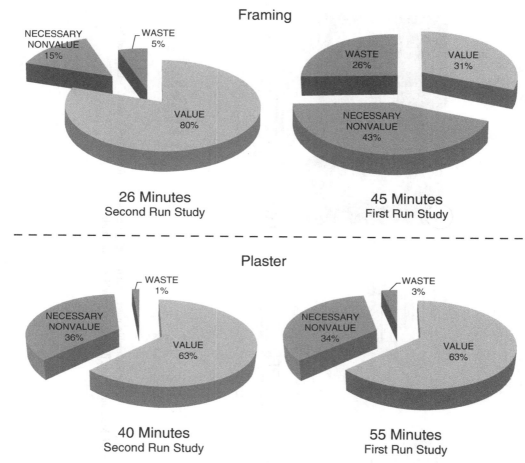

FIGURE 11.17 First Run Studies method based on typical activity times. © DPR Construction, DPR SPW Drywall Group–SoCal.

team members (described in Chapter 8). The percentage of material submittals approved without need for resubmittal (4), and the percentage of RFIs answered without need for resubmittal (5), both of which are quality metrics, were also updated and charted weekly. The number of tasks planned and completed as planned (PPC) and the trend compared to the 80 percent goal (6) were also reported visually each week. The hundreds of people working on the project in the co-location space could see the progress they and other subteams were making toward designing and constructing the project within its budget.

Dashboards

In a practical way, metrics form the basis of project management and define the actual tasks that a project manager and the project teams undertake. This is the principal reason that metrics are more

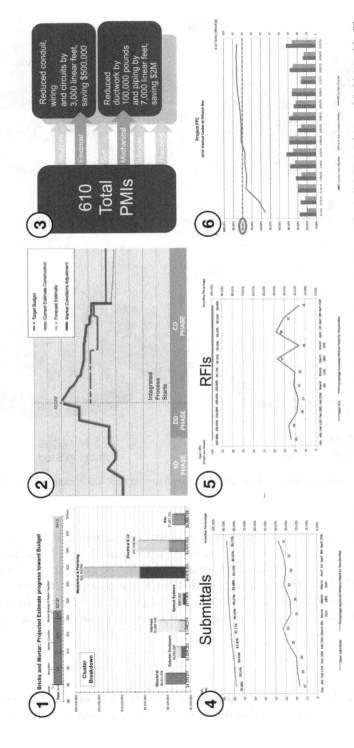

FIGURE 11.18 UCSF Mission Bay Hospitals project weekly reports for leadership meeting and posting on the wall for "visual control." Licensed by The Regents of the University of California on behalf of its UCSF Medical Center; illustration provided by the authors.

than a report card and more than a way to determine incentives, bonuses, and corrections. When a project team has enough metrics and collects the data about those metrics in a consistent and low-overhead way, they can be aggregated to provide insight into the process of the job.

One way to think of production metrics is like the dashboard of a car. The car's dashboard gives the driver feedback on the state of his vehicle and, crucially, allows him to change the way he is driving *before* catastrophe strikes. With no speedometer, he might speed straight past a waiting policeman. With no gas gauge, he would run out of gas before he realized there was any danger. Worse problems arise with no engine temperature indicator, no oil, engine, or air bag lights, and so forth. With no dashboard, the only way a driver can know if he is making a mistake is after he experiences the disastrous results; he has had no advance warning and no way to adjust his behavior ahead of time.

This dashboard analogy is also useful in understanding the development of effective metrics. The information displayed on a car's dashboard is only a small fraction of the information collected in a modern vehicle. Moreover, the reason that today's cars are safer, more reliable, and more comfortable is because of thousands of smaller improvements in the suspension, chassis, engine, and components—all done at a level of detail where they can individually and measurably improve their individual performance. In the construction industry, the opposite is often the case. We use aggregated performance metrics like overall schedule compliance (think speedometer), but often without measuring the thousands of small improvements that would result in significant improvement.

Consequently, successful construction projects must also have a useful dashboard that provides real-time information that the team can use to adjust its delivery process midstream. In fact, most modern construction projects already track a variety of metrics. With the widespread adoption of design and construction software at every level of a project, numbers are easily available and can be included in spreadsheets, charts, and graphs. Yet simply extracting numbers for the sake of putting them on paper does not inherently help the team improve. The kind of metrics that are tracked, the level of detail at which they are tracked, how they are interpreted, and what the team does in response to them make all the difference.

11.4   INTERCONNECTIONS

The problem with all process flow diagrams such as the Simple Framework is that they are abstractions and cannot account for a dynamic reality. For example, it looks as though the last thing a team should do is to establish metrics to measure value. This couldn't be further from the truth because the entire process might start here, as it did for the Packard Foundation. The facility owner and the IPD team should define or validate value as one of the first initiatives of their new integrated organization. This will likely occur as they are negotiating the IPD contract. The owner, users, and delivery team members must agree on what value is and how to measure it before moving forward. The IPD team should also establish metrics and KPIs to assess how well it is doing overall and with its processes. "Keeping score" allows people to learn from their failures and see opportunities to improve. Modeling and simulation are a combination of methods and technology that allows teams to predict the performance of alternatives, all in the service of making better decisions. Framing decisions and making good predictions are the stuff of collaboration and both require metrics. Metrics are the currency of collaboration. A team cannot manage production well in any phase of a project without looking at metrics for value-adding work versus waste. Without metrics and measurement and without translating client and project objectives into production metrics and controllable factors, a team's chances of delivering a high-performance building are significantly reduced. The effort must start at the beginning and continue through operation, as is required for Net Zero Building certification.

11.5   REFLECTIONS

A new approach to designing, building, and operating facilities must be based on a robust information infrastructure and a culture of collaboration around specific facility performance goals so that project teams maintain their focus on the critical issues as they work out the details. To accomplish this, teams need to have tools to navigate. Through measurement and control, project teams can manage projects to achieve client and project goals. But teams will also need to change.

Making every facility a highly sustainable high-performance facility calls for a departure from business as usual. It will require the collective engagement of owners, designers, and builders to overcome aspects of today's practice that make it unlikely that a facility will be as good as it could be. For example, the culture of focusing on the form of a facility almost from the start of a project without properly defining and prioritizing its goals and expected performance limits will most likely set the project team off in the wrong direction. The culture of focusing on first cost to select design options omits the reasons why the facility is designed and built in the first place. Hence, unless first cost prices all future costs and values correctly—for example, the cost of air pollution is included in the cost of a material that requires a lot of energy generated by coal-fired power plants—the predominant focus on first cost undervalues life cycle concerns, which are, of course, determining whether a facility is sustainable or not. The culture of engaging professionals from various disciplines sequentially leads to suboptimization from the moment the owner says "go" and makes it unlikely that a facility model that could be used over the life of a facility is set up. Finally, not measuring actual facility performance and comparing it to the predictions of the design team limits learning, the adoption of innovations, and long-term risk management.

11.6   SUMMARY

In navigation, you need to know where you are going and you need to measure your progress in order to steer your ship to port. It is the same with projects. You need to establish your goals and then measure your progress toward them. But what should you measure? Measurement (project controls) should provide information that allows you to adjust the project's controllable factors, that is, assert control over the processes that affect project outcome. Once you have determined your goals and what to measure, you can use Deming's Plan-Do-Study-Act (or Spear's Predict-Test-Adjust) cycle to continuously improve the work processes to achieve better outcomes.

NOTE

1. We are indebted to Roberto Arbulu of Strategic Project Solutions (SPS) in San Francisco for alerting us to this very useful distinction.

REFERENCES

Ballard, G., & Zabelle, T. (2000). Project definition. *White Paper #9*, Lean Construction Institute.

Design Process Innovation (2015). *Wecision—collaborative decision models*. Retrieved June 19, 2015, from https://wecision.com.

Drucker, Peter F. (2008). *Management Revised*. New York: HarperCollins.

Herrera, J., & Munoz, A. (2015). First Run video studies: Driving continuous improvement. Presentation at the 17th Annual Lean Construction Institute Congress, Boston, MA. Retrieved from http://www.leanconstruction.org/media/docs/2015-congress/presentations/TH18- Herrera.pdf.

International Living Future Institute. (2015). Living building challenge: Two part certification. Cascade Green Building Council. Retrieved February 16, 2015, from http://living-future.org/living-building-challenge/certification/certification-details/two-part-certification.

Laine, T., Hänninen, R., & Karola, A. (2007). Benefits of BIM in the thermal performance management. *Proceedings of IBPSA Building Simulation, 2007*, 1455–1461.

Pennanen, A. (2004, April 3). *Workplace planning: User Activity Based Workplace Definition as an Instrument for Workplace Management in Multi-User Organizations*. Dissertation for the degree of Doctor of Technology, Department of Architecture, Tampere University of Technology, Finland, Haahtela-kehtys Oy, Helsinki. Retrieved October 7, 2010, from http://www.haahtela.fi/main/Workplace_Planning.pdf.

Schwartz, S. H. (1992). Universals in the content and structure of values: Theoretical advances and empirical tests in 20 countries. *Advances in Experimental Social Psychology, 25*(1), 1–65.

Senge, P. M. (2006). *The fifth discipline: The art and practice of the learning organization*. New York, NY: Broadway Business.

Spear, Steven J. (2009). The high-velocity edge: how market leaders leverage operational excellence to beat the competition. New York: McGraw-Hill.

Suhr, J. (1999). *The Choosing by Advantages decision making system*. Westport, CT: Greenwood Publishing Group.

Thomson, Derek S., and Simon A. Austin (2006). *"Using VALiD to understand value from the stakeholder perspective."*

CHAPTER 12

Visualizing and Simulating Building Performance

"Vision is the art of seeing what is invisible to others."

—Jonathan Swift

12.1 WHAT ARE SIMULATION AND VISUALIZATION?

How do we communicate what a building will look like before it is built? How can team members ensure that a building will perform according to its intended use before it is built? The answer to both these questions lies in virtually designing, constructing, and operating a building through the use of building information modeling (BIM). BIM can display design information as we would see it in a finished building, deepening understanding and reducing interpretive errors. Through simulation of a building's behavior, BIM enables exploration and optimization across multiple dimensions of cost, quality, and schedule. This results in better performance predictions, which are vital to enable productive collaboration among the multiple disciplines and create an optimal workflow (Kindler, DeLuke, Rhea, & Kunz, 1994).

Visualization represents a product in a form that is meaningful to a diverse group of stakeholders. Visualization creates understanding of what the product looks like and how it will function. For example, a rendered three-dimensional (3-D) model of a building represents how it will look and can be used to communicate and develop a common understanding about the building's features.

Simulation is the prediction of the behavior of a Product, Organization, or Process (POP) or their combination based on analyzing abstract POP models. Because of the effort involved in a simulation and the advantage of considering many scenarios, computers are used to perform the calculations. For example, an energy model that predicts energy use of a building under changing occupancies, seasons, and so on is a simulation model that can be used to predict actual energy use of the building.

FIGURE 12.1   The professor takes us inside VDC. Courtesy of CDReed.

Simulations also enable exploring design alternatives. Michael Schrage (2013), in his seminal book *Serious Play,* calls for companies and teams to use prototypes and simulation to innovate and provide insights into generating value for the customer. He argues that prototypes and simulations are necessary for predicting behavior of any system and that even simple simulation models are imperative to good decision making. Simulation, and the insights it provides, he adds, changes how teams interact and innovate.

Virtual design and construction (VDC) is the use of multidisciplinary performance models of building projects, including their products (facilities), organizations, and work processes for business objectives. Our professor, who helped orient us in the Preface, decodes VDC in Figure 12.1.

The professor is explaining that we must use computers and realize that design and construction determine a large portion of life cycle costs. Every project organization must include all stakeholder perspectives to understand the business objectives and translate them into project goals that are quantitatively or qualitatively measurable. Then the team must create an integrated organization and implement virtual and Lean work processes to simulate and test many alternatives to select optimal solutions. Finally, the team must integrate their design solutions virtually so they can build the integrated systems that make up a high-performance building.

VDC brings together BIM and production management principles within a collaborative team to build models of the product, organization, and process to predict the performance of the building and the project team in terms of cost, quality, schedule, energy use, life cycle cost, and so on. Not only is the resulting building optimized, the processes for creating the building are optimized, as well (Kunz & Fischer, 2012).

FIGURE 12.2   Indian fable of Blind Men and the Elephant. Courtesy of CDReed.

12.2   WHAT DOES SUCCESS LOOK LIKE?

Visualization is successful if the entire team and all relevant stakeholders have the same understanding of what is being built and how it is being built. Simulation is successful if it accurately predicts building and team behavior and leads to optimization of project solutions. Moreover, they are both successful if they enable the team to communicate more effectively and make better decisions.

Unfortunately, success eludes many teams. More often than not, building performance and the value a client derives from the building is poorly represented, and thus poorly understood by team members. In the ancient Indian fable of "The Blind Men and the Elephant," depicted in Figure 12.2, a group of blind people are asked to describe an elephant. A woman touches the leg, and says the elephant is like a pillar; a man touches the tail and says the elephant is like a rope; still another touches the tusk and says it is like a strong pipe. The moral of the story is that the world is viewed differently depending on an individual's perspective. Although everyone was touching the elephant, they were each touching different parts and walked away with a fervently held, but incomplete understanding of the actual elephant.

Similarly, when engaged in a large and complex construction project, it is difficult to see the whole. Most project team members struggle to see the project from any point of view other than their own. Without a holistic view of the project, optimization of design or construction is almost impossible.

The problem is even more severe with users and owners. Too often, during the final walk-through an owner or user will exclaim with dismay that "she thought she was getting something else or this is not what she was expecting." This is a reflection of the reality we all face. What we communicated to the owner and thought she understood is not what she thought it would be. Two-dimensional renderings or drawings are simply not adequate to create a mutual understanding of the eventual reality. Without visualization and simulation, we are looking at the project through the hands of the blind men in the Indian fable. With visualization we can develop a common understanding among the project participants and inform stakeholders to allow them to make the right decisions at the right time.

Three-dimensional visualizations, scaled models, accurate renderings, 4-D simulations, estimates based on model quantities, and prediction of energy performance based on models enhance the understanding of the wider team and play a role in building the same multifaceted version of reality for everyone. Rapid feedback presents an unprecedented opportunity to incorporate the knowledge and experience of all stakeholders while ensuring that the design stays true to the values and goals of the owner. These visualization and simulation tools also allow teams to make informed decisions on hard trade-offs so that value is maximized within the constraints of the project, and to accurately predict the impact of decisions early in the process. With visualization and simulation, we are no longer blind and can see the project in its entirety, and as it actually is.

The Utility of Architecture

Architecture is the art and science of creating spatial experiences for people, and working with light to aid their activities. The occupants' activities are based on a particular function. How that function is served, and how people experience the space is the utility of architecture.

This last statement holds a juxtaposition, because function is objective, and peoples' experiences are subjective. A language to communicate the utility of architecture needs to convey both a confirmation that factual needs are met and project the sense that people will be pleased occupying the space. This language is visualization and simulation.

A technical drawing (e.g., floor plan, section) is a form of visualization to describe function. Renderings show the exterior or interior of the building and help the owner and the construction team imagine its appearance.

Architecture is a relatively new profession. The word originated from the Greek word *arkhitekton*, which means "chief builder." Through ancient times and the Middle Ages, master builders who were artisans and stonemasons had the role of imagining and communicating the vision of a new building. It wasn't until modern times that paper and pencil became ubiquitous, enabling the rise of design drawings by many professionals. Even in ancient times, visualization was necessary to bring stakeholders to the table and agree on what the designed space would look and feel like. Clearly describing something that needs to serve a purpose for many people and create an environment for individual well-being isn't simple. Art and architecture have been going hand-in-hand in the past centuries to achieve this through visualization.

Today, computers can create precise 3-D models and photorealistic renderings to simulate the look of the designed space. Computers can be powerful tools in this process, but many criticize their use because the precision of the renderings can limit subjective interpretation.

As computers are evolving, simulation of occupant movement and other functional aspects of buildings are becoming more mainstream, along with an artistic use of computer renderings to allow subjective imagination. Regardless of the techniques, visualization and simulation will remain important to describe the complex layers of the utility of architecture.

Adam Rendek, LEED AP, BIM/Engagement Manager, DPR Construction.[1]

12.3   HOW CAN THIS BE DONE?

We have explained that a high-performing building should be buildable, operable, usable, and sustainable. In order to understand the performance of the building and the project delivery process we must be able to predict whether the building is indeed buildable, operable, usable, and sustainable. Virtual design and construction (VDC) methods can be used to predict the performance of buildings for these performance criteria. To test whether the building is usable and meets the aesthetics and the beauty that the client desires a project delivery team can build 3-D models that are rendered and also use the same models to simulate the performance of the building in terms of its usability. To check whether the building is buildable the team can perform constructability analysis using 3-D models or build a 4-D model to see how the building can be built most efficiently. Among other types of analysis, the team can also analyze energy use based on the 3-D model and predictions for building occupancy and use.

In the following sections, we describe various visualization and simulation techniques to predict building project team performance across the dimensions described above. Specifically, we address tools and methods to:

- Understand the design;
- Simulate its function;
- Analyze sequencing and logistics;
- Simulate building use;
- Model costs; and
- Optimize the design and construction solutions.

12.3.1   Understanding the Design through Visualization

Visualization refers to the realistic 3-D representation of the building form. Using detailed and accurate 3-D models, the team is able to communicate more clearly and effectively with each other and with the owner. Many owners have little design and construction experience and cannot understand complex 2-D drawings. Too often, one hears owners say: "It may be what I asked for or approved, but it is not what I wanted." Three-dimensional models are vastly easier to comprehend, because they look like what the building will actually be. Because they are easier to comprehend, they allow the owner to better understand the building being designed and to provide designers with feedback during the design process. With visualization, we can hear owners say, "This is what I thought you were designing and it's what I wanted. Great job!" Figure 12.3 is a digital rendering created by Devenney Group of a patient floor nurses' station. The photograph taken of the finished product, from the same vantage point, is almost identical, as it should be.

This rendering allowed the project stakeholders to understand the space before it was ever built. Virtual design and construction tools, such as BIM, are critical to creating a consistent understanding of future reality across the entire team.

FIGURE 12.3   Rendering of the fourth-floor nurses' station/lobby of the Sutter Eden Medical Center Hospital. Courtesy of Sutter Health; courtesy of Devenney Group Ltd. Architects.

FIGURE 12.4   Stakeholders seeing their space in 3-D in the Penn State Architectural Engineering Center CAVE. Courtesy of DPR Construction.

Three-dimensional models can also be visualized in a computer-assisted virtual environment (CAVE), where the model is rear-projected onto multiple screens, which surround the user, forming a small room. CAVEs offer an immersive experience that can give an owner an even better idea of the space that will be built. Figure 12.4 shows the CAVE at the Penn State University Campus, which was used to demonstrate an interior of a courthouse to the project participants in order to create a shared understanding of what it would feel like to be inside the courthouse.

Cross-Pollination between Industries Enables Understanding Design

In recent years, a strong development emerged connecting virtual design using BIM and the video game industry. As computer hardware and video cards are becoming more powerful and more sophisticated, the real-time rendering of the 3-D environment in video games is becoming extremely compelling to the human eye. In a video game, the environment is fixed and highly pre-detailed. In virtual design, the 3-D environment is continuously changing as the design is progressing. Yet the recently developed rendering technologies by the gaming industry are becoming so efficient that they allow for the rapid loading of building information models and rendering with accurate lighting and material representation. This way, design teams and building owners can review the latest design options in an almost photorealistic and dynamic presentation. This is combined with a game-like navigation, which was actually adopted years ago in BIM.

Representing lighting and the realistic appearance of materials and surfaces is computation intensive. That is why about 10 years ago video games and rendering software were primarily based on simulating photorealistic conditions as opposed to calculating the true behavior of light and its reflections on materials. Stronger computers changed that. Real-time calculation based on the physical properties of light is possible today resulting in very convincing renderings of building models.

Virtual reality headsets are becoming the media to present photorealistic and dynamic 3-D environments. An earlier example of virtual reality—the computer assisted virtual environment (CAVE)—has already showed promising advantages of an immersive simulation. This will become even more seamless with the next virtual reality headsets.

Adam Rendek, LEED AP, BIM/Engagement Manager, DPR Construction.

12.3.2 Simulating the Function of the Design

Simulation builds on the 3-D model to predict how the envisioned form will function. Architects and designers often design the form of a facility by drawing on their experience to develop particular forms, for example, the layout of a floor of offices and conference room for a commercial building. These are then test fit and adjusted against the building owner's and users' performance goals. Understanding and validating whether or not a particular form does indeed satisfy the required functions has been an elusive pursuit, and the actual behavior of the building may not be clearly understood until it is actually built.

In almost any other area of modern engineering, it is standard practice to predict how a product will behave and function *before manufacturing it*. For example, a semiconductor designer knows how much power a computer chip will use, how much heat it will produce, and how it will fit into a computer before it is manufactured (Lam, 2005). Automobile companies predict the gas consumption of a car, cell phone makers know how long a battery will last, and so on. Yet in the AEC industry we often make design decisions without understanding how they will affect building performance. Some argue that it is difficult to anticipate how a building will be used and that it is therefore difficult to create accurate predictions of future performance. This is correct, but an imperfect tool is better than no tool at all. Even tools that provide qualitative results lead to better-informed decisions. One of the most significant benefits of simulation is to enable a discussion concerning the parameters affecting performance and providing insights that allow better decisions. Even imperfect tools can accomplish these purposes, and, when combined with the experience and knowledge of a multidisciplinary project team, can significantly improve outcomes.

With BIM-based simulation, teams can effectively predict everything from energy performance to occupancy, functioning of business processes, and other performance parameters. The expected behavior of the building can be simulated and then compared with the actual performance once measurements are available. For example, a hospital team can simulate patient flow to determine how long patients will have to wait before being seen by a doctor, or how long their walk would be to the nearest available patient room. Among many other issues, BIM can also be used to evaluate the building's compliance with the prescriptive description of clearance requirements under the Americans with Disabilities Act (ADA) (Han, Kunz, & Law, 2001).

Simulations allow the team to carry multiple design options forward for comparison. For example, long- and short-span steel, precast and timber structural systems could be kept in play along with appropriate mechanical, plumbing, and electrical systems. Distributed and central mechanical approaches, use of space, energy, and natural light can all be calculated and analyzed. Through simulation, the team can build virtually before building physically, testing many different prototypes and options to make the right design decisions based on as much hard evidence as possible—not just human judgment alone.

As computing power becomes exponentially cheaper and faster, teams are able to develop semantic models, that is, models that represent an actual situation as closely as possible, and test fairly complex scenarios that once were only available to supercomputing labs. Computers can model and evaluate thousands or even millions of options and, under various criteria, ensure that the best possible design is reached.

Simulating the Safety Performance of the Building

It is also important to incorporate operational knowledge, that is, how the facility will be run, into the design phase because creating a design that can be operated is, of course, a key task for the integrated design teams, too. Health and safety of the facility users and operational efficiency are key focus areas for testing the operational fitness of a facility. Agent-based simulations show how occupants would use a building and provide useful insights into the usability of a building layout (Höcker, Berkhahn, Kneidl, & Borrmann, 2010). Similarly, traffic simulations test whether a road network accommodates the expected traffic (Wedel, Schunemann, & Radusch, 2009) and factory simulations allow design teams to study the efficiency of a particular factory layout (Kühn, 2006).

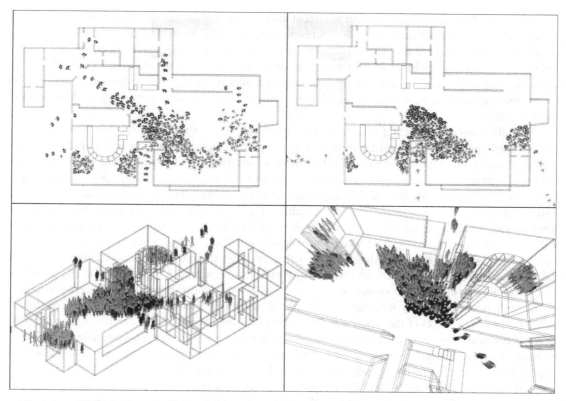

FIGURE 12.5 Simulation of nightclub evacuation. Illustration provided by Xiaoshan Pan, PhD.

Figure 12.5 is a screen shot of a simulation of the evacuation of a nightclub in Rhode Island that suffered a devastating fire in 2003.[2] The top row shows clustering at some of the exits as the evacuation progressed while other exits were only lightly used or not used at all. The bottom row shows an agent-based simulation of the evacuation of the nightclub (Law et al., 2006).

12.3.3 Understanding Sequencing and Logistics by Simulating Construction (4-D)

Four-dimensional simulation connects the building elements in the model with information from the schedule to reveal what will be built when, where, by whom, and with what. In other words, 4-D is a 3-D building model that represents the status of the building at any moment in time. The result is an animation of the construction sequence, which illustrates the actual schedule and helps the team identify time and space conflicts (Collier & Fischer 1995).

The bulk of the cost of any project is almost always centered on actual construction in the field. A major task for construction planners is determining the sequence of construction activities to allocate resources appropriately, and avoid overlapping sub-trades. Field construction offers the best opportunity for productivity gains, and 4-D modeling can optimize construction sequences and logistics. Teams can address concerns such as time and space conflicts, availability of laydown areas, access for equipment installation, and potential safety hazards before construction even begins (Haymaker & Fischer, 2001).

Four-dimensional modeling effectively communicates the desired construction sequence to all project participants. A good construction manager is able to envision the flow and sequence of construction as a result of his deep professional experience, but the tools to communicate this understanding have been lacking. Traditionally, construction sequences have been communicated through Gantt charts and Critical Path Method (CPM) schedules, which are inadequate because they do not allow the entire team to see the consequences of the schedule in terms of time and space conflicts. Many problems go undetected because traditional methods result in different perceptions about how the work will actually be installed in the field. Four-dimensional modeling allows teams to understand construction sequences and develop multiple scenarios that can be explored before construction begins, so resources are optimally allocated, and workflow between various trades is optimized at the outset.

There are two distinct 4-D modeling techniques, which are defined based on the level of granularity of the sequence that is being investigated. A "macro" level 4-D animation communicates the flow of construction on the level of the whole site. It is typically a simulation that encompasses many months or an entire year. This model is used for overall site logistics analysis and understanding issues such as road closures, emergency access routes, and communication with non-construction stakeholders (e.g., a hospital that is staying openduring a renovation or expansion).

On the PAMF Mountain View Center project the general contractor created a macro level 4-D model to explain the sequencing and logistics plan with stakeholders such as doctors, community team members, hospital staff, as well as project team members.

An example of the macro level 4-D model from the PAMF Mountain View Center is shown in Figure 12.6. The screenshot shows the progress of the exterior skin and glazing system work. It was critical to analyze this work using the 4-D model to coordinate the use of the crawler crane and maximize availability of laydown areas. The model was used extensively to communicate with the contractors on the project. The 4-D model was created at the 100 percent DD (design development) stage and updated at the 50 percent stage of CD (construction documents) and again after the 100 percent CD models were created. The macro-level model revealed important job logistics issues like car parking, impact to adjacent buildings, and how to plan major deliveries. It also effectively demonstrated the overall sequencing to non-construction stakeholders.

A "micro" 4-D model is an animation that shows the sequence of work for a much shorter period of time (four to six weeks). This type of model is used by the crews working in the field to understand issues such as laydown areas, and imminent time or space conflicts (Staub-French & Khanzode, 2007). An example of using a short-term schedule with BIM is the KanBIM system (Sacks, Barak, Belaciano, Gurevich, & Pikas, 2013), which combines the pull planning and Last Planner® System of making short-term commitments with the use of BIM. The experimental use of this system on a large residential project in Israel showed positive results with the site team reporting that the use of this system improved their ability to visualize work and reduced the waste in terms of looking for work. The screenshot in Figure 12.7 shows a drywall crew's assignments on a floor of a residential building using KanBIM.

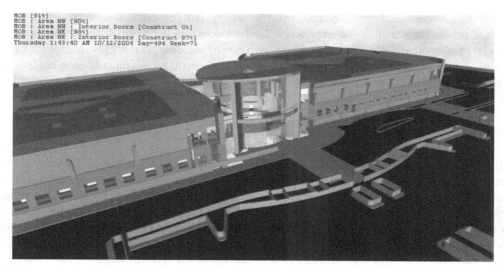

FIGURE 12.6 Screenshot of the construction sequencing of the PAMF Mountain View Center project. Courtesy of Sutter Health; courtesy of DPR Construction.

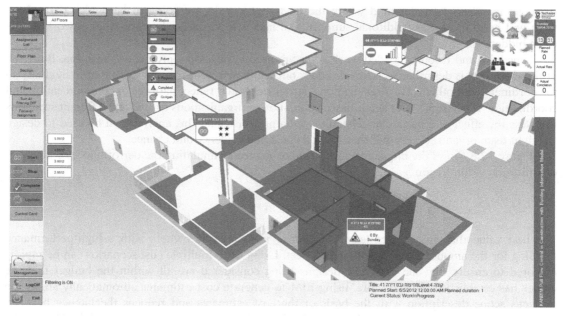

FIGURE 12.7 Drywall crew assignments in KanBIM. © Rafael Sacks, Associate Professor, Virtual Construction Lab, National Building Research Institute, Technion—Israel Institute of Technology.

Four-dimensional models can be developed by anyone with access to a schedule and a BIM, but they must reflect the construction experience (and wisdom) of the team. If the professionals who are directly responsible for delivering and constructing the building do not provide input for the 4-D scenarios, the resulting model will be of little use. Furthermore, the model must be continually updated with new schedule information, which is the only way for the 4-D model to remain relevant and useful to the project team.

12.3.4   Simulating the Work of the Organization

Creating a project organization from a diverse group of companies and individuals is an art. On most projects it is left to the senior most representatives of each company or to organizational design consultants. The project organization may be based on who is available, not who is best suited for the project. Team members often haven't worked together before and no one can really know if the team that got assembled is the right team to handle the work that is in front of them. If the project flounders, more experienced personnel are added to "shore up" the failing areas. The project organization that results is more a product of circumstance than design. Although this is common, it is not inevitable.

Organizational design research offers insight into the VDC method of designing project organizations. The Virtual Design Team research at Stanford (Levitt, 2012) and subsequent development of a commercial simulation technology (SimVision®) can help a project team simulate the work of the team members to understand where the likely bottlenecks are going to be, which part of the team will be in most need additonal support, and also help visualize the coordination work that is often hidden.

Schedules and resource-loaded work plans are used to develop a project organization. But most of the work that is described in schedules does not really represent the real work that individuals do to coordinate with others. The organizational structure should reflect patterns of communication and coordination. SimVision® uses agent-based simulation of organizations to evaluate different scenarios to assess the effect of a particular schedule on the project team. This analysis can be used to design the right team organization rather than rely on anecdotal evidence and chance.

Figure 12.8 shows the impact on several subteams from simulating the performance of a project organization in SimVision®.

12.3.5   Modeling the Cost of Construction

The best value that any project team can deliver is to provide owners with a high-performance building for the amount they are willing to invest. Using BIM, multiple cost scenarios can be rapidly explored to ensure that the design or designs being considered are all within the budget that the owner has established. Furthermore, using BIM to generate cost estimates automatically aligns the project's scope description with the basis of the cost estimates and reduces the latency between design iterations and cost feedback dramatically, shortening the preconstruction schedule. This results in a leaner approach, minimizing wasted effort and maximizing value (Tiwari, Odelson, Watt, & Khanzode, 2008).

Model-based cost estimating is the process of integrating the designer's 3-D model with cost information from the estimator. Object attributes in the 3-D model relevant for cost estimating are

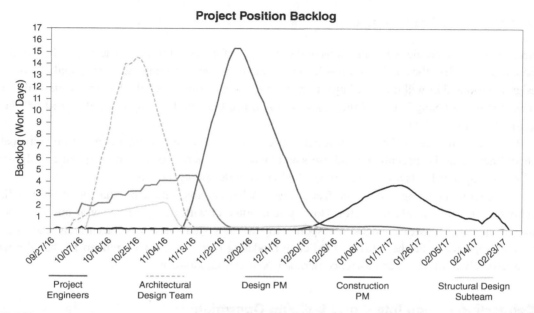

FIGURE 12.8 Simulating the performance of a project organization. Courtesy of ePM LLC.

associated with cost information from the estimator's database, so the resulting cost estimate is much more accurate than simply compiling a rough estimate from 2-D drawings. This process is faster, more reliable, more auditable, and far less likely to be incorrect due to omissions and other human error.

Using the 3-D model for cost estimating also allows teams to provide rapid cost feedback for potential design options or changes. Manually estimating is slow and cumbersome, and almost never as thorough as a model-based estimate can be. Using the model, a team can make more informed decisions because it is getting nearly real-time and regular input on constructability and cost feasibility of design alternatives.

In order to truly integrate cost and design, estimators and designers must work together to develop a BIM that can be understood by estimating software. All project participants involved in the process of realizing the design, including supply chain providers, must provide cost feedback through the model in order to ensure the most accurate results. This is easier said than done; not only must software be compatible, but estimators must be thoroughly trained on software and run test cases to ensure that the model-based estimates are accurate and trustworthy.

The model-based estimating process requires significant investment in formalizing cost information and its connection to BIM and in setting up tools and training. Without the backing of senior management, such an overhaul of the estimating process is extremely difficult if not impossible. The process must also be supported by the team leaders and the owner—their performance expectations in terms of speed and accuracy of cost estimates are essential for a transition to BIM-based estimating. Successful implementation of this process is as influenced by the organizational and contractual setup as it is by the actual software solutions.

12.3.6   Modeling Energy Use

Energy models are predictive simulations of the energy performance for a given design of the building. These models utilize the combined knowledge of team members, allowing the team to make informed design decisions that will have an impact on the energy performance of the facility for its entire life cycle. Energy modeling is one of the key ways to ensure that the building being designed will be a high-performance building.

For energy models to be as accurate and useful as possible, input must come from principals, subcontractors, and especially the end users and facility manager. Information on specific equipment and real occupancy loads are crucial to developing accurate energy models.

Yet, energy modeling is not a perfect science. A big variable—human behavior of the building occupants—is often an elusive factor. How systems interact with occupant behavior and usage and how occupants' behavior changes in response to various systems is not yet accurately understood (Maile, Bazjanac, & Fischer, 2012). Predictive modeling methods must be refined further through more research at labs such as Lawrence Berkeley National Laboratory.

Connecting Design Intent and Building Operations

Through the design and construction of a project, the energy modeling process always aims at the future. It is, by its nature, predictive. At its best it lets the design team identify what's important to focus on, gauge if they are meeting their goals, and ultimately understand how much energy their project will really use. Throughout this process, the energy model gets tweaked and updated as input information becomes more defined and design decisions are solidified. The model is constantly being adjusted to be more like the building being built.

Once the project is complete and occupants move in, the role of the energy model can be flipped. The actual building can now be constantly adjusted to be more like the model. This makes perfect sense—after all, the as-built energy model is a representation of the design intent, so it is the perfect tool with which to run the facilities.

Submetered data with end-use energy runs from a calibrated model can be used to verify energy performance and serve as a valuable diagnostic instrument for optimizing ongoing operations. This energy use comparison can be visualized by a digital dashboard. The expected performance levels change automatically with weather conditions, time of day, and other factors, while the building's energy information system provides real-time energy use.

This kind of information is most effective when it provides an easy to understand message that can influence future action. For example, if more energy is used than expected, a dial moves into a red zone. A user can tell at a glance what is responsible and can drill down into the data to get more information. This allows facility managers to identify when energy trends deviate from expected patterns, suggesting that equipment may need maintenance or adjustments. Such impacts on operational behavior are powerful for managing energy use for the life of a building after design and construction.

Matt Grinberg, PE, BEMP, EDAC, LEED AP[3]

Aligning Expectations for Energy Use

While working at Stantec, Matt Grinberg and I led a multiyear research project exploring how the connection between BIM and net zero energy modeling can be optimized. Our work had the benefit of building on the complex energy modeling work Stantec performed for the National Renewable Energy Laboratory (NREL) Research Support Facility. To our surprise, we found that the real challenge of optimizing the BIM and energy modeling connection was not related to technology, but rather to deficiencies of communication between the stakeholders that are involved in the work. Designers, energy modelers, and building owners often have misaligned expectations of what needs to be communicated or brought to the table at coordination meetings.

It is still hard and resource-intensive to properly predict the energy use of a net zero energy building design today, yet the software industry has been making significant progress in the last years and the technology issue will be ultimately resolved in the near future. However, the misalignment between stakeholders is based on differences of work cultures, and it needs to be addressed with properly designed processes and workflows that enable everyone's—including the owner's—uniform understanding of the goals and the steps to achieve the goals. To that end, we developed a simple workflow tool that shows who is responsible to deliver what kind of information at which coordination meeting. The tool also shows the level of development of the exchanged information to eliminate misunderstanding based on differing assumptions of what a particular facet of data represents.

With DPR, we are looking at how the implementation of BIM for facilities maintenance and operations can be combined with organizational development. Clear goal setting and coordinating processes across various disciplines are important to establish effective energy standards and operate buildings at high efficiency.

Adam Rendek, LEED AP, BIM/Engagement Manager, DPR Construction.

12.3.7 Creating Simulation Models to Choose the Optimal Solution

Recent advancements in cloud computing make it possible to run massively parallel computer resources to solve problems of optimization. Professionals are already accustomed to storing, sharing, and accessing data through cloud-based services. Virtually all software providers are enabling their software to operate on and with cloud-based data. But the cloud can also be used as a computation center, making massively parallel computing available for everyone willing to pay a few cents per hour of computing. Cloud computing makes the Multidisciplinary Optimization (MDO) methods that have been key to improving the design of airplanes (Obayashi, 1998) and other manufactured products available for construction project teams.

For example, in a recent experiment at Stanford's Center for Integrated Facility Engineering, (Welle et al., 2011) studied the daylighting performance, that is, the amount of daylight available in offices and its impact on the heating and cooling load of the building, of the thousands of design alternatives. These alternatives were identified by the project team by combining various glazing and shading options for the upper and lower levels with the various orientations of three wings of a major office building. With about 200 hours of setup time and using 4,000 computers (called nodes) in the cloud for 12 hours, the design team gained insight into every possible alternative. Since there were thousands of feasible combinations of different glass and shading devices for the upper and lower floors of the west, south, east, and north-facing façades of the three wings of this office building, it was impossible for the project team to determine the best alternative from intuition only or from analyzing a handful of alternatives.

In another CIFE research study, Forest Flager analyzed the steel structure of a roof using an MDO method that quickly generated 12,800 design options for the design of roof trusses compared to 39 options generated by conventional practice and was able to demonstrate that the steel weight could be reduced for the steel structure—without losing any of the performance criteria—from 1,414 metric tons to 1,146 metric tons for each of the two sides of the roof, resulting in savings of $5 million (Flager, Adya, & Haymaker, 2009).

Before the advent of parallel processing systems, the design team could only have analyzed a few alternatives. It would have been unthinkable to explore the performance of all the feasible design alternatives. But using parallel processing in the cloud, the research team was able to evaluate thousands of alternatives for only $5,000 in computing costs (plus the setup or preparation time). Using this approach, project teams will no longer be limited to exploring a handful of design alternatives. They will be able to optimize a design from many perspectives to understand the trade-offs and make better-informed design decisions faster.

Leveraging cloud computing in this way builds on the following technologies.

- *Parametric BIM.* While some projects—mostly projects with complex geometry—are already using parametric BIM (Shah & Mäntylä, 1995; Lee et al., 2006), this technology is still fairly new, but provides the foundation for large-scale automation and optimization of design. In contrast to representational BIM tools, which are predominantly used today for tasks like clash detection and which simply represent a building design in a BIM requiring professionals to change design parameters manually as other design parameters change, parametric BIM tools allow professionals to set up a logical model of the design so that a change in a few parameters leads to automatic adjustments of other design parameters. Parametric BIM will enable a design to be studied more fully and possibly optimized through cloud computing.

- *Parallel computing.* Decomposing a design or analysis problem into parallel streams of tasks allows multiple computers (or nodes in the cloud) to execute them in parallel thereby dramatically shortening the time required for design and analysis. Leveraging parametric BIM, parallel computing shortens the time to develop a new design option or analyze a design option dramatically—from hours to seconds—for problems that can be parallelized. As the daylighting study above shows, cloud computing makes parallel computing affordable to everyone because it does not require the investment in dozens or hundreds of computers to obtain the computing power and time needed.

The combination of parametric BIM, parallel computing, and cloud computing enables the application of MDO for building projects, which in turn allows project teams to explore far more design options (thousands versus a few dozen that are feasible at best today) for many more criteria (dozens versus a handful) to find the best design possible. As the following case study illustrates, cloud-based MDO will be a key tool to optimize the performance of buildings and infrastructure and design truly sustainable buildings. In several cases, there have been improvements of about 20 percent over the best design developed by professional project teams with today's leading methods when the project teams employed cloud-based MDO practices.

12.3.8 Life Cycle Cost/Carbon Footprint Optimization

A design-build team was asked to find the optimal design for three or four multifamily housing buildings for an integrated live-work campus (Del Monte, 2012). The owner was looking for a design that minimized the carbon footprint and the 30-year life cycle cost of these buildings. These criteria were somewhat new to the design-build team that was more accustomed to developing designs that minimized the construction cost. The owner was willing to consider the construction of three or four buildings with any orientation. The owner also provided a list of acceptable façade options and was flexible with respect to the shape of the floor plate as long as the buildings provided the required square footage and maintained code-required distances to the site perimeter and between each other. These options, including feasible increments in building length and width and number of floors, created a design space with 146 billion possible alternatives.

After defining the design space in terms of variables for which the owner wanted to explore the expected building performance, the design-build team developed a base design and two design options that improved on the life cycle cost performance and the carbon footprint performance respectively (Figure 12.9). Note that it is impossible to know whether these are three really good solutions or average or poor solutions. It is also impossible to know whether spending time and money on developing a fourth design would pay off.

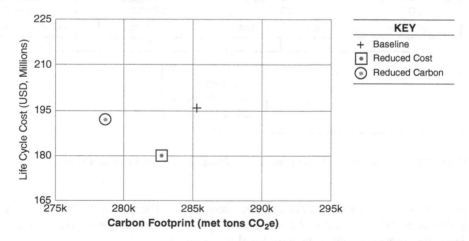

FIGURE 12.9 Life Cycle cost versus carbon footprint for three designs.[4] Courtesy of Forest Flager and John Basbagill, developed in collaboration with the Beck Group, Dallas, TX.

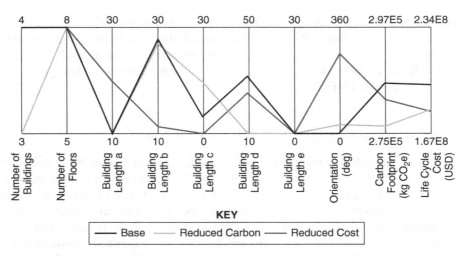

FIGURE 12.10 Comparison of three designs with independent and dependent variables. Courtesy of Forest Flager and John Basbagill, developed in collaboration with the Beck Group, Dallas, TX.

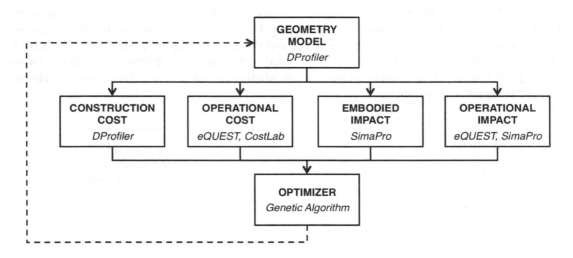

FIGURE 12.11 Modeling, analysis, and optimization tools. Illustration provided by Forest Flager.

Figure 12.10 shows a comparison of the three designs with the settings of the independent variables (number of buildings, number of floors, building footprint parameters a to e, and orientation) and the dependent variables (life cycle impact and life cycle cost).

Meanwhile, the research team set up the MDO process using the same BIM and analysis tools the design-build team was using (Figure 12.11) (Flager et al., 2009).

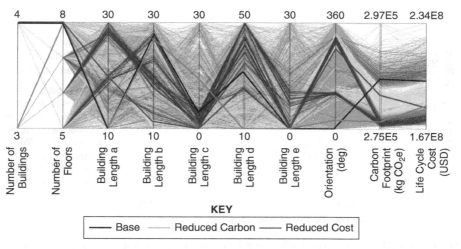

FIGURE 12.12 Result of the MDO process. Courtesy of Forest Flager and John Basbagill, developed in collaboration with the Beck Group, Dallas, TX.

Each design alternative was represented with a BIM. The MDO environment allowed varying the independent variables (number of buildings, number of floors, footprint parameters, and building orientation) so that the performance of a wide range of feasible design alternatives could be calculated quickly, since the parameters from the BIM drove the analysis. By using optimization techniques to limit the exploration of design options the MDO environment was able to find a near-optimal design by considering "only" 21,360 design options (from the 146 billion possible options), which was, of course, dramatically more than the design-build team could have ever considered with today's methods. When these options were graphed it became clear which parameters drove high-performance, that is, low carbon footprint and low life cycle cost (dark / blue lines in Figure 12.12). The owner and the design-build team appreciated this insight into the factors driving high-performance to guide the development of the design because it showed which parameters were particularly important in creating a set of high-performing buildings.

Comparing the cost and outcomes of the "manual" typical BIM-based design process and the MDO process, the typical process used about 25 percent fewer hours (about 160 vs. 210) although about 70 percent of the hours used on the MDO process were spent on setting up the MDO software environment. For a similar second problem or project, the set-up cost would be significantly lower. The best design option found through the MDO process outperformed the best solution found with the typical process by about 10 percent in terms of life cycle cost and 3 percent in terms of life cycle impact. Figure 12.13 shows life cycle cost versus carbon footprint for the 21,360 design alternatives, with the three base designs shown per the legend.

Similar approaches are also being researched to create construction schedules (Dong, 2012). Cloud-based MDO simulation dramatically increases the speed with which a project team can consider a large number of alternative designs and alternative schedules and improves the team's capability to develop near-optimal solutions and understand which parameter settings lead to high-performing building designs and which parameter settings limit the best-possible performance.

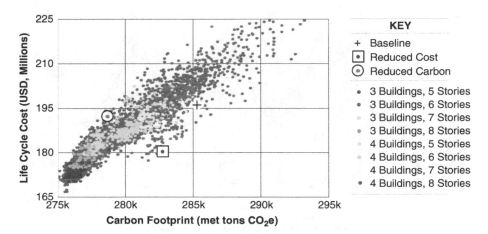

FIGURE 12.13 Life cycle cost versus carbon footprint for 21,360 design alternatives. Courtesy of Forest Flager and John Basbagill, developed in collaboration with the Beck Group, Dallas, TX.

Big Data in Visualization and Simulation

The basis of visualization and simulation is becoming more complex than ever before. Skilled professionals were able to rely on their knowledge and experience to interpret and integrate a wide range of objective and subjective input information. They did this in the form of creating drawings for decades, but, since the beginning of the digital age, people have been increasingly exposed to more data than what they can make sense of.

The source of input is also multiplying, as sensors in new buildings create a large amount of raw data that needs to be collected, stored, categorized, and ultimately turned into useful information. Because of these factors, the building industry is confronted by the transforming effects of big data.

The goal of gathering big data for building projects is to enable the visualization and simulation of their future use and performance with a high level of confidence. However, making big data meaningful to building owners and project teams is a challenging task.

Innovation is needed in this area, which could rely on cross-pollination with different sciences and disciplines. Scientists of weather prediction, traffic simulation, disease prevention, and economy have been dealing with a large amount of input data for a longer period than professionals in the building industry. The data models and methods used to see patterns and draw meaningful conclusions in these areas could serve as examples to follow for architects, builders, and building owners in the future.

Adam Rendek, LEED AP, BIM/Engagement Manager, DPR Construction.

12.4   REAL-LIFE EXAMPLES

12.4.1   PAMF Mountain View Center

The Camino Medical Group, now part of the Palo Alto Medical Foundation (PAMF) and Sutter Health, needed a new medical office building in Mountain View, California. Construction began in January 2005 under general contractor DPR Construction, architect Hawley Peterson and Snyder Architecture, and mechanical engineer Capital Engineering Consultants. The owner, along with the architect, engineers, and contractor, selected the mechanical, electrical, plumbing, and fire protection subcontractors on their ability to collaborate using 3-D and 4-D tools.

The team wanted to focus on limiting reciprocally interdependent work during construction, with the objective of avoiding rework and maximizing productivity. The project superintendent, Ralph Eslick, consulted the foremen for each of the three MEP trades and the superintendent to determine the work they committed to complete, how it should be sequenced, and how it would flow through the project site. They broke the site down into quadrants, and then identified the installation sequence within each of the quadrants.

Figure 12.14 shows the sequence of the ductwork and wall framing and ductwork for the second floor northeast quadrant for the project. Clockwise from top right: (1) installation of full-height "priority wall"; (2) medium-pressure ductwork; (3) both medium- and low-pressure ductwork; and (4) exhaust, along with the other ductwork and fire sprinkler. The sequence, BIM, and 4-D simulation were developed based on the detailed input from the trade foremen on the project.

As a result of the 4-D tool combined with disciplined use of the Last Planner® System, the superintendents were able to spend less than five hours over a three-month period dealing with field issues, compared to two to three hours a day on a comparable project. Only two of 233 requests for information were related to a field conflict; there were zero change orders due to field conflicts, and only one recordable injury for 203,448 work hours. All of the trades finished their rough-in work ahead of schedule, and the mechanical subcontractor estimated that productivity improved by 5 percent for the piping work and 30 percent for the sheet metal work.

12.4.2   Sutter Health Eden Medical Center Cost Estimating

The Sutter Health Eden Medical Center (SHEMC) project team implemented model-based cost estimating on the $320 million, six-story, 130-patient bed replacement hospital in Castro Valley, California. The team was using target value design, had an integrated form of agreement (IFOA) with 11 signatories, and committed to implement Lean and integrated practices.

The Eden Medical Center set up cross-functional groups of architects, engineers, and estimators for the general contractor, and BIM engineers to automate the cost-estimating process on the project and generate cost estimate updates on design changes every two weeks. This early integration resulted in significant time savings, enabling one estimator to generate an updated estimate from

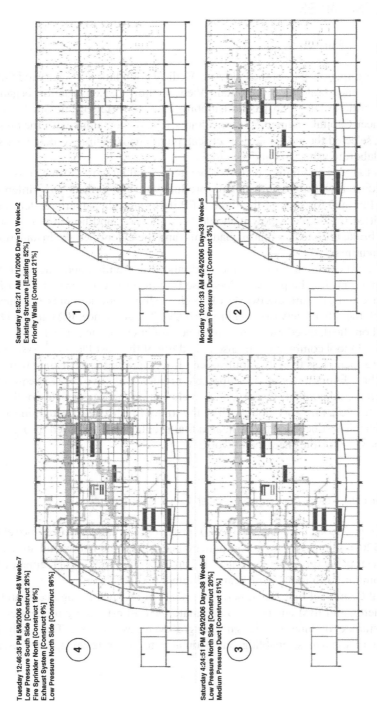

Saturday 8:52:21 AM 4/1/2006 Day=10 Week=2
Existing Structure [Existing 52%]
Priority Walls [Construct 21%]

1

Monday 10:01:33 AM 4/24/2006 Day=33 Week=5
Medium Pressure Duct [Construct 3%]

2

Tuesday 12:46:35 PM 5/9/2006 Day=48 Week=7
Low Pressure South Side [Construct 26%]
Fire Sprinkler North [Construct 19%]
Exhaust System [Construct 9%]
Low Pressure North Side [Construct 96%]

4

Saturday 4:24:51 PM 4/29/2006 Day=38 Week=6
Low Pressure North Side [Construct 20%]
Medium Pressure Duct [Construct 51%]

3

FIGURE 12.14 Screenshots from the 4-D interior construction simulation of the PAMF Mountain View Medical Center. Courtesy of Sutter Health; courtesy of DPR Construction.

the 3-D model in just two days—about one-fifth of the time for traditional estimating (Tiwari et al., 2009).

Figure 12.15 shows the model-based estimating (MBE) process followed by the SHEMC project team. The elements in the BIM were linked to component assembly data in the estimating software via an intermediary system to provide rapid cost feedback to the team as the design kept evolving. For example, the one-hour wall in the BIM was mapped to the one-hour wall assembly. The quantity parameters were brought in from the BIM and used to calculate the labor, material, equipment, and other costs from the assembly data to produce an estimate.

The team concluded that in spite of the challenges in available technology and process workflow, the advantages of model-based estimating far outweighed the up-front time and effort required to enable the process. They found that by automating tasks like quantity takeoff, the team was able to use the time saved to allow them to advance the design and improve constructability by modeling more of the project. The estimators produced the graph shown in Figure 12.16, which reflected the state of design. Their premise was that the higher the percentage of cost data coming directly from BIM, the more developed the design had become.

12.4.3   SHEMC Visualizing Design

At the SHEMC project, the team used BIM extensively in the modeling and coordination process and evaluated the progress of design completion in biweekly design review sessions with all hands at the SmartBoards. The team developed its designs in BIM at the fabrication/installation-level detail. Engineers simulated energy use and daylighting, comparing performance metrics to targets. Every inch of space in the patient rooms had to be taken into account, which meant that the team had to coordinate installations within very tight tolerances. Models were repeatedly coordinated until clashes between systems were eliminated or "soft enough" so that they could be resolved during installation.

The owner's senior project manager, Digby Christian, asked designers and builders how they could be certain there would not be a problem unless the information was coordinated in the model. For instance, how did they know that exterior skin embeds, which weren't modeled, weren't conflicting with MEP risers, which were modeled? It became self-evident that the embeds should be modeled. Eventually, all but four types of components were modeled because the core team, with Digby in the lead, agreed to release contingency to pay for the additional modeling.

Figure 12.17 shows a design review session in progress at SHEMC. The designers and trade contractor modelers and managers were going through the coordinated BIM at the site office along with other disciplines to understand the progress of design. This meeting was an all-hands-on-deck meeting that occurred every other week during the design phase.

12.4.4   SHEMC Seeing Accuracy of Installation

At SHEMC the team performed laser scanning to see if they had indeed built what they had modeled. Once the utilities were installed, the construction team laser-scanned and compared them to the BIM to make sure they were installed per the model. They continued this practice through interior construction. All the walls and above-ceiling utilities were laser-scanned before they were enclosed with wallboard. The native point cloud files were turned over to the hospital facilities department to

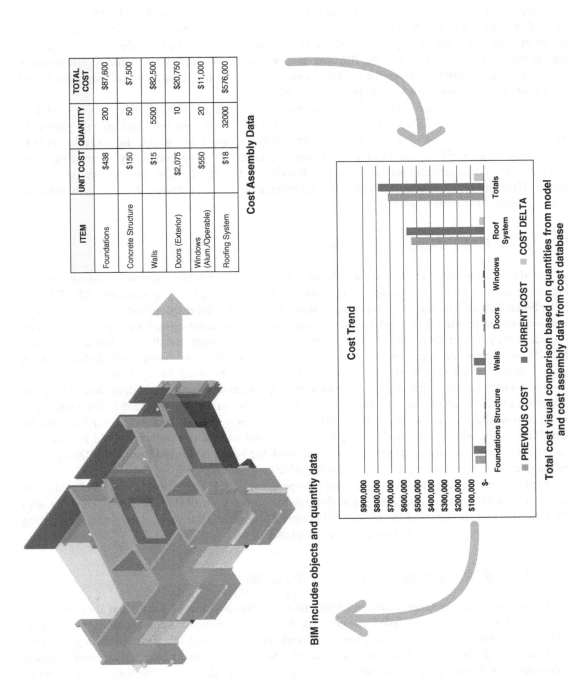

Cost Assembly Data

ITEM	UNIT COST	QUANTITY	TOTAL COST
Foundations	$438	200	$87,600
Concrete Structure	$150	50	$7,500
Walls	$15	5500	$82,500
Doors (Exterior)	$2,075	10	$20,750
Windows (Alum./Operable)	$550	20	$11,000
Roofing System	$18	32000	$576,000

BIM includes objects and quantity data

Cost Trend

Foundations Structure Walls Doors Windows Roof System Totals

■ PREVIOUS COST   ■ CURRENT COST   ■ COST DELTA

Total cost visual comparison based on quantities from model and cost assembly data from cost database

FIGURE 12.15 Model-based estimating (MBE) process. © DPR Construction.

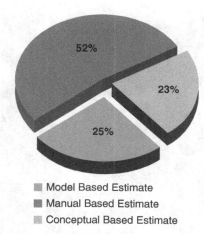

Model Based Estimate
Manual Based Estimate
Conceptual Based Estimate

FIGURE 12.16   Today's estimate source. Courtesy of Sutter Health; courtesy of DPR Construction, by Simon Eldridge.

FIGURE 12.17   Collaborative design review using BIM. Courtesy of Sutter Health; courtesy of Ghafari Associates, LLC.

FIGURE 12.18   Laser scanned point cloud. Courtesy of Sutter Health; courtesy of DPR Construction.

serve as a true as-built record. Figure 12.18 shows the laser-scanned point cloud image of the walls and above ceiling utilities at SHEMC.

Figure 12.19 shows the laser-scanned point cloud image superimposed on the BIM to check the accuracy of installation of the walls and above-ceiling utilities at SHEMC.

12.5   INTERCONNECTIONS

Visualization and simulation using computer models of integrated product, organization, and process information are key methods to predict how a building and project team will perform. They allow the project team to build a common understanding and are key to the project team's decision making process. Simulation and visualization rely on integrated information. You can only visualize and simulate if you are able to integrate the information about the product, organization, and process and represent that in a computer model. Simulation and visualization also require an integrated organization so that all the concerns of the various stakeholders are represented in the models used to predict the performance of the building. Simulation and visualization are also key methods in representing the desired behavior of a high-performing building.

12.6   REFLECTIONS

This chapter described the use of visualization and simulation tools to support integrated project delivery and highlighted the main uses of these tools we see in practice today. We anticipate that these tools

FIGURE 12.19 BIM superimposed over laser point cloud. Courtesy of Sutter Health; courtesy DPR Construction and Ghafari Associates, LLC.

will be used more widely and for more performance goals over the next few years, including life cycle assessment, CO_2 footprint, maintenance costs, and more. As discussed in the section on optimizing design using simulation, the recent advances in cloud computing, the availability of cheap computing resources, and the ability to describe a design problem to a computer make it possible to generate thousands of design options and explore those options against a set of criteria so that the integrated project delivery team can always see whether it is indeed designing and building the high-performing building the owner envisions.

Simulation methods will become increasingly sophisticated as our ability to describe problems to the computer advances and the availability of massively parallel cloud computing keeps improving. In addition, there are many buildings where the actual building performance is being tracked using thousands of sensors that track everything from occupancy to energy use and water use. The wide availability of data captured through these sensors helps calibrate the models used to predict building performance so that they become increasingly accurate. We believe that our industry is ready to adopt simulation and visualization methods to deliver truly high-performing building that we all can be proud of.

12.7 SUMMARY

Visualization is a process of revealing information to make it more easily understood. Two-dimensional plans, elevations, and sections may contain sufficient information to describe a structure, but a 3-D digital model, or a physical mock-up, convey that information more effectively. In addition, because the digital model is continuous and interactive, the user can explore the entire model, rather than be limited to the views and sections drawn by the designers.

Visualization can also affect understanding of relationships and possibilities. For example, a graph of numbers can show relationships, trends, and predictions that would not be apparent in just a list of numbers. Similarly, visualization of a sloping drainage pipe in a model is much more likely to reveal hard and soft conflicts than a line in plan view with a note to slope the pipe.

Simulation allows the project team to examine the planned structure as it will be used or constructed. The simulation is the digital prototype of the building and allows the team to test different alternatives and practice building the structure.

Simulation also creates the opportunity for digital optimization. Designers can rarely consider more than a handful of alternative solutions. But computers, particularly with the advent of parallel processing, can quickly consider thousands of options and rank them according to one or more programmed values, such as energy efficiency. And systems being developed will use simulation of a building's performance to create a design best suited to supporting the building's functions.

NOTES

1. After graduating from the Budapest University of Technology and Economics with a master's degree in architecture and building engineering, Adam Rendek moved to San Francisco to pursue his career in design and construction. He has worked on numerous large projects both in Europe and the United States, including the Palace of Arts in Budapest and the UCSF Mission Bay Hospitals project in San Francisco. He is the recipient of the 2006 FIABCI Prix d'Excellence. In addition to his professional work, he taught a lecture series on digital presentation and BIM at UC Berkeley. He collaborated with the Lawrence Berkeley National Laboratory and consulted with IDEO, Adobe Systems, Autodesk, and Graphisoft on product development for the AEC industry. He also worked on a research project that explored optimization opportunities in the energy modeling workflow and developed solutions for building life cycle management and operations. He published several papers and co-authored a book on BIM and energy modeling and presented at major conferences in the United States and Europe. He is currently working on the BIM implementation of a large public organization on behalf of DPR Consulting. His responsibilities include client stakeholder engagement as well as the development of technology implementation strategies.

2. The Station nightclub fire (2012). Retrieved March 27, 2012, from http://en.wikipedia.org/wiki/The_Station_nightclub fire

3. Matt Grinberg is a mechanical engineer and registered PE who has focused on energy simulation and sustainable building design for over 10 years. He has lead projects and helped produce award winning designs, including LEED Platinum projects such as the Windrush School and the UC Davis Graduate School of Management. He was involved with the Research Support Facility for the National Renewable Energy Lab in Colorado, performing energy analysis and helping the team integrate sustainable options into the net-zero design. ENR named him as one of its "Top 20 Under 40" recipients in 2014. Active in the green building industry, Matt has served on the organizing committee for the International Building Performance Simulation Association and participated in the Northern California USGBC Chapter. He has

worked on numerous LEED submittals including official LEED reviews for the USGBC. As part of a collaborative R&D effort, he co-developed a workflow map tool to help align design teams and energy modelers on projects. He has published several papers on the topic and presented at major conferences in the United States and Europe. Matt is currently leading product strategy and design at a San Francisco based technology firm.

4. This chart and the four following were developed by Forest Flager, John Basbagill, and Mike Lepech at CIFE, in collaboration with the Beck Group, Dallas, TX.

REFERENCES

Collier, E., & Fischer, M. (1995). *Four-dimensional modeling in design and construction.* CIFE Technical Report No 101, Stanford University, Stanford, CA.

Del Monte, Rick. (2012). The Beck Group. Advanced BIM Applications. Presentation at the Annual Industry Advisory Board meeting, Center for Integrated Facility Engineering, Stanford University, October 18, 2012, Stanford CA.

Dong, Ning. (2012) "Automated Look-ahead Schedule Generation and Optimization for the Finishing Phase of Complex Construction Projects." PhD diss., Stanford University.

Flager, F., Adya, A., & Haymaker, J. (2009). *AEC multidisciplinary design optimization: Impact of high-performance computing* (pp. 1–8). CIFE Technical Report Number 186, Center for Integrated Facility Engineering, Stanford University, Stanford CA.

Han, C. S., Kunz, J. C., & Law, K. H. (2002). *Compliance analysis for disabled access.* In W. McIver Jr. & A. K. Elmagarmid (Eds.), *Advances in digital government* (Vol. 26, pp. 149–162), *Advances in Database Systems.* New York: Springer US. Retrieved from http://dx.doi.org/10.1007/0-306-47374-7_9

Haymaker, J., & Fischer, M. (2001). *Challenges and benefits of 4D modeling on the Walt Disney Concert Hall Project.* CIFE Working Paper 64, CIFE, Stanford University, CA.

Höcker, M., Berkhahn, V., Kneidl, A., & Borrmann, A. (2010). Graph-based approaches for simulating pedestrian dynamics in building models." *eWork and eBusiness in Architecture, Engineering, and Construction, Proceedings of the European Conference on Product and Process Modeling*, K. Menzel & R. Scherer (eds.). Boca Raton, FL: CRC Press, Taylor & Francis Group, 389–394.

Kindler, C., DeLuke, R., Rhea, J., & Kunz, J. (1994). *Development and demonstration of an agent-oriented integration methodology.* CIFE Technical Report No. 97.

Kühn, W. (2006). Digital factory: Simulation enhancing the product and production engineering process. *Proceedings of the 38th Winter Simulation Conference, ACM,* 1899–1906.

Kunz, J., & Fischer, M. (2012). *Virtual design and construction: Themes, case studies and implementation suggestions.* Center for Integrated Facility Engineering (CIFE), Stanford University, Stanford, CA.

Lam, W. (2005). *Hardware design verification: Simulation and formal method-based approaches.* Prentice Hall Modern Semiconductor Design Series. Upper Saddle River, NJ: Prentice Hall PTR.

Law, K. H., Latombe, J.-C., Dauber, K., Pan, X., & Peng, G. (2006.). *Computational modeling of nonadaptive crowd behaviors for egress analysis.* Retrieved March 27, 2012, from http://eil.stanford.edu/egress/

Lee, Ghang, Rafael Sacks, & Charles M. Eastman. (2006). "Specifying parametric building object behavior (BOB) for a building information modeling system." *Automation in construction 15*(6), 758–776.

Levitt, R. (2012). The virtual design team, designing project organizations as engineers design bridges. *Journal of Organizational Design, 1*(2), 14–41.

Maile, T., Bazjanac, V., & Fischer, M. (2012). A method to compare simulated and measured data to assess building energy performance. *Building and Environment*, 56, 241–251.

Obayashi, Shigeru. "Multidisciplinary design optimization of aircraft wing planform based on evolutionary algorithms." In IEEE International Conference on Systems Man and Cybernetics, vol. 4, pp. 3148–3153. INSTITUTE OF ELECTRICAL ENGINEERS INC (IEEE), 1998.

Sacks, R., Barak, R., Belaciano, B., Gurevich, U., & Pikas, E. (2013). KanBIM workflow management system: Prototype implementation and field testing. *Lean Construction Journal*, 2013, 19–35. http://www.leanconstruction.org/media/library/id9/KanBIM_Workflow_Management_System_Prototype_implementation_and_field_testing.pdf

Schrage, M. (2013). *Serious play: How the world's best companies simulate to innovate.* Boston, MA: Harvard Business Press.

Shah, Jami J., and Martti Mäntylä. (1995). *Parametric and feature-based CAD/CAM: Concepts, techniques, and applications.* New York: John Wiley & Sons.

Staub-French, S., & Khanzode, A. (2007). 3D and 4D modeling for design and construction coordination: Issues and lessons learned. *ITcon, 12,* 381–407.

Tiwari, S., Odelson, J., Watt, A., Khanzode, A. (2009). *Model based estimating to inform target value design.* AECBytes: Building the Future. Retrieved from http://www.aecbytes.com/buidingthefuture/2009/ModelBasedEstimating.html

Wedel, J. W., Schunemann, B., & Radusch, I. (2009). V2X-based traffic congestion recognition and avoidance. *10th International Symposium on Pervasive Systems, Algorithms, and Networks (ISPAN),* 637–641.

Welle, Benjamin, Haymaker, John, and Rogers, Zack (2011). ThermalOpt: A Methodology for Automated BIM-Based Multidisciplinary Thermal Simulation for Use in Optimization Environments. Technical Report Nr. 200, Center for Integrated Facility Engineering, Stanford, CA, available at http://cife.stanford.edu/sites/default/files/TR200.pdf (last accessed on February 4, 2013).

CHAPTER 13

Collaborating in an Integrated Project

"Some people think architecture is about the genius sketch; I don't. Great architecture is a collaboration among a lot of people over a long period of time."

—Joshua Prince-Ramos

13.1 SO WHAT'S THE PROBLEM?

Look at the website of almost any firm in the design and construction space and it will hawk the firm's collaborative culture. Phrases abound like, "We have always been collaborative," "Our firm is known for our collaborative culture," and "Collaboration is in our DNA...." But is it?

The design and construction industry is also known for being litigious, claims oriented, and siloed. In the Norton Rose Fulbright annual litigation trends survey, 53 percent of the firms in engineering and construction reported that they had at least one arbitration commenced against them during the prior year (Norton Rose, 2014). This was the third highest of any industry surveyed. Parties are required to stay in their siloes. One of the most influential of all contract documents is the American Institute of Architect's Document A201 (2007), which in Section 4.2.4 requires the owner and contractor to communicate through the architect, requires that all communications to trade contractors flow through the contractor, and requires that all communications to consultants flow through the architect. And there is an unfortunate kernel of truth to the saying that "people come to projects as friends and leave as enemies." Maybe we aren't as collaborative as we say we are.

But if we look back at projects that were truly successful, to projects that everyone loved doing, how would they be described? From our experience, we know that these were the projects that were magical. Where everyone worked together and had fun. And yes, we would use the term *collaborative*.

In this chapter, we ask, what is this magic called collaboration, and more importantly, how do we make it happen reliably?

13.2   WHAT IS COLLABORATION, REALLY?

Collaboration is a community of people working together to achieve a common goal. In a project, the community is mostly defined by the immediate participants: designers, contractors, trades, vendors, and the owner. But in a larger sense, the community may also include end users, facility managers, and regulators. In its broadest sense, the community is anyone that has a material effect on project outcome or who derives value from the project. Projects, especially complex projects, advance more smoothly when collaboration includes stakeholders and regulators, as well as the project participants.

Working together implies an engagement among participants who are not only attempting to execute their work well, but are also supporting the success of others. Each party may have its own talents and resources, but they are all being utilized for the common good. Everyone has the same compass, pointed in the same direction. The project participants have a clear understanding of why the project is being undertaken, what they are attempting to do, and how they are doing it.

Perhaps collaboration is best defined by example. If you consider an Amish barn raising, everyone in the community understands the need for a new barn and what the barn must accomplish. The individuals in the community offer their individual talents—some are carpenters, some are roofers, some lay out the barn, and some bake the bread to feed the workers. But everyone is doing his or her best to help everyone achieve the common goal. They collaborate.

The failed Apollo 13 lunar mission also demonstrates collaboration. After an oxygen tank exploded, the crippled spacecraft needed to return to earth. As astronaut Gene Lovell famously reported: "Houston, we've had a problem." But it wasn't just the astronaut's problem, or the rocket designer's problem—it was everyone's problem. One specific problem was the limited capacity of the carbon dioxide scrubbing system, which was insufficient for the trip home. The lunar lander did have CO_2 scrubber cartridges, but they were the wrong shape and would not work in the command module that housed the astronauts. The team had a clear goal—keeping the astronauts alive. The NASA ground team swarmed the problem, with everyone working together, proposing ideas, rapidly prototyping and eventually developing a makeshift solution that the astronauts then built. This was just one of the many crises the team overcame, working together while owning the entire problem. They collaborated.

13.3   WHAT DOES SUCCESS LOOK LIKE?

The work hums. No one looks at a problem and says, "It's not my job." Problems are dealt with organically almost like an immune system attacking an infection. The participants determine who is necessary to solve the problem and then "swarm" to a solution. It is not only clear that they know what they are trying to accomplish—they are committed to doing so. Everyone is sharing his talents for the common good.

Conversations abound. The participants engage in what Peter Senge refers to as "dialogue" (Senge, 2006). Not just communication, or argument, dialogue is a joint discussion—even a debate—whose goal is to develop a better solution than either speaker could alone envision.

Commitments are being made and kept. Plan percent complete (PPC) metrics show that predictions are accurate and accomplished. The commitment log reflects a high degree of follow-through. Planning is a group activity, with those responsible for executing work taking responsibility for coordination and scheduling. Everyone owns the constraint log.

Mistakes get made. But rather than burying them—or assigning blame—they are quickly brought to the surface where the project team can address and resolve them. Decisions are jointly made to achieve the project objectives.

And it is fun.

13.4 HOW CAN THIS BE DONE?

13.4.1 Trust and Transparency

Trust is the fundamental element of collaboration, but it is not a blind or naïve trust. It is the *authentic trust* discussed by Fernando Flores, which is a conscious choice developed through interactions and conversations between people (Flores, 2013). The choice to trust creates a relationship between the person trusting and the person being trusted. It is not a state of being, but an action.

> We make decisions to trust. We make promises and tacit commitments. We see them through. We come to have expectation of others, and we respond to the fulfillment or frustration of those expectations. Trust isn't something we "have," or a medium or an atmosphere within which we operate. Trust is something we do, something we make. Our mutual choices of trust determine nothing less than the kind of beings we are and the kinds of lives we will live together (Flores, 2013).

Trusting teams are more effective and encourage team members to take risks. Trust facilitates information sharing and trust enhances productivity (Robbins and Judge, 2012). In *The Speed of Trust*, Stephen Covey (2006) argues strongly that developing integrity and trust is fundamental to high-performance organizations. Without trust, efficiency plummets and a firm suffers a "trust tax" that increases compliance cost and reduces productivity. Although other factors are important, trust is the one thing that makes collaboration really work (Prusak, 2011).

> Build communities of trust. When people trust one another, they are more likely to collaborate freely and productively. When people trust their organizations, they are more likely to give of themselves now in anticipation of future reward. And when organizations trust each other, they are more likely to share intellectual property without choking on legalisms (Evans & Wolf, 2005).

Trust is built by developing relationships, treating people fairly, and meeting your commitments. Flores (2013) proposed that coordination and commitment was accomplished through conversations for action. In its simplest form, conversation for action consists of four separate speech acts:

- Request or offer,
- Promise or acceptance,
- Declaration of completion, and
- Declaration of satisfaction.

These speech acts enable collaboration by clearly identifying the commitment made, establishing conditions of satisfaction, and acknowledging whether they have been satisfied. This avoids inadvertent communication breakdowns, as where the promisor through misunderstanding provides an outcome that doesn't fully address the promisee's needs, thus creating a perception of unreliability.

Trust reflects the team's perception of competency. The team needs to rely on the information and recommendations of its members. No matter how open and honest team members are, their recommendations won't be trusted if their ability to provide accurate information and recommendations is doubted. The perception of competence is a combination of prior history and reputation, but it is also based on the accuracy of information provided to the team. Competence is tested and proven through continual interaction.

Trust is also tied to similarity and transparency. It is easier to trust people you know and who are similar to you. This identification-based trust is based on a mutual understanding of others' intentions and appreciation of their needs and desires (Robbins and Judge, 2012). If people do not know each other well, they are slow to trust, creating a knowledge "transfer barrier" (Hansen, 2009). In contrast, hoarding information plants seeds of distrust that grow into defensive reactions. What starts as "prudent information control" spirals into team dysfunction.

Transparency is difficult for those who have interacted in internally competitive situations. If winning a negotiation means that the other party loses, then guarding information is an effective, and even necessary, strategy. When you are trying to create cooperation, however, this strategy is self-defeating. In fact, one of the most intriguing examinations of cooperation, Robert Axelrod's *The Evolution of Cooperation*, used the Prisoner's Dilemma to show that cooperation, as long as you were also being cooperated with, is the optimal negotiation strategy for many common and complex situations (Axelrod, 2006). Because cooperation is a predicate for collaboration, his findings have at least four implications for integrated project delivery (IPD) projects.

- Overall project value is maximized if the parties choose a collaborative strategy.
- Collaboration requires knowing what signals the other party is sending, that is, whether a provocation has actually occurred or whether they intend to collaborate. Thus, clarity and transparency are essential.
- Collaboration is improved if you increase the frequency and duration of interactions.
- Payoffs affect collaboration. Thus, risk/reward systems can be used to increase collaboration.

The Prisoner's Dilemma

The Prisoner's Dilemma is an abstract formulation of some very common and interesting situations in which what is best for each person individually leads to mutual defection, whereas both parties would have been better off with mutual cooperation. In the Prisoner's Dilemma, two thieves have been captured and then separated by the police. Each prisoner is told that if he will betray the other, he will receive a light sentence. But if they both betray each other, they will both get the same sentence. If neither betrays the other, then they will both get a more moderate sentence as shown in Table 13.1. In all cases, the effectiveness of a strategy—betray or remain silent—depends on the strategy adopted by the other prisoner.

The question posed by Axelrod was what strategy achieved the best outcome if the game was repeatedly played? The answer, derived from the international competition he hosted, was that cooperation—as long as the other party cooperated, was the most successful strategy.

TABLE 13.1 Prisoner Dilemma Strategies and Outcomes

Prisoner A's Strategy	Prisoner A's Sentence	Prisoner B's Strategy	Prisoner B's Sentence
Betray	0	Remain Silent	5
Betray	5	Betray	5
Remain Silent	3	Remain Silent	3
Remain Silent	5	Betray	5

It is very difficult to develop a mutual understanding of intention without face-to-face interaction. Moreover, research shows that face-to-face communication regarding attitudes and feelings is only 7 percent in what is said, 38 percent in how it is said, and 55 percent in body language (Covey, 2006). This information is lost if the parties aren't physically present. The perception of sincerity, a key factor in trust, is diminished if people only interact virtually. One of the reasons that co-location is so important is that it provides the opportunity for frequent, direct communication.

13.4.2 Clear Goals and Values

The chapter on IPD teams discusses the importance of goals in motivating and managing teams. In this chapter, we look at goals and values from a different perspective.

If we are going to collaborate, what are we going to collaborate about? To collaborate effectively, we need to know why we are building, what we are building, and how we are building it.

"Why" is a profound question in collaboration because it gives purpose. If "why" is to enable learning, the school we build must respond to this need. If "why" is to reduce illness or promote wellness, the hospital we build must enable healthcare professionals to be more effective. Understanding "why" gives individuals the common purpose necessary for collaboration.

In *Building a Collaborative Enterprise*, Adler, Heckscher, and Prusak (2011) emphasize the power of shared purpose.

> In focusing on the fourth alternative—a shared purpose—collaborative communities seek a basis for trust and organizational cohesion that is more robust than self-interest, more flexible than tradition, and less ephemeral that the emotional, charismatic appeal of a Steve Jobs, Larry Page, or a Mark Zuckerberg.
>
> *****
>
> This shared purpose is not an expression of a company's enduring essence—it's a description of what everyone in the organization is trying to do. It guides efforts at all levels … from top management's business strategy, to joint planning by the company's unique labor-management partnership, right down to unit-based team work on process improvement (Adler et al., 2011).

"Why" also provides guidance. Project members, whether in cross-functional teams, ICE sessions, or small meetings must make thousands of decisions. Clear values and goals provide the criteria for these decisions. For example, if a team is using Choosing by Advantages to evaluate alternatives, the goals and values provide the categories for comparing advantages.

"Why" also allows synchronization. If members of a team all understand the project goals, they can work in parallel to reach the same destination. Although communication between members is still important, a clear understanding of values and goals enables working toward the same goal during the periods between meetings.

There are many different ways to identify and document project values. One approach, described below, is to have a values workshop that jointly develops the key project values. Project management can then refine the values, establish goals consistent with the values, develop metrics to measure achievement, and establish targets. Moreover, the values can be used as criteria to guide project decisions throughout the life of the project. Other approaches involve deep discussions among stakeholders and participants resulting in conditions of satisfaction, charters, or other declarations of values. The key is to create a common understanding and a common purpose throughout the virtual organization.

Values Workshop

One of the authors uses a values workshop to both develop values for the project and to demonstrate principals of team facilitation/leadership. The workshop participants are grouped in tables of five to seven participants, preferably with a diversity of backgrounds and organization, and given a stack of blank cards. They are then instructed to:

- Select a table facilitator.
- Without speaking to anyone, write down the five most important project values, one to each card.
- The facilitator then leads the team in doing the following.
- Each participant hands his or her cards to the person to the left.
- The person receiving the card stack, reads each value, and explains what he or she believes the value means, preferably with examples.
- The facilitator asks the other team members what they understood the value to mean.
- The facilitator then asks the writer what he or she intended.
- The team discusses the value until writer and readers come to a clear understanding of the value.
- This continues through all cards until complete.
- The facilitator then asks whether any cards are essentially the same. After discussion, the cards are grouped into stacks of "same" cards and stapled together.
- The facilitator then leads a debate concerning the relative importance of each value until the team can assign a rank of first to nth of their stacks.
- While the process is under way, the workshop facilitator constructs a swim lane made of painter's tape on a wall. The top will be populated with unique values and the left will be arranged by rank, first on top.

- Each table will come to the wall and arrange their card stacks by value and rank. As they do so, they will explain what they meant by a value. See Figure 13.1.

- Each table, in turn, will come to the wall and start placing their cards. This will generate another discussion regarding similarity and difference and what is meant by a value. The workshop facilitator will manage the discussion using questions.

- After all cards are in place, the facilitator will lead an analysis of the swim lane, seeking to distinguish outcome from process values and to better rationalize differences and similarities.

- Finally, the workshop facilitator will explain the potential dysfunctions in team decision making, and how team size, individual idea creation, reflective listening, debate, depersonalized analysis, quantification, and visual management were used to improve decision outcome and how the participants can use similar techniques in project teams.

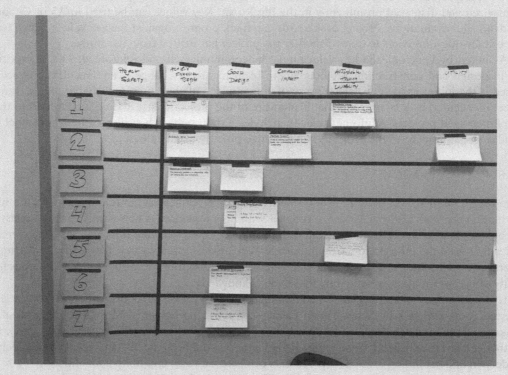

FIGURE 13.1   Swim lane populated with value cards. © Howard Ashcraft.

Although successful teams have used different approaches to defining and documenting values, we have observed that teams that spend the effort to understand the project values are significantly more focused—and more successful—than teams that skip this step.

Patrick Lencioni described management's responsibility to ensure that the entire organization understands the key values. This cannot be achieved by stringing buzzwords together into a marketing or mission statement. He recommends a rigorous approach based on deeply analyzing and simultaneously answering six questions.

1. Why do we exist?
2. How do we behave?
3. What do we do?
4. How will we succeed?
5. What is most important, right now?
6. Who must do what?

But management's responsibility does not end with creating clarity; they must continuously repeat their message (over-communicate), and reinforce clarity (Lencioni, 2012).

An IPD project has several tools for reinforcing values. On many projects, the key values are summarized and displayed prominently in the Big Room. Key values can be used in A3 or Choosing by Advantages evaluations. Project dashboards can track progress against the key values. Some teams use periodic questionnaires to assess whether the project has drifted from its course.

As noted by Lencioni, the key values must be communicated and reinforced. This is especially true in design and construction projects where team members change over time. Little is accomplished if the initial group develops a sense of community, collaboration, and shared values, but the project is executed by a later group that is unaware of the project values and not committed to them. In well-managed projects, reinforcing clarity includes a thorough on-boarding of new personnel that includes a commitment to the project's collaborative processes and values.

13.4.3   Conversation

There are three modes of communication between people. Unilateral communication occurs when information is broadcast from sender to receiver with no interaction. A news broadcast or a construction bulletin is a unilateral communication. Information, often valuable information, is delivered to parties that need the information. Communication can also be for advocacy. The sender is attempting to persuade the listener to adopt the sender's viewpoint. In both of these communication modes, the quality of the information is limited by the sender's knowledge.

IPD uses unilateral and persuasive communication, but it relies on conversation. In conversation (referred to as *dialogue* by Senge), both parties are active participants. Moreover, the goal of the exchange is not to persuade, but to build. Using the viewpoints and intelligence of each party, they together can be more insightful and creative than either would be alone. Debate, even spirited debate, can deepen understanding if the debate remains centered on ideas and not personalities.

Not everyone is adept in conversation. Facilitators—whether consultants or existing personnel—can guide persuasion to conversation. Using structured decision tools can depersonalize the debate and focus conversation on the issues and not the people. Increasing the frequency of interactions—through co-location or other methods—allows for spontaneous conversation.

13.4.4   Ownership of the Whole

In a collaborative project, the participants are committed to overall project success. In traditional projects, a trade could properly state that it did everything possible if it faithfully executed the plans and specifications for its work—even if the project did not meet its goals. But in an IPD project, the participants' profit is tied to project outcome. Only by achieving the overall goals will the individual succeed. This is succinctly summarized in two of Sutter Health's Five Big Ideas: Collaborate, really collaborate, and optimize the whole.

In practical terms, ownership means that the individuals are concerned with the entire project. A problem is a problem for all.[1]

Although this is a relatively simple concept, it is hard to overcome traditions and traditional thinking. Often, it is better to demonstrate the difference between local and project optimization through simple examples. One common example is the use of resources. If trades are asked whether they have the right number of scissor lifts, front-end loaders, or similar equipment, they will usually consult their estimates and confirm that they do. But if the question is how many scissor lifts or front-end loaders the project needs, the number may be quite different than just summing the number of scissor lifts in each trade's estimate.

Similar discussions can be had concerning which firm should execute work. When should design end and detailing begin? Should hangers be placed by one firm or three? Should trades or laborers clean up the work space? Should handoffs between trades benefit the trade or the project? How should work be performed to optimize project, rather than individual outcomes?

Ownership of the whole often results in solutions coming from parties that had no responsibility for the problem. In the Cook Children's Rosedale Office Building Project, a program goal was to create flexible spaces that could be reconfigured to meet future needs, and the team had chosen to use moveable walls for these spaces. Fire protection was provided by an automatic sprinkler system. As the project was being laid out, one of the participants noted that the fixed sprinkler locations would have to be re-plumbed and re-balanced if the walls were moved. His suggestion, adopted by the team, and shown in Figure 13.2, was to put the sprinkler heads on flexible hose to permit later readjustment. He wasn't the fire sprinkler contractor, or a mechanical engineer—he was employed by Haworth, the furniture and fixtures IPD team member.

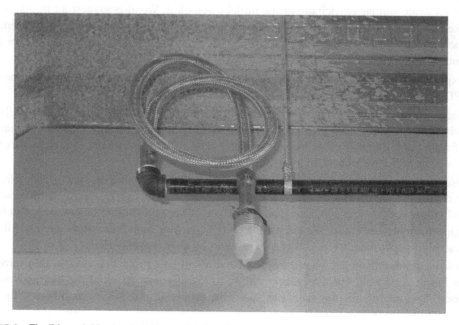

FIGURE 13.2 Flexible sprinkler head. © Howard Ashcraft.

13.4.5   Pull Planning

Pull planning was developed to improve the reliability and flow of work. But it has another outcome, as well. When parties come together to jointly work on developing a project schedule, they must necessarily take responsibility for the whole. Moreover, they get to immediately see the effect of their efforts on others and the project. In more than one instance, we have heard individuals engaged in a pull planning session exclaim, "So that's what you do" or "So that's why you need that." It builds a level of understanding and community.

13.4.6   On-Boarding

Although we know on-boarding is important, we often leave it to chance. This is a serious mistake because continuous on-boarding is one of the most cost effective techniques for enabling a high-performance project.

Most firms on-board incorrectly. The first several days that a new employee spends with an organization is a crucial, never-to-be-repeated, opportunity for creating a connection between the new employee, the organization, and the work they will do together. Yet this opportunity is wasted addressing administrative or human resource matters, computer training, and explaining work rules. This critical time should be used to communicate the mission of the organization, its values, and how the new employee/participant can contribute to the project. The same approach should be used with project on-boarding. With this in mind, we recommend that on-boarding:

1. Focus on why the project exists. For instance, in a hospital project you could focus on how the project will allow physicians and nurses to better heal people. You might want to have a nurse or someone similar participate in the initial discussion. We also suggest an interactive values exercise, such as we do with project teams. You might then compare the values generated to the project's mission statement/values matrix. We want to assure that everyone on the team understands what the organization is trying to achieve.

2. Explain what an IPD project is, why IPD is being used, what the differences are between an IPD project and a traditional project.

3. Focus on how the person can contribute to the project. What experience and skills do they bring? What about their personality will allow them to help a team? Have them share stories of something they did contributing to a team. Have them speculate (with some direction) on how they will have to act differently.

4. Use team exercises. Choose something useful, perhaps focusing on how a task will be done and get the team to analyze the task and suggest improvements. Draw the distinction between individual success and project success. Work on continuous improvement. This can be a good opportunity to use simulations, Last Planner™ coaching, and "teaching moments" for how to lead high-performance teams. Learning should be an active experience.

5. Have team members list what they have liked about past projects and what they haven't liked or hasn't gone well. Have a group exercise to identify how to avoid the bad and enforce the good. You might role-play scenarios. Seed the groups, if possible, with others who have real

IPD experience and can tell "war stories" about how the projects went. It would be good to have someone emphasize the "fun" aspect.

6. Take care of the administrative details.

7. Revisit with teams after a short period of time to determine if they are sliding back to conventional behavior and to learn whether they think the project is being run as was described in the on-boarding. Based on this information, adjust and make changes.

On-boarding can be made more efficient if you have a common set of materials, videos, and the like to reuse whenever a new group needs to be on-boarded. Common on-boarding materials also improve message consistency. Videos of earlier training sessions (edited to the relevant material) can be engaging and efficient. Not everyone is a good on-boarding leader. You should probably have a single group doing on-boarding or at least "train the trainers."

13.4.7 Cultural (Corporate) Considerations

Another challenge is that the teams are created from individuals from different companies with different management systems, incentive packages, and corporate cultures. When firms join an integrated project, they bring these differences with them. In this way, IPD is similar to a corporate merger. When mergers are successful, the individual firms are forged into a single seamless organization with common goals, common procedures, and a common culture. But mergers often fail because of cultural incompatibility. The failed AOL/Time Warner merger[2] is a classic example of complementary capabilities stymied by different cultures (Mahavidyalaya, 2012). The success of IPD teams will reflect how well these cultural issues are addressed.

A firm's culture reinforces the behavioral norms and beliefs that may have attracted like-minded employees to join that firm. This combination of self-selection and reinforcement stubbornly resists change. Culture can strongly affect performance because most employees act in accordance with their corporate culture. In some instances, this strengthens the team. In other instances, the differing firm cultures will hamper team effectiveness. And if project and corporate culture differ, employees are placed in the awkward position of complying with inconsistent norms.

Although project culture can be developed, successful teams are more easily created if their employing firms have cultures compatible with project values and with each other. Owners can improve compatibility by requesting qualifications or proposal responses from self-assembled teams of firms, rather than assembling the IPD team from individually selected firms. Firms that have chosen to work together, particularly if they have prior experience doing so, are more likely to have compatible cultures. Another useful strategy is to use rolling team selection where each key firm added to the team participates in the interviews and selection of additional team members. This allows considering how the new team member will fit into the existing mix. The contract should also have the option of replacing a party that does not meld well, and the entire procurement strategy should consider how well a firm will fit into the IPD team as well as their subjective competence and competitiveness.

If you can't choose partner firms on compatibility—and likely some of the IPD member firms will not have supportive cultures—you will need to spend more time building a project culture. This will never be perfect because team members will be caught between two cultures, but it will be better than allowing them to bring their differing cultures to the common workplace.

The first step to creating a project culture is to have one. The project leadership must clearly express their values and expectations from the onset, and then must act in accordance with those values. The leadership must demonstrate the required behaviors, recognize and praise congruence, and promptly correct deviance from norms.

Any action that builds or reinforces a common identity helps create project culture. Project logos and signage should replace individual company identification. Group events such as barbecues, community volunteering, fundraisers, project-supported sports teams, or any activity that intermingles individuals without reference to their employer helps to promote group identity. Informal lunch areas where workers can eat together gives an opportunity for community growth. Recognizing achievements, even informally, strengthens the sense of project community. Jointly working on developing project norms and procedures can also strengthen commitment to those norms.

Co-location offers opportunities to enhance project culture. The co-location site provides a physical space separate from an employee's firm and a demarcation point between corporate and project cultures. The organization within the co-location site can reduce the effect of corporate cultures and support the project identity. Workspaces should be organized by cross-functional unit, not by employer. You shouldn't be able to tell who an employee works for without asking.

Project-sponsored joint activities build community. The project might sponsor and mentor a school or class, jointly work on a Habitat for Humanity™ or similar project, field a team in a recreational sports league, or have "family day" barbecues with activities. It doesn't matter what the event is, provided it is inclusive and is undertaken at the project level.

Boot camps can start building culture from project inception. The military has long known that separating people from their environment and then forcing them into intense, joint activity, creates group identity. In a similar, if less rigorous way, moving team members to a physically separate place, reorganizing them across corporate lines, and then engaging them in demanding training or work exercises, starts the transformation to a project identity.

Whenever possible, teams should be built from participants who share similar cultures, but if that is not possible, the project leadership should take positive steps to create a project culture that supplants the participants' firm cultures.

13.4.8 Declaring Breakdowns

In traditional Lean theory, the production workers are empowered to stop a production line if they discover a defect. If a defect is observed, the line workers activate an "Andon" signal that alerts their immediate supervisor that there is a problem. If it can't be fixed immediately, the line is stopped to prevent creating further defects. Once the cause of the defect is fixed, the line can restart. A root-cause analysis is undertaken to determine why the defect occurred, the team then proposes a solution to prevent the problem from occurring, tests the solution, and, if successful, restarts the production line.

At first blush, it seems that it would be easier and less wasteful to continue production and correct (or discard) the defective parts. But in practice, continually analyzing and fixing the causes of defects leads to better overall quality, reduced rework, and higher throughput.

While this process is well understood in Lean manufacturing, it is often overlooked in Lean processes. If the Lean processes are not working correctly, the frontline workers must be empowered to "declare breakdown"—in effect, to press the Andon switch for processes. The IPD team leaders must then respond, fix the problem, and undertake steps to eliminate the dysfunction at its root cause.

This raises an important point about collaboration. It isn't about being nice. Communication in an IPD project should be respectful, but it should be the truth, the whole truth, and nothing but the truth. Even if it is uncomfortable, the parties need to say what needs to be said. To enable this level of openness, the IPD project must be a safe harbor for communication, without recrimination or blame.

In IPD, the processes—collaboration, dialogue, reliable promising, honesty, transparency, co-location, integrated concurrent engineering, virtual design and construction, target value design, pull scheduling, and similar processes—are the "assembly line" of ideas and information. In almost every project, especially under severe schedule pressure, team members will revert to traditional behaviors that will undermine the IPD process. When this happens, the other team members need to declare a breakdown, to figuratively pull the Andon switch and refocus the team members on why, what, and, most directly, how they are supposed to behave.

13.4.9 Agile and Scrum

Many of the best ideas for how to improve the design and construction process have been inspired by how other industries have solved similar problems. As noted in the discussion of culture, the information about corporate mergers can be useful to the design of an IPD project because the conjunction of many firms has aspects of a corporate merger. Similarly, the inception of a project is somewhat similar to product development, and there are lessons that can be learned from how the software industry has addressed collaboration.

Software development is plagued by many of the types of problems seen in the design and preconstruction phases of a project. Traditionally, these projects were managed using management and scheduling systems, such as elaborate product specifications and discrete tasks organized in "waterfall" Gantt or Critical Path Method (CPM) schedules. Quality was enforced through testing, executed after the software was "finished." Despite all of the attention on prescription, logical planning, and testing, software projects were notorious for being over budget, late, and often did not meet the expectations of the users.

About 20 years ago, groups of programmers started experimenting with flexible approaches built around small cross-functional teams that were self-organized and managed. Work was quantized to fit the teams and time boxed. Rapid prototyping was used to explore solutions and test their effectiveness. The smaller batches were tested by the product team as they were developed, rather than after the project was completed.

The changes worked. The result was the development and adoption of flexible approaches such as Agile and Scrum, which at their core were lean processes executed by small cross-functional teams focused on meeting the needs of their customers. Although some of the concepts need to be translated to fit the design/preconstruction phase of a project, their experiments and successes serve as hints for those seeking a better way to deliver infrastructure projects.

Agile is not a practice. It is a quality of the organization and its people to be adaptive, responsive, continually learning and evolving—to be agile, with the goal of competitive business success and rapid delivery of economically valuable products and knowledge. Although people speak of agile practices, the concept is to be agile, rather than to do Agile (Larman & Vodde, 2008). The nine principles of Agile software development are remarkably similar to the recommendations for efficient, Lean design.

- Deliver something useful to the client; check what they value.
- Cultivate committed stakeholders.
- Employ a leadership-collaboration style.
- Build competent, collaborative teams.
- Enable team decision making.
- Use short, time-boxed iterations to quickly deliver features.
- Encourage adaptability.
- Champion technical excellence.
- Focus on delivery activities, not process-compliance activities.

Three points of Agile are worth mentioning in a design context. To be agile is to be close to the client. To avoid queuing issues, an agile team works on specific deliverables in specific time limits (time boxing or sprints), reducing variability by adjusting batch size, another Lean concept, and increasing throughput. This allows reasonable utilization without gridlock. Finally, the software, not the documentation, is the goal. Thus, the output of structural design is the building, not the prints or even the model. From this viewpoint, the designer's client is the contractor and design should be "pulled"; that is, design should provide the information requested by the contractor to allow it to achieve the designer's intent. Finally, testing is not a culmination phase, but is built into every step of the progress. Thus, quality assurance (QA)/quality control (QC) is not a separate phase, but is continuous throughout design.

Scrum is a specific adaptation of Agile built around small, self-managed, cross-functional teams. Scrums are usually composed of around seven people from different backgrounds and with different responsibilities. Moreover, the scrum does not have a "leader" per se, but is mentored by a person with excellent subject matter and interpersonal skills (scrum master) who provides experience, training, and resources. The team selects the projects it will accomplish within a sprint from a prioritized list (product backlog) and then determines the best method for accomplishing the goals.

One of the visible elements of Scrum is the daily scrum meeting where each person reports three things he or she is working on, what they have accomplished since the last meeting, what will be accomplished by the next meeting, and any blocks or impediments they are facing. The entire meeting takes 15 minutes or less—there is no discussion within the meeting. As the meeting breaks up, if parties have ideas or suggestions, they can talk in smaller groups. The scrum's time is not used to discuss issues that are only relevant to one or two nor is anyone allowed to monopolize the conversation. In some scrum meetings, management is not allowed to attend. The meeting is to enable the team to operate better, not to allow management to grade or guide the activity. Each day, the team members update their estimate of the amount of time required to complete their current task in the sprint backlog. The remaining hours in the sprint are calculated and the progress recorded in the sprint burn-down chart. The team's goal and its progress are completely transparent.

Just like Agile, Scrum has underlying values that resonate in a design and construction environment.

- Commitment— Be willing to commit to a goal. Scrum provides people all the authority they need to meet their commitments.
- Focus—Do your job. Focus all your efforts and skills on doing the work that you've committed to doing.
- Openness—Scrum keeps everything about a project visible to everyone.
- Respect—Individuals are shaped by their background and their experiences. It is important to respect the different people who make up a team.
- Courage—Have the courage to commit, to act, to be open, and to expect respect.

The design and construction industry has benefited from techniques such as Lean, prefabrication, cross-functional teams, and BIM that were commonly used by other industries long before they were adopted by the architecture, engineering, and construction (AEC) community. Agile and Scrum move the focus of efficiency from documentation to the creation of value. Short, time-boxed sprints limit scope creep, reduce gold plating, and increase focus on producing useful deliverables. Design quality is built into the process and teams are empowered to achieve results the client values.

13.5   REAL-LIFE EXAMPLES

13.5.1   MaineGeneral New Replacement Hospital

MaineGeneral is the leading provider of health care services in Central Maine. In 2009, it decided to replace two existing facilities with a 640,000-square-foot new regional hospital costing approximately $322 million. Besides the normal complexities of acute healthcare facilities, the MaineGeneral team faced several unique challenges. Weather in Maine is difficult with severe winters that limit when construction can be under way. Snowfall at the MaineGeneral site averages 71.3 inches per year. In addition, the project would be the largest project in Maine, effectively absorbing the entire commercial labor force. Conventional wisdom held that skilled labor would have to be imported from Boston and New York to handle the sophisticated construction packages. But MaineGeneral, as a regional hospital with close community ties, was committed to keeping labor local. Finally, this was a major project for MaineGeneral, which did not have recent experience in building a large hospital. Faced with these challenges, many executives would have chosen to stick with traditional approaches. But MaineGeneral's CEO, Chuck Hays believed that if it was going to undertake this project, it should use the best tools available. After studying the options, he committed MaineGeneral to IPD.

The State of Maine uses a Certificate of Need process before allowing a healthcare system to construct a new hospital. As part of this process, MaineGeneral had to obtain four separate estimates of cost and schedule. Thus, the comparison of planned against actual outcomes are based on very solid estimates of cost and schedule using a construction management at risk approach.

Despite their inexperience with IPD, the MaineGeneral team fully embraced collaboration. The contractor was a joint venture of Robins & Morton and H. P. Cummings Construction Company. The design was led by SMRT of Portland with heavy support from TRO (formerly TRO Jung|Brannen) of Boston. Several of the local trades, who normally competed against each other, formed joint ventures to have sufficient capacity to execute the project (over 90 percent of the trade dollars remained in Central Maine).

The team co-located during the design and preconstruction phases, which is discussed in the co-location chapter. They also finished under the target cost, while simultaneously adding many value enhancements. In addition, there were only five change orders, all requested by the owner to cover new or changed scope. These are impressive results, but the signature accomplishment was the acceleration in schedule achieved through collaboration.

Schedule Acceleration

Schedule was a significant issue in MaineGeneral. General conditions and financing costs were approximately $1,600,000 per month. Early completion would also result in an earlier first patient date and earlier revenue. There were difficulties that affected schedule, but the team still completed construction in less than 25 months. This equals a construction finish rate of 25,600 square feet per month and completion 9 months ahead of schedule. This is the fastest delivery of a large acute care hospital by the contractor team (R&M is a very experienced healthcare contractor). In the team's opinion the schedule acceleration could not have been achieved outside the IPD delivery process.

To illustrate the drastic improvement, not long before the beginning of the MGMC project, R&M completed a Texas hospital project of 480,000 square feet in 25 months, an average monthly completion rate of 19,200 square feet. Based on several other large, 300,000+ square feet greenfield projects it has successfully completed in the past, R&M determined that 15,000 to 20,000 square feet a month is an aggressive goal for a large health care project.

At MGMC, Robins & Morton proposed a goal of approximately 20,000 square feet per month, with an additional two months added for extreme weather and manpower concerns in this rural area. This estimate equated to a 33-month overall construction timetable and was the basis for the IPD contract, which R&M and its joint venture partner HP Cummings (HPC) were awarded.

The MGMC project immediately faced several significant challenges that would, in the traditional design-bid-build world, usually lead to delay claims, cost overruns, large change orders and contentious relationships. MGMC's financing was delayed and instead of breaking ground in April 2011 as originally planned, the project's start was delayed until August. Missing the peak construction season in Maine leads to huge inefficiencies produced by difficult soil conditions. IPD allowed the site work subcontractor performing cost plus work to proceed without waiting for what would have been a change order for well over a million dollars.

The next challenge faced was the discovery of unforeseen boulders below grade while driving piles. Again, on a traditional delivery approach the pile contractor would file a claim and weeks or perhaps even months would be consumed in negotiating a fair price. With IPD, Robins & Morton was able to work closely with the structural engineer, material testing company, and pile contractor to not only prevent any delay in the schedule but actually finish the piling ahead of schedule and under budget!

IPD not only created a good foundation for overcoming problems, it also provided an environment where a realistic and information-rich schedule could be produced. Early in the process, the major contractors were brought on board on the front end of the design nearly a year before construction started. This allowed the contractors to understand the logic and all the critical path activities. Durations could be and were changed as the design developed. By the end of the preconstruction period, not only was the schedule very well thought out, but it was developed and committed to by everyone of the construction team members without worries of future claims or traps.

Also important was the utilization of "Lean" construction principles to continually improve the efficiency of the construction process. On traditional jobs, every subcontractor works only to make their scope of work as efficient as possible. Efforts were made to remove constraints at all levels. This included everything from material procurement to workflows. With IPD, the partnership between owner, constructor, and designer was able to accept that the overall project could be more efficient if occasionally minor trades were less efficient, realizing that it sometimes costs a "dime to save a dollar." With this understanding, means and methods were designed into the project to aid workflow efficiency. Optimization carried over to the design and owner responsibilities as well. Their tasks both were pulled to meet the pace of construction; urgency was no longer just the GC's focus. It was the entire team's responsibility to manage their pacing within the project flow.

MGMC owners, architects, engineers, and contractors were all "co-located" in a single office on site. Questions, concerns, challenges, and opportunities could be dealt with immediately. Almost daily, a situation would arise that could be solved quickly by having on hand all the resources necessary to immediately respond. This prevented schedule delays and provided ultimate value for the owner. The owner was literally watching the work in real time, and the constructors were able to adjust at once if owners wanted something done differently.

IPD allowed intense communication, buy-in, and continuous improvement to have an accurate and realistic schedule designed to optimize the whole project. Robins & Morton firmly believes that the gains made in the schedule would not have been so great if the project was under another form of contract.

Draw It Once

Another source of efficiency at MaineGeneral was the team's commitment to "draw it once."

The usual approach to design and construction is to have the designers prepare materials that are transferred to contractors and fabricators who redraw (physically or digitally) the information to meet their construction and fabrication needs. The MaineGeneral team recognized that this was inefficient, reduced quality, and absorbed time. They instituted a "draw-it-once" policy that required close collaboration between designers and trades.

The "draw it once" process called for the mechanical and plumbing contractors to model and produce the construction documents instead of the architect/engineer (A/E). The documents were produced using each trade's fabrication software, with the modeling team working under the direct supervision of the project engineers. In fact, for the majority of the construction documents production phase, modelers working for the subcontractors were co-located at the A/E offices. Once the model reached the permit set level of development, 2-D plans were produced, reviewed, and stamped by the engineer of record. The modelers then continued the level of development to the fabrication level, coordinated with other trades, and then sent directly to fabrication.

The draw-it-once approach had challenges due to lack of modelers and incompatibility among the different software programs. But in the end, the "draw-it-once" approach accomplished its ultimate goal of shortening the project schedule by overlapping the design and coordination efforts. It also gave the mechanical contractor more control of their product deliverable, and therefore a greater level of cost control. The close collaboration between fabricators and designers also resulted in a better understanding of what was being built, reducing expensive field changes and adjustments.

One of the best examples of the design/fabrication collaboration was in steel design. When the structural design team was offered the opportunity to work with a steel fabricator within the IPD framework, the steel group committed to eliminating all the inefficiencies of the design to erection process. The overall methodology of the design and detailing of the structural steel was examined, conventional practices challenged, and opportunities explored for using the capabilities of everyone in the design and construction team.

The first opportunity presented itself during the schematic development of the hospital's structure. Various methods of lateral bracing were discussed as options: cross-bracing between columns for low cost and moment frames for best utilization of floor space and future flexibility. The moment frames also offered additional options in their connection types: traditional field-welded, field-bolted end plates, field-bolted top bottom flange plates, and a proprietary system "SidePlate." Having a fabricator (Cives) who could translate the output of the design software into their estimating software allowed the team to generate multiple analysis models using the different types of bracing systems, including those generated by SidePlate staff. These analysis models were sent directly to the fabricator for comparison pricing, without any drawing preparation required. The results were presented to the project management team (PMT) for the rational selection of a bracing system that provided the best compromise between project cost and facility utilization.

The next place identified for streamlining was the typical in-house drafting process. By eliminating in-house drafting by sending design sketches directly to the steel fabricator, the team was able to capitalize on the unique relationship provided by IPD. The fabricator used the sketches to verify constructability during design and suggest value-engineering ideas throughout the detailing process. The A/E, SMRT and TRO‖JB, then inserted the fabricator's details into the A/E Revit model to verify potential conflicts, which also provided a pre-checking of the information used in the shop drawings. This pre-checking significantly improved the shop drawing review process, shortening turnaround time and greatly reducing the amount of corrections that usually accompany the review process.

The intimate designer-constructor relationship coupled with the use of an integrated design-construction schedule resulted in the realization that the fabricator would have all the information needed to begin work six weeks earlier than originally planned. This resulted in a lean, but fully coordinated, set of structural drawings. One change in practice was that the nonstructural details were eliminated from the structural set and everyone had to learn that "it's in the model." But the net gain of six weeks was worth having to refer to the model. The close collaboration between the structural engineer and the steel fabricator/erector also resulted in eliminating welding on all the moment connections and instead using bolted paddle plates. Fabrication speed was increased. In addition, elongated shear connection plates, or slap plates, were used for the steel connections, eliminating the need for the erector to tilt and rotate beams into position, saving significant time "hanging" the steel beams. This approach to design and erection was worked out incorporating input from the design team, construction managers, steel fabricator, steel erector, and the owner. A real IPD win.

Value Additions

One of the goals of the MaineGeneral process was to improve the value of the building without exceeding the cost target. For example, the project was planned to achieve LEED Silver, but was awarded LEED Gold. As the project proceeded, the team was able to add value to the project and still complete early and under target.

The initial design of the project only provided mechanical penthouses for three of five building wings and the target cost assumed three penthouses. But the project team recognized that additional penthouses would improve the life span of the mechanical equipment and aid maintenance. The team was able to find other opportunities to reduce cost and thus fund the additional penthouses.

The Chiller Plant Ice storage system is a similar story. The system produces ice during times of low energy rates and uses the ice to cool the building, offsetting the cost of additional electric chillers when energy rates are higher. But this was considered an "optional" feature not required to operate the hospital. But because of the tight integration between the designers and contractors, they were able to estimate costs with sufficient accuracy to allow the IPD team to add the system when it was still possible to do so. If they would have had to wait for the savings to be realized, it would have been too late to add the system.

The ability to add value at the right time testifies to the trust this team developed. In traditional projects, even if there are projected savings, the non-owner parties don't want to release the savings because they are concerned that they will need the savings as additional contingency against future problems. But in this case, the team transparently shared information leading to confidence concerning the accuracy of estimates. Moreover, they knew that if something went wrong, they (including the owner) were watching out for each other and would jointly solve problems as they occurred. Knowing they were in a collaborative environment allowed them to add value to the project.

13.6 INTERCONNECTIONS

Within the Simple Framework, collaboration is both an enabler and an emergent outcome. It is an enabler because collaborative teams more effectively achieve the goals of integrated organizations and integrated processes. It is also an outcome because tools such as simulation and visualization, ICE, and production management are aimed at helping the project team work better together—to collaborate.

13.7 REFLECTIONS

There are many pressures leading to collaboration. Digital systems such as BIM work best when the information is jointly created to jointly agreed standards. The MaineGeneral "design-it-once" approach is an example of how sophisticated use of technology requires collaboration. Sustainability also requires collaboration as achieving high-performance buildings requires—as the Simple Framework shows—integration throughout the entire design and construction process. And Lean similarly requires the early involvement of builders as well as designers. The upshot is that the requirements of modern projects will demand higher levels of collaboration than we have traditionally seen.

Project delivery methods will have to respond to the need for collaboration. Traditional design-bid-build is incapable of doing so. Construction management at risk with preconstruction services can partially achieve collaboration. But it is only with fully integrated project delivery approaches that we can achieve the level of collaboration future projects will require. The upshot is that the industry will move toward more vertically and horizontally integrated solutions or to solutions, such as IPD, that create project-specific virtual integration.

13.8   SUMMARY

Collaboration is the heart and soul of IPD. When teams are collaborating, every step in IPD works better. When they are collaborating, incredible achievements become possible. But collaboration can't be taken for granted. Great collaboration needs:

- A deep level of trust among the team members built on honesty, integrity, and demonstrated commitment. This allows honest communication and positive risk taking.
- Respect for the knowledge and skills of all members. Passionate debate is positive if the speakers respect each other and are committed to jointly achieving the best solution.
- A clear understanding of project goals and values. Every team member should know why the project is being built, how it is being build, and what is being built.
- A sense of community. Each member should feel they are part of a larger and worthwhile whole that justifies taking ownership of overall project success.

NOTES

1. The joint ownership of the project should be distinguished from legal liability or legal partnership. In an IPD project, the parties retain their traditional legal liabilities to third parties. But through profit dependent on project outcome, they have a financial interest in overall project success.
2. On paper, the merger between Time/Warner and AOL (America On line) was a match made in heaven. Time/Warner had an immense content library. AOL had unmatched digital distribution capabilities and a large client base. But the merger was a disaster due, in large part, to the cultural differences between the "suits" of the traditional company (Time/Warner) and the "geeks" of the Internet (AOL).

REFERENCES

Adler, P., Heckscher, C., & Prusak, L. (2011, July–August). Building a collaborative enterprise. *Harvard Business Review*, *89*(7–8), 94–101.

American Institute of Architects. (2007). AIA Document A201™, General Conditions of the Contract for Construction.

Covey, Stephen MR. (2006). *The speed of trust: The one thing that changes everything*. New York: Simon and Schuster.

Evans, P., & Wolf, B. (2005, July–August). Collaboration rules. *Harvard Business Review*, *83*(7–8), 96–104.

Norton Rose. (2014). Fulbright litigation trends survey report. Retrieved from http://www.iam-media.com/files/Norton%20Rose%20Fulbright%20Annual%20Litigation%20Trends.pdf.

Flores, Fernando. (2013). *Conversations For Action and Collected Essays: Instilling a Culture of Commitment in Working Relationships*. CreateSpace.

Hansen, M. (2009). *Collaboration: How leaders avoid the traps, create unity, and reap big results*. Boston, MA: Harvard Business School Publishing.

Larman, Craig & Vodde, Bas. (2008). *Scaling lean & agile development: thinking and organizational tools for large-scale Scrum*. Pearson Education India.

Lencioni, Patrick. (2012). *The advantage: Why organizational health trumps everything else in business*. Hoboken, NJ: John Wiley & Sons.

Mahavidyalaya, S. S. (2012). Cultural dimension analysis of AOL-Time Warner Merger. *Journal of Applied Library and Information Sciences*, *1*(2), 39–41.

Prusak, Larry. (2011). The one thing that makes collaboration work. *Harvard Business Review*, July. Retrieved from: https://hbr.org/2011/07/one-thing-that-makes-collaboration (last accessed on October 18, 2016).

Robbins, S. and Judge, T. (2012). *Essentials of Organizational Behavior*. Boston: Prentice Hall.

Senge, Peter M. (2006). *The fifth discipline: The art and practice of the learning organization*. New York: Broadway Business.

Solomon, Robert, and Fernando Flores. (2003). *Building trust: In business, politics, relationships, and life*. Oxford: Oxford University Press.

CHAPTER 14

Co-locating to Improve Performance

"With four people you can create one very strong energy, but if you can get 65 people working together, and swinging together, that's a whole other kind of energy."

—Chuck Mangione

14.1 ASPIRIN FOR INTEGRATION

Co-location is one of the most potent tools of the integrated project. Like aspirin, which is incredibly useful for a large variety of ailments, co-location solves many difficult problems. Co-location:

- Radically reduces latency (time to make a decision).
- Improves accuracy of communication through direct discussion and feedback.
- Increases creativity through collision of different perspectives and ideas.
- Provides designers with a fuller understanding of design consequences and alternatives.
- Supports a common understanding of values, goals, and project status.
- Improves project management by making work visible.
- Strengthens relationships among team members.
- Supports decision-making approaches like integrated concurrent engineering (ICE) and swarming.
- Improves virtual communication that follows co-located work.
- Provides a location for visual management controls.

Despite the enormous advantages of co-locating, some teams or team members are reluctant to co-locate, arguing that it is too expensive, particularly on smaller projects. Rarely does this argument compare the costs of co-location with the additional costs and reduced value the project will incur if it does not co-locate. From this perspective, the more relevant question is whether a project can afford not to co-locate.

Co-location does create practical issues that must be considered in developing a co-location plan. But whatever level of co-location is adopted, the goal is to obtain the maximum benefits that budget and personnel considerations allow.

14.2   WHAT IS CO-LOCATION, EXACTLY?

Co-location is the collaborative execution of work by the key project team members in a single location. In most instances, the single location is a physical place that has been created or modified for use by the project team. This may be augmented by virtual collaboration tools, but virtual collaboration is not co-location.

> The Integrated Center for Design and Construction (ICDC) really gave us the opportunity to enhance collaboration by integrating the different firms in one place around one problem. The key here is really two words: collaboration and integration.
>
> Stuart Eckblad, VP Major Construction, UCSF Medical Center

Co-location is also referred to as the "Big Room" in reference to Toyota's practice in placing multidisciplinary teams in a physical location to enhance communication and creativity. The Big Room or "co-lo" can take a wide variety of forms scaled to meet the needs of specific projects and teams. On smaller projects, the co-lo can simply be a conference room dedicated to the project where a smaller team can jointly meet to work together, and which has no more technology enhancement than whiteboards and wall space for displaying project goals, status, and other relevant information. At the other end of the scale, a co-lo has a fully integrated information technology (IT) structure, open workspaces organized around project function, workstations tied into the integrated IT infrastructure, phone and videoconferencing, large-scale monitors for displaying models, and breakout conference rooms/work areas for subteam use. Although these co-los look vastly different, they both serve the identical purpose of improving communication, enhancing creativity, and enabling collaboration.

14.3   WHAT DOES SUCCESS LOOK LIKE?

All of the key project members are working in the co-lo. Although production work may be occurring outside of the co-lo, the co-lo is where planning occurs. "Planning," in this context, refers to every intellectual action that must be accomplished to enable physical realization of the project. Design decisions are being made in the co-lo, but so are logistics and production planning, construction simulation, scheduling, and cost management. What you see in the successful co-lo is execution of the processes described in this book.

The work in the co-lo reflects the stage of the project, with early design exploration at commencement, and production planning in later stages. But in both cases, the co-lo is where the team works jointly to achieve the project goals.

Looking around a moderate size co-lo, you would see project information all over the walls: status updates, organizational diagrams, prints and renderings from models. There might be status boards showing where the project stands—and what improvement is needed. Some of the people in the co-lo have shirts with firm logos—but many do not, and the group you see gathered around a monitor has some of each. All you can really tell is that they are working on the mechanical systems.

The morning started with a huddle, where everyone explained what they were planning to do that day. As the huddle broke, a few groups coalesced to discuss some aspect of the work outlined during the huddle.

There are neighborhoods of workstations with people updating models, drawings, working on cost projections, but the neighborhoods are functional and multidisciplinary. A small group is gathered around a larger monitor and jointly exploring issues and opportunities in the 3-D model.

Somewhere in the co-lo are the project "leaders," but it would be hard to spot them. Yet there is no sense of disorganization, everyone seems to be doing real work, together. In fact, you are left with the sense that you are observing a community at work. And the laughter you heard from the corner isn't a bad sign, either.

14.4   HOW CAN THIS BE DONE?

Chapters 9 and 13 discuss in greater detail how to improve collaboration and develop high-functioning multidisciplinary teams. A lot of what is described in those chapters happens in the co-lo. This chapter focuses on the design and management of a co-location space and presents recommendations for co-location. It also discusses strategies for partial co-location and the use of virtual collaboration tools. Finally, we will touch on some real, human problems that must be considered when co-locating.

14.4.1   Start Early

> If I were to do something differently, I think it would be to jump ahead and ... set up the systems a little earlier.
>
> Cindy Lima, Executive Director, Mission Bay Hospitals Project, UCSF Med. Center

One of the mistakes that teams make is to start co-locating later in the design/preconstruction phase. Sometimes this is because the team is looking for a good space; in other instances, the team has been brought together gradually and significant amounts of work are done before the team is fully assembled. And then it gets hard to overcome the tendency to stay in existing offices. But teams that eventually co-locate and begin to see the advantages recognize that they lost significant opportunities and efficiencies by starting later.

14.4.2   Self-Limiting

Co-locating can be a significant expense. You need to create an appropriate working area. You need to transfer people to the co-lo and start organizing their work. You need to spend time in meetings planning how the project will proceed. You are nervous about not getting started in "real work" because of a tight schedule. You think about what you have to do individually and know you can be more productive on that work in your own office, surrounded by your normal tools and resources.

Stop! Co-location isn't designed to improve your utilization or the utilization of your firm's employees. It is designed to improve the progress and flow of the project. An hour spent in the co-lo making decisions that release work for others may have a far greater effect on the project than eight hours spent working individually.

Rather than thinking about what co-location will cost, or how it will impact individual utilization, think about what it will do for the project. Before you start self-limiting, ask: How much damage will that do to the project? How much additional waste will be caused? What opportunities will be lost? If everyone's profit is tied together through the IPD agreement, can you afford not to maximize your co-location opportunity?

14.4.3 Physical before Virtual

Almost all co-location facilities use some level of virtual collaboration tools. These can range from simple phone and videoconferences, to web conferences, or to fully interactive videoconferences and collaborative environments, such as Bluescape walls. They all serve a purpose and expand the reach of co-location by involving those who can't be physically present.

But there are three major lessons about virtual collaboration:

1. Physical collaboration is always better than virtual collaboration, especially if you need to engage the talents of the entire team and achieve creative outcomes.
2. Virtual collaboration works better *after* you have developed a face-to-face relationship. Once you have gotten to know your team members well, you can more effectively interact with them virtually. So start with your team together, even if you know you will eventually have to separate. Create a physical community before launching virtually.
3. The virtual tool should be chosen based on the type of interaction you need.
   o Phone calls rarely work well when you have more than two or three people on the phone. Moreover, so little context is transferred on the phone that it is best limited to exchanging basic information, like scheduling meetings or asking for confirmation of an event.
   o E-mail is a bad medium for discussing anything. E-mail has no context and is easily misunderstood. In addition, to get a full understanding of a subject, you often need to plow through long e-mail threads that are hard to decode. Finding information buried in e-mail is often difficult. E-mail is good for organizing and documenting decisions or serving as a wrapper for sending documents with an explanation.
   o Messaging systems built into a collaboration system improve on e-mail by organizing the information into discussions. They can also have robust tagging, organizing and searching capabilities. But the context issues are similar to e-mail.
   o Web conferences are good for presenting information, such as analyses and reports, but not as good for discussing or debating the information.

○ Videoconferences, especially in HD and in combination with web conferencing, provide a better level of interactivity and offers some visual context. If you are going to be creative, you want to have as much context available as possible.

Patterns and Context Matter

Communication researchers have long known that the bulk of human communication is in context, not the words spoken. In *Honest Signals,* Alex Pentland (2008) surveys the existing research and concludes that nonverbal clues are the "honest signals" we trust. Moreover, groups can come to decisions just using these "honest signals" without speaking. No wonder physical collaboration is superior. Pentland's team at MIT has taken this understanding to a new level. They "wired" people with monitors that recorded the frequency of interactions, together with information concerning affect. No substantive information was recorded. But based on the patterns of communication, they could predict which teams would be successful (Pentland, 2012).

14.4.4 Optimize the Physical Layout

Organize by Clusters, Not Firms

Co-location is designed to improve the level of communication and creativity within the team by creating multidisciplinary interactions. But they should be related to a common focus. Those responsible for mechanical systems should be located in the same area, regardless of firm affiliation. Using the mechanical example, you should have the mechanical engineer, the mechanical contractor, the owner's facility manager, and perhaps a vendor in adjacent workstations. Related clusters, such as an electrical group, might be immediately adjacent to the mechanical group. Ideally, the people who are supposed to interact should be in physical sight of their colleagues.

In a smaller co-lo, the groups may be merged into one related cluster, such as mechanical, electrical, plumbing, and fire (MEPF) protection. In a large co-lo, there may be more division by function, with the larger (in this example MEPF group) forming a "neighborhood."

Workstations can—and should be personalized—but tall dividers should be avoided because they interfere with line of sight communication.

Lots of Wall Space

Visual management requires lots of wall space for communicating with the team and for working together. Information about team status, planning, and goals is displayed on walls. Planning activities, such as pull-planning, occurs on walls. Brainstorming occurs on drawing spaces that can be used collaboratively, such as large whiteboards or drawing paper that can be pinned or stuck to wall surfaces as the team explores concepts and alternatives.

One team that was "wall space challenged" developed rolling walls that could be moved around the co-loco. The movable walls were an "A-Frame" design. They had usable surfaces on both sides and were mounted on industrial-size casters. They were made from materials that were already on site. When a team needed a wall surface, it pulled a movable wall from where they were stored and into a cluster area.

Another wall-challenged team created a sliding pull-scheduling wall. Each wall segment reflected a block of time, such as a week. As the project moved forward in time, the oldest wall segment was erased, removed, and transferred to the most future position. This created a "perpetual" calendar in a relatively small space that was always looking six weeks into the future.

Technology can also be used to expand wall space. SMART Boards and Bluescape walls are examples of interactive digital spaces that allow the users to collaboratively display and manipulate information. The Bluescape software allows simultaneous interaction, in multiple locations and on multiple devices on a massive digital plane that can be expanded or contracted to shift focus from detail to overview and back. Systems such as these make interaction with the information easy, thus inviting engagement of the project or cluster team with the project information.

But there is also an advantage to information that is always available. Although we all have computers with scheduling and task reminder capability, most of us occasionally put Post-it ™ notes on our monitors to "really" remind us of something we need to do. Information that is visible has greater impact than information that is out of sight, regardless of how easily it could be retrieved.

Small Breakout Spaces

Team members need to be able to spontaneously interact, and this requires some common areas where they can meet. These can be small conference rooms or open areas. In one co-lo, the small breakout spaces were circular tables in each "neighborhood" where team members could move to when they needed to discuss an issue. In another co-lo, there were small breakout spaces with couches, chairs, and a large monitor that could display information from the BIM. Teams could move to the breakout space to jointly evaluate the design or work in the BIM.

Respect Differences

The open office environment is wonderful for many people, but others work better in a quieter setting. Some people need music to concentrate; some need silence. These are legitimate differences that should be respected as long as the solution isn't a return to a closed office environment. White-noise generators may be a way of dampening sound while maintaining visual connectivity. Others may be fine with headphones (especially if they concentrate better with music) that block noise and keep their sound from affecting others.

Right Technology

Your project budget may allow use of every new technology on the market. But should you use all of it? Lean processes are about using the right technologies. Pull planning is an interactive process that is best done with "stickies" in the hands of the people making the commitments as this drives the conversation necessary for collaboration. Understanding the design intent, in contrast, may require extensive use of visualization technologies, all the way to immersive three-dimensional (3-D) virtual imaging, such as the digital immersive showroom (DISH) used by Disney to visualize design information. The type of tool will affect how the team interacts. Carpenters and mechanics know that the work moves more smoothly with the right tools. The same is true for collaboration.

14.4.5 Managing Co-location

Allow Time to Work

Most of the emphasis on co-location concerns collaborating, meeting, discussing, debating, deciding—but you need to do your own work, too. Project management should schedule open periods that allow people sufficient time to do their independent work.

Schedule the Space

A weekly schedule should be posted in the co-lo and available through whatever collaborative digital tools the team is using. As noted previously, it should allow time for individual work. It should also be regular and should respect the time of the participants. If you are using a schedule that segregates days by function (e.g., structural/foundation groups meet on Tuesdays), then the days should stay the same each week to allow people to plan their work. And overlapping areas should be adjacently scheduled to efficiently use people's time. The upshot is that project management should plan who is in the co-lo facility and when they should be there.

Project management should also get the commitment from everyone to attend when they are scheduled. Having almost everyone necessary for a decision wastes the time of everyone who made the effort to attend. For the same reason, scheduled meetings should start and stop precisely when scheduled with no one drifting in late or leaving before the work has been completed or the scheduled ending has arrived.

But don't schedule just to schedule. Calendar software can set meetings that repeat infinitely. While a regular schedule is useful, having meetings that aren't required is not. To avoid clogging everyone's calendars with meetings that may not occur, or may not be necessary, it may be better to only schedule meetings for a few weeks at a time.

Huddle Daily

Every schedule should start with a short daily huddle. These are also referred to as "stand-ups" because they are often done with the team standing. This avoids multitasking during the meeting and tends to shorten the meeting duration. The outline for a daily huddle is to:

- Identify yourself unless that is obvious to the team;
- State the work you plan to be doing that day; and
- Note any particular problems that you are struggling with.

The huddle is not a time for extensive discussion. After the huddle is over, if a team member needs to discuss an item with another, or if someone has information or a suggestion regarding a problem, it occurs in a separate conversation after the huddle. This avoids the whole team having to take the time to listen in on a conversation that is not relevant to them.

Huddles are very effective tools to enhance coordination and avoid waste by making work apparent. If someone is doing work that isn't needed or is working on an issue that is of lower priority than another task, you discover it that day, not weeks later.

For those who can't physically attend the huddle, it may be worthwhile to include them by videoconferencing.

Very Specific Agendas

Meetings in the co-lo—actually all project meetings—should have specific agendas. This allows people to know what is being discussed and whether they should be present. If decisions are going to be made, a good practice is to list the decisions as an agenda item to make sure that the right decision makers are present and the necessary preparatory work has been done. For meetings that have follow-up, another good practice is to create the agenda for the next meeting as the last agenda item of the current meeting. This creates continuity between meetings and alerts the project team about what work needs to be done before the next meeting.

Minutes/A3s

Collaborative does not mean disorganized. Meetings should have specific agendas and there should be minutes that track the decisions and commitments made. There should be accountability. The minutes should record who is responsible for an issue and should track the commitments made. The minutes should also follow the same format in all meetings and should be quickly posted and readily accessible by team members. A commitment log, which records the commitments made, can be used to ensure accountability. A3s can also be used to record decisions.

Only Right People in Meetings

One downside to co-location is "meetingitis," an excessive amount of time spent in meetings that aren't effective. One of the biggest time wasters is to have people present who have no interest in the subject matter or to have people absent that are necessary for the meetings, either because they have needed knowledge or expertise or because they are a decision maker for the items being considered.

In facilitated meetings, the facilitator often begins the meeting by asking if there is anyone not present that is necessary for the purpose of the meeting. If the answer is yes, the follow-up question is whether there is anything useful that can be accomplished without that person. If the answer is no, or relatively little, the meeting is adjourned and rescheduled. Another question is whether there is anyone present that the group agrees does not need to attend. If there is, the person is excused from the meeting. If there is any doubt whether the right people are present, these questions should be asked.

Respect Time of Smaller Players

Sometimes you do need to have people present, but not for the entire meeting. You can respect the smaller players by discussing their items at the beginning of the agenda and allowing them to leave, or scheduling a specific time within the meeting for their attendance.

Be Practical and Tactical

Projects can be viewed as a flow of dependent decisions. Being practical, you need to have done sufficient work prior to a meeting to be able to reasonably address the issues. Being tactical, you should prioritize issues on their ability to release work that affects the project progress. There is no point addressing interior artwork when you need to be solving civil site problems. In this respect, meeting agendas should reflect the "pull" in the schedule. They should also allow sufficient time between meetings to accomplish work. For example, if MEP costs are addressed one week, structural/envelope costs might be addressed the next so that the MEP group can advance their work without having to spend all their time to report on costs every week.

Park Items

Have you ever been in a meeting where someone hijacks the agenda to discuss something that was not on the agenda, but very important to them? Or where someone comes up with an intriguing idea that will take the discussion way off track? These should be recorded in a "parking lot" and then addressed at the end of the meeting, if time permits. If not, and if they are sufficiently important, they should be placed on the agenda for another meeting.

ICE, ICE, ICE

Major problems and opportunities should be attacked with intensive multidisciplinary sessions. Integrated concurrent engineering (ICE) is a discipline for engaging all necessary resources and decision makers in a high-energy burst of analysis, rapid prototyping/simulation, and decision making. ICE sessions should be encouraged during the project.

Use Models

Very few people, even design professionals, can effortlessly translate two-dimensional drawings into a mental three (or four!) dimensional image. Project stakeholders rarely have this ability.

3D-models are easier to understand and reduce the likelihood of misunderstanding within the team. Models can also be augmented with schedule and cost information allowing the team to visualize the construction sequencing or to compare quantities—and to some degree costs—associated with different alternatives.

Immersive 3-D visualization, using 3-D visors or entering 3-D visualization rooms, now allow users as well as team members to "enter" the model and experience functionality. Moreover, research is under way that would allow the person in the model to not only experience the model visually but to interact with the model adjusting parameters and experiencing the modification.

Models are the prototype building. They allow the team to understand what they are building, test and refine it, and virtually build it multiple times before anyone steps into the field. Research has also shown that meetings that use models effectively are much more productive than those that don't; one study showed that, in meetings with models, participants resolved 75 percent of the issues during the meeting, whereas in meetings without models, no issue was resolved (Liston, 2009).

14.4.6   Strategies for Partial Co-location

By now you are hopefully convinced that co-location is valuable and have an understanding of how co-location should be organized. But you may have skimmed through the prior material thinking that it really doesn't apply to your "smaller" project. It does, but you may need to modify the approach. Before looking at the modifications, however, we should examine the arguments against co-location.

Two Common Objections to Co-location

The Opiate of Utilization

> Initially, we didn't bring our entire team out here. Our architects are very interactive with our entire office, and we all are very used to bouncing ideas off on each other all the time. . . . When we started in the Big Room, we started with a partial presence, and that didn't work at all.
>
> Laurel Harrison

The first argument is that "I'm more efficient in my office." If you only focus on your individual utilization rate, this is probably true. In your office, you have work available from a variety of projects, you have your resources immediately at hand, and people aren't dropping in to ask you questions. And if you (and everyone else who works out of the co-lo) are more efficient, then you might think the project is more efficient. Unfortunately, no.

Niklas Modig and Pär Ahlström, in *This Is Lean*, directly address the conundrum that individual utilization leads to a lack of flow. Viewed from the viewpoint of the work, individual utilization results in large blocks of time in which the flow is not progressing. There are several reasons for this.

If you are switching between projects in order to maintain your utilization, you are probably batching work from the project in order to create a sufficient inventory of work to assure that you stay busy. Once you have worked through that stack, you then move to another project stack. Utilization is high. But work done by one party on a project releases work to be done by others. So by batching work you are creating a delay in releasing other work on the project. You are efficient, but the project is inefficient. And if the profit of your company is tied to the efficiency of the project, as it is in an IPD contract, then you are undercutting your firm's profit in the name of "efficiency."

There is another pernicious effect of utilization. As utilization of a resource goes up, throughput goes down. Think of a freeway. As the number of cars increases, the speed of the cars starts going down until they move at a crawl, if at all. And if there is any variability in speed (no matter how hard they try, people drive at somewhat varying speeds), the system begins to pulse in going faster and slower. We have all experienced this on the freeway where you speed up, only to have to slow down shortly after. Even worse, the drop-off in throughput is not linear. As utilization of a resource exceeds 50 percent throughput starts dropping off exponentially. This is why the freeway has a traffic jam when it is only 75 percent utilized. The upshot is that if everyone is perfectly utilized, the project will come to halt.

We Can't Afford It

The second argument is that co-location is too expensive. It is expensive to create a space, outfit it with appropriate technology, and have people travel to be in the space. But what is the cost of not co-locating? How much waste is caused by having teams that are poorly coordinated? How many great ideas are lost because there are no "water cooler" discussions? How much rework must occur because teams worked on issues out of sequence or went in a direction that differed from what the team needed? How much of the waste in design and in construction is because we don't co-locate? Without some estimate of what is lost by not co-locating, you can't actually say it is "too expensive."

14.4.7   Maximizing Co-location Value

Rather than arguing why a smaller project shouldn't co-locate, the discussion should focus on how much we can stretch to invest in co-locating and then how can we gain as much benefit from co-location using the resources we can apply. The next sections discuss some of the strategies teams have used to "stretch" their co-location budget.

Rotating Co-Location

In the Autodesk AEC Waltham project, the contractor, Tocci Building Company, moved into the architect's office (Kling Stubbins, now a part of Jacobs) during the design/preconstruction period. As the architect was designing, the contractor was detailing in the same model. The key trades also spent time in the architect's office providing their input into design. Once construction was under way, the model was shifted to Tocci's office, and the architect, supported by its consultants, moved in with Tocci. Questions were immediately answered, design issues confronted, and decisions made jointly. There was no separate co-location space or infrastructure, but the purposes of co-location were mostly achieved.

Other teams have taken similar approaches with early joint efforts occurring in one of the parties' offices and then moving to a co-location facility or to another firm's office. This approach can also be coordinated with a systems focus—for example, moving to the mechanical engineer's office during development of the mechanical and electrical systems.

Pulsed Co-location

Instead of having continuous co-location, the team could co-locate every other week, or one week per month depending on the pace of the project. These pulses of co-location would be similar to ICE sessions and should focus on making decisions that will release work. Certain team members may meet between co-location sessions and virtual tools, such as video conferencing, can also be used to bridge between co-location sessions. Project collaboration applications or portals can also be used to keep some communication flowing between sessions.

Office Hours

Everyone doesn't have to be in the co-lo all the time. It may be sufficient to establish a schedule of when certain members will always be present. Thus, the entire team might meet on Monday to discuss the week's planned work, but on Tuesday only the parties concerned with structural systems are in the co-lo. Wednesday is devoted to mechanical/electrical systems, and so on. If an "office hours" approach is used, it is imperative that participants treat their attendance as a critical obligation. If they are present only when it is convenient, the system will fail.

Offsite Co-location

Co-lo facilities are generally at or very close to the project site. This is particularly important during construction. But it is less important during the design/preconstruction period. In some instances, teams have decided to co-locate where it is most physically convenient for the team, and then move closer to the site as construction begins.

The Co-located Telecommuter

Why can't you do your other work from the co-lo? Many firms already allow or even encourage telecommuting from home. If you can do work on multiple projects from a home office, there is no reason you can't do the same work from within the co-lo. If having the structural engineers continuously available is valuable to the project, there should be no objection to their working on other projects while housed in the co-lo.

Paying for Value

If there is value in having people in the co-lo, then perhaps the team should be willing to absorb some of the lost utilization that may occur. So far, the authors haven't seen an example where this was done, but it might be a solution in some cases.

Virtual Augmentation

As noted previously, virtual collaboration tools alone are not co-location. But once a team has developed strong relationships, videoconferencing, Bluescape and similar techniques can provide some of co-location's benefits. Virtual collaboration techniques are limited, however, and should be considered an augmentation of co-location, not a replacement.

14.4.8   Real Problems

In the partial co-location section of this chapter, we discussed some of the negative effects of high utilization. But the reality is that some parties, particularly professionals, need to maintain a reasonable level of utilization to be profitable. This is not a significant problem on large projects, because their time is fully utilized on the project. But in smaller projects, the team members must realize that the need for high utilization draws professionals out of co-location and into their offices. This problem needs to be openly confronted and addressed, possibly with one of the strategies discussed in the partial co-location section.

Another real problem relates to the effect co-location can have on the personal lives of the participants. If everyone lives in a metropolitan area, co-locating to another part of the city may have little effect on the participants. But if the project is distant from a party's offices, then it may be difficult for their employees to work from that location, especially if it means extended absences from home. This may put stress on the employee and the employee's family as they struggle with parenting, schools, or other family obligations. This is a book about integrating project delivery (IPD), not human resources, but it may be necessary to make some accommodations to allow valuable employees to contribute to the co-location effort without completely sacrificing their family life.

14.5   REAL-LIFE EXAMPLE

14.5.1   MaineGeneral Replacement Hospital

Background

MaineGeneral is the leading provider of acute health care services in Central Maine. Traditionally, it operated out of two separate facilities, which resulted in substantial operational inefficiencies. In addition, the facilities were aging and needed to be refreshed or replaced. MaineGeneral decided to consolidate their facilities into a new, modern, 640,000-square-foot hospital with an anticipated cost of $322 million. Chuck Hays, MaineGeneral's CEO studied various options for delivering the new hospital and chose IPD.

No one in Central Maine had ever done an IPD project. The principal architect, SMRT from Portland, Maine, had a good relationship with MaineGeneral and with a Boston architect, TRO (formerly TRO Jung|Brannen). TRO had some IPD experience on a smaller project in Connecticut. The team selected a construction manager that was a joint venture of Robins & Morton, a national-level healthcare contractor from Birmingham, Alabama, and H. P. Cummings, a smaller but well respected Maine contractor.
Neither contractor had done an IPD project. The team then selected the trades, who were almost entirely Maine based, and the additional consultants. They had the players, they had an IPD contract, but did they have a team?

Process and Examples

One of the first steps the project team took was to develop a plan for how they were going to collaboratively execute the project. After surveying the approaches taken by other projects, particularly early IPD hospital projects, the project management became convinced that co-location was critical to transforming individually competent participants into an effective IPD team. Under the IPD agreement MaineGeneral was using (three-party Hanson Bridgett form), the Project Management Team (PMT) had the authority to invest in opportunities for improvement. They decided to lease a building close to the Augusta project site that could accommodate the key members for working and meeting sessions.

The space was outfitted for team use including installing a high-bandwidth Internet connection. Videoconferencing was also installed to allow the Boston architect, and others, to be involved in meetings when they could not physically attend. An added benefit was that the space was large enough to be used for patient room mock-ups and similar physical modeling.

The co-lo was the focal point of communication and collaboration. Immediately on entering the space, you were notified of project status through use of a "traffic light" for budget and schedule. Over target was red, concerned was yellow, and under target was green. The immediate focus on project outcome, combined with the open office arrangement, was an immediate reflection that "we are all in this together." Everything was project-oriented. If you called the co-lo, the receptionist answered, "Good afternoon, New Regional Hospital Project, can I help you?" The project team referred to it as the "co-lo" with its own address and logo, shown in Figure 14.1.

Early exercises took place at the co-lo site to collaborate on the schedule. With design team members side by side with the subcontractors for each trade, a lean process of determining schedule was able to take place. Figure 14.2 shows the first Big Room.

A major undertaking like this project requires a tremendous amount of planning with the user groups. User group meetings are formed and necessary for the planning process. The Big Room integration concept lent itself to this process greatly. Members of the construction team attended each user group meeting, with real-time feedback on logistics and cost impacts. The users themselves, being at the co-lo, saw the budget status and schedule milestone goals and more readily understood and supported these aspects. As a result, the decision-making process was more effective.

While the user group planning meetings took place, the collaborative design and contractor trade teams formed project implementation teams (multidisciplinary teams). These teams, one of which is shown in Figure 14.3, met at the co-lo on a regular basis to develop the design together, with feedback on cost from the contractor, and input from the owner. This Big Room integration process allowed the team to work together to find the best value for the project. If the cost rose on an item, then the

FIGURE 14.1   The MaineGeneral co-location site signage. Courtesy of MaineGeneral Alfond Center for Health Integrated Project Delivery Team.

FIGURE 14.2 A schedule and workflow process meeting at Co-Lo 1. Courtesy of MaineGeneral Alfond Center for Health Integrated Project Delivery Team.

collaborative team worked to balance the project by reducing cost elsewhere. By reviewing scope, cost, and schedule jointly with all parties, decisions were made that each party supported.

The co-location concept continued into construction, although the co-lo site was moved closer to the project for convenience. Safety meetings, weekly superintendent IPD/Lean meetings, joint scheduling, and continued Project Implementation Team ("PIT") meetings were all held at what was now known as Co-Lo 2.

The job site trailer complex was set up with a central area that replicates the co-location site with owner's reps, designers, and contractors intermingled throughout the room. A large conference room with movable partitions and full videoconferencing capability was included to continue the ability to have user group meetings with the entire project team nearby and maintain the integrated spirit. The large conference room also served as a central location for project information such as floor plans, schedules, etc.

Challenges

The co-location effort faced two major challenges and an accommodation issue. The first problem was that the architects were not located close to the project site. TRO, which also provided engineering services, was located in Boston. SMRT was in Portland, but that is still a three-hour drive from the project site.

FIGURE 14.3 A multidisciplinary/"project implementation team" meeting at Co-Lo 1 with videoconferencing. Courtesy of MaineGeneral Alfond Center for Health Integrated Project Delivery Team.

The team overcame the distance challenge by committing to regularly work—although not full time—in the co-lo. The team also installed high-quality videoconferencing to connect Co-Lo 1 and later Co-Lo 2 to the architects' offices. During design development, a sheet metal detailer moved into each architect's office to facilitate rapid coordination between the heating, ventilating, and air-conditioning (HVAC) detailers and the mechanical designers. Effectively, there was a mini co-lo occurring within the architect's offices.

The other major challenge was technical. The team was committed to working jointly in the model, but integrating the different software platforms proved challenging. In addition, because there was a delay installing the high bandwidth connection to the co-lo, the participants had to keep the initial model development on their own IT platforms.

The open office layout promoted collaboration amongst the project participants, but proved, in this location, to be noisy. In part, the participants had to adapt, but the team also moved the meeting areas further from production to reduce noise from meetings.

Outcomes

Despite the challenges, co-location served as an important catalyst, bringing parties into close physical proximity, enhancing communication and bridging between entities accustomed to working more independently. Co-location of the owner, design, and construction team is, without a doubt, a key to developing the culture of shared decision making necessary to move the parties beyond simple collaboration to a sense of unity of purpose with alignment of interests to achieve success. The MaineGeneral IPD project team acted as a single enterprise, with all parties focused on the same goals.

The MaineGeneral team had significant accomplishments. The project schedule was reduced by over nine months in comparison to the multiple schedules created by competing contractors during the Certificate of Need process. With financing costs of $1 million a month and general conditions of $600,000, this was a substantial savings. The design and construction periods overlapped by 18 months, greatly accelerating the schedule. Working together closely built the confidence that allowed the construction team to move into the field while design had not even been started for some areas. The team also was able to substantially increase the value of the project by funding, from projected savings, an ice chiller system, additional mechanical penthouses, and a rainwater reclamation system. The team was also able to improve the project's LEED certification from Silver to Gold. And the project was delivered below target cost with additional profit provided to the team.

14.6 INTERCONNECTIONS

Co-location is directly connected to IPD teams and collaboration. It is also an element of creating an Integrated Organization and should be supported by Integrated Information. Working in multidisciplinary teams allows the Integrated Processes that lead to Integrated Systems. Co-location touches on virtually every aspect of IPD.

14.7 REFLECTIONS

All roads lead to co-location. Whether your goal is sustainability, Lean design and construction, or effective use of building information modeling (BIM), you will need to collaborate. In the future, someone may crack the code of virtual collaboration and develop a system that is as effective as direct physical contact. But until then, high-performance teams will require significant amounts of direct, physical interaction. And the best way to bring multidisciplinary teams together is through co-location.

14.8 SUMMARY

Co-location is an exceptionally valuable collaboration tool. Bringing multidisciplinary teams together improves communication, enhances creativity, and develops the personal relationships that collaboration requires. Although there is a monetary cost to collaboration, its value greatly outweighs the investment. However, if you can't fully co-locate, there are strategies for obtaining value from partial co-location or co-location enhanced by digital collaboration tools. But we also have to recognize that co-location puts demands on firms and their employees that need to be accommodated in a successful project.

REFERENCES

DPR Review. (2011, Spring–Summer). Q&A: Building a great Big Room. pp. 1–4.

Liston, K. (2009). A Mediated-Interaction Approach to Study the Role of Media Use in Team Interaction. Ph.D. dissertation, Stanford University, CA.

Modig, N., & Ahlström, P. (2015). *This Is Lean*. Stockholm, Sweden: Rheologica.

Pentland, A. (2008). *Honest signals: How they shape our world*. Cambridge, MA: MIT Press.

Pentland, A. (2012, April). The new science of building great teams. *Harvard Business Review*, *90*(4), pp. 60–69.

CHAPTER 15

Managing Production as an Integrated Team

"Being busy does not always mean real work. The object of all work is production or accomplishment and to either of these ends there must be forethought, system, planning, intelligence, and honest purpose, as well as perspiration. Seeming to do is not doing."

—Thomas A. Edison

"Don't mistake activity with achievement."

—John Wooden

15.1 WHAT IS INTEGRATED PRODUCTION MANAGEMENT?

The question is how do team members define and break down the work into manageable chunks that match available resources, move the project toward goals, and advance the project meaningfully, productively, and safely? The answers can be seen in a real-life example of a hospital completed in July 2013 in Southern California. Success came from completely integrating production of the hospital.

15.2 WHAT DOES SUCCESS LOOK LIKE?

The superintendents, general foremen, and foremen planning production for the Temecula Valley Hospital had a clear vision of what success would look like. Success, they believed, was not measured by a typical "good project, but by achieving everything that was possible."

In the successful project, the crews would install their work only once at the right time and in the right sequence to make good on the promise their foremen made in the weekly production planning meeting. Hand-offs between trades would be without surprises and as promised. Crew members would have the right skills and equipment to do their job safely. Crews would be working from documents, which are clear, accurate, and correct. There would be no outstanding design questions or changes being considered when the crew began work in a particular area. The work area would be clean and unobstructed. The crew would be aware of hazards and would have taken appropriate measures to protect themselves and anyone else who might be in harm's way. The crew would understand the quality expectations and would have developed a work plan to produce nothing less than what was expected. The right material would be ready and close at hand without obstructing other workers. The crew would have been given adequate time to do their work correctly. The crew leaders would share problems as soon as they emerged and announce completion of their work as soon as they knew when that would likely be.

15.3 HOW CAN THIS BE DONE?

Value is created for owners/clients when their building is constructed or transformed. The planning to make that happen is necessary non-value-adding from the Lean perspective. Figure 15.1 shows the elements of integrated production management and their relationships, starting from the creation of value (1). The arrows show the flow of information from strategic planning (5), the most distant from value creation, and prerequisite for all other planning. Establishing project milestones is above the line to distinguish this work from production management. Steps 1, 2, 3, and 4 are the methods, processes, and practices for that, all of which are essential for delivering valuable, high-performing buildings in fact, as opposed to intent.

Figure 15.1 shows the continuous cycle of production management. If we want flawless execution, we must plan production very well. If we want to plan production effectively, we must do a very good job of scheduling production. If we want to schedule production, we need to know how well we are doing so we can continually improve performance. If we want to schedule production, we must also create a sound plan for what should happen in the master schedule. If we want meaningful milestones to steer the project, we must incorporate what we've learned in our continuous improvement process.

15.3.1 Execute Work to Produce Value

Value is created by executing only and exactly the work needed by downstream customers. To do this, all the required elements must be in place: Preceding work is properly installed; safety awareness is high and precautions are in place; space is clear; the right-sized, trained work crew is ready; the correct materials are at hand; proper equipment is available; information in the form of quality criteria is clear and understandable. The amount of work, the "batch," must match the size and composition of the crew; both must be adequate to support delivery of the project within the target cost.

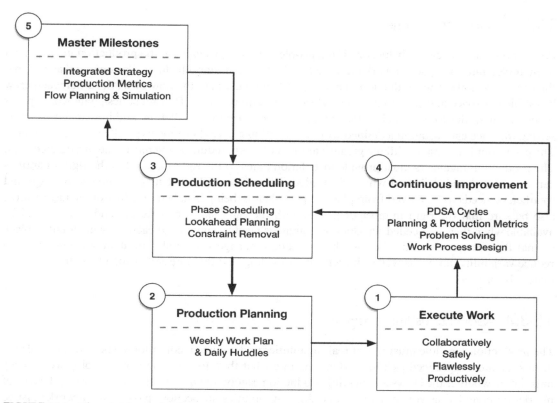

FIGURE 15.1   Blueprint for production management.

15.3.2   Production Planning

Safe, high-quality, and timely execution of work is only possible by planning production each and every day. The last people to plan work, in most cases the foremen, are called "last planners" in the Last Planner® System (LPS). Production supervisors and the last planners meet every day to report what their crews have accomplished and raise concerns about anything that might disrupt the flow of work. All the information they need is at their fingertips: up-to-date design documents from the architectural and engineering drawings, building information models, fabrication and shop drawings, clear quality and completion criteria, and installation instructions. The current work plan reflecting progress up to meeting time, material deliveries, inspections, and corrective work are reported and discussed in the daily check-in. Clarity and honesty are imperative. Even if a problem is caused by their company, the last planners must openly disclose the issues. Without accurate production information, the trades cannot plan their work to integrate smoothly with the work of others. Promising to "do my best" is inadequate because it does not permit work to be released. The only useful responses are "yes, I can get that done or "no, I don't have what I need, and this is why."

15.3.3 Production Scheduling

Production planning can only be done if the people planning production, the production management team, understands the pace of work required to achieve the master schedule milestones and agrees that they are achievable. The production management team must plan the flow of work so that each crew has sufficient space and time to install its work safely and properly. They must designate work spaces based on quantities to be installed, the equipment necessary for installation, and how much the crew setting the pace can install in a defined work area within a period of time, typically measured in days. The pacesetter is the trade with the greatest amount of work, requiring the most time in a defined area. The production managers must provide installation dates so the right material in the right quantities can be ordered and delivered when it's needed, but not so early that it is exposed to damage and weather, or needs to be moved from place to place before being installed. Production managers must also be sure that the right number of people with the right skills, experience, and training will be available when they are needed. Production managers need to know what each crew must put in place so that the next crew can do their work. Production managers must also think about how they can reduce variability in flow as well as buffer it by providing slightly more rather than less crew capacity, materials, or time.

15.3.4 Continuous Improvement

The production schedule must reflect real capabilities under current conditions. Those responsible for it must create a rapid feedback loop to incorporate what they are learning from ongoing work. They must know whether the crews are meeting production, safety, and quality objectives. This is the role of metrics for completing work as planned. Productivity, injuries, inspections passed, and rework must be measured. Work must be standardized to remove variations that mask root causes. The team managing production must look at the results to see problems that they wouldn't see otherwise, especially in the sequencing and pace of production. Once managers and production supervisors see their problems, they need a process to understand root causes. Asking "why?" five times is one. Swarming over a problem from different perspectives and expertise is another. Operations can be improved by doing First Run Studies of operations and asking crew members to redesign them as they watch a video of themselves at work. After each change, the team must accurately measure to determine whether the adjustment was effective and, armed with the new information, repeat the process.

15.3.5 Master Milestones

Master schedule milestones establish the targets for the production scheduling. Working backward from delivery, the team defines major deliverables for construction, fabrication, and procurement, permits, and design. All the project participants must be heard together. Inevitably, team members draw on their and others' past experience, especially when they intend to apply new methods. Ideas should be heard and different scenarios considered. Debate should be encouraged.

The master schedule is not planning at a granular level. The Critical Path Method, often used for documenting the master schedule, can be useful for understanding how work should be sequenced, provided the activities are not so detailed that the logic network becomes a maze of interdependencies.

Scheduling specific work months and years in advance without knowing what the actual opportunities and constraints will be is potentially damaging and certainly wasteful, at least with today's largely manual planning and scheduling tools. The master schedule is a living document that must follow production planning and scheduling, rather than dictate how the work will be performed.

15.4 REAL-LIFE EXAMPLE

From the beginning, the Temecula Valley Hospital (TVH) team understood that the project would not be business as usual. Bill Seed, then the Universal Health Services staff vice president of facilities, made it clear that the team he selected would have to be committed to Lean integrated project delivery (IPD). He was convinced that this approach could yield much better outcomes, as it had for Sutter Health and other hospital owners across the United States.

The first step for the new team was to learn what Lean production is and how it should be done in construction. Bill Seed encouraged them to listen to one of their team leaders, George Zettel, for whom Lean construction had become a mission. Bill also asked Kristen Hill, a licensed architect and Lean construction consultant, to work with the team during design. In the second year of the project, the team formed their own community of practice (CoP) to improve understanding and application of Lean practices. In particular, the Lean coaches and the CoP focused everyone's attention on eliminating waste in the three areas where 70 percent of time is lost: production, waiting, and movement.

What terms and vocabulary did the team learn and use? Bill Seed and the Lean construction experts provided a vocabulary for the IPD team to use in describing both the challenges they faced as well as their solutions. Designers and builders together learned at least five terms, as follows:

- *Conditions of satisfaction (CoS).* What would make the customer satisfied? Bill Seed initially described these as "measures of success" and later used "conditions of satisfaction," the more common Lean term. In the first meeting of all the IPD partners, the team heard what these were for Universal Health Services and discussed what they meant. They later learned that every customer needed to know their CoS and that everyone was a customer in a network of commitments.

  Waste. Anything that does not directly provide something the customer could use and is willing to pay for is waste. In this case, the customer was willing to pay for making the hospital facility and grounds functional in every respect. Design, estimating, planning, and project management were, at best, supporting activities and not value producing themselves. The Lean experts encouraged everyone to recognize the seven wastes in mass production first described by Taichi Ohno, the Toyota chief engineer credited with inventing the Toyota Production System (TPS) (Ohno, 1988). For the first time, managers, superintendents, and foremen learned that doing work before it was needed was overproduction and drove the other wastes of waiting, excess inventory, unnecessary movement of people and materials, unnecessary effort, rework, and overprocessing of information and other hand-offs. The team also learned that unused human potential was a very big eighth waste. The Lean coaches talked about how Taichi Ohno taught new employees to see waste by drawing a circle and asking them to stand inside of it for an entire day to observe and report the waste they saw. Team members learned that observation was the only way to really understand how and why waste was occurring (Liker, 2004).

- *Pull planning.* Team members learned how to involve everyone responsible for achieving a deliverable in planning every step starting with the last task required for reaching the objective and working back to the first one. Rather than a few people trying to do all the thinking and creating a Critical Path Method schedule in a computer, every voice could be heard, and a team with sticky notes could create a better plan.

- *Last Planner® System.* The Lean coaches explained that trade foremen were the last planners in a series and that they held the key to reducing the swings in production by work crews. These last planners knew what conditions were on the ground, whether they had clear and accurate information, the right people, the material specified, and whether the trade in front of them was likely to produce what their crew needed (Howell, 2011). The Lean coaches told the team that they needed to implement a different planning process called the Last Planner® System that empowered the last planners to plan the sequence of work for a phase, create weekly work plans of tasks ready for their crews, to look ahead to identify roadblocks in the weeks ahead, and meet every day to make whatever adjustments in their plan that were necessary to help their crews (Ballard, 1994).

- *Plan-Do-Study-Act.* The Lean coaches also explained how individuals and the team could continuously improve by following the method taught in 1950 to Japanese scientists and industry leaders by W. Edwards Deming. *Plan* is establishing the objectives and process to achieve a targeted improvement and thereby creating a more complete and accurate specification. *Do* is following the process to make a product while at the same time collecting data. *Study* is analyzing the actual results and comparing against the expectations set in the Plan step. *Act,* called *Adjust* by some, is determining corrective action to take for significant differences between planned and actual results based on root cause analysis (Latzko & Saunders, 1995).

15.4.1   How the Team Saw the Challenge

How did the team grapple with the challenges they saw? After some education, project leaders recognized twin challenges: make work flow and eliminate waste. Anything that did not contribute to delivering exactly the facility UHS specified, meeting quality requirements safely, on time, and within budget should not be done. The work to install material and system components should flow.

For most team members their new understanding of Lean ideas and the Last Planner® System created a platform they did not have before to frame and solve the challenge of workflow (Ballard, 2000). They had a vocabulary for taking control rather than responding to failures. The team could operate within the PDSA cycle, beginning with hearing the last planners' ideas for sequencing the work through clearly defined work zones on each level of the hospital. Rather than the general contractor's project managers and superintendents determining activity durations, the trade general foremen were able to do that in a way that would avoid the trade stacking, which almost always results in lower quality, safety hazards, and lower productivity.

The team now understood that they should spend time seeking out problems that could force work crews to stop-and-start or hopscotch from area to area, completely undermining flow. The team could begin to see waste that they had previously accepted as the way things have always been done. Once they saw people waiting, moving material unnecessarily, and redoing another person's work in the job site office, they could reduce or eliminate those wastes.

Team members also realized that they had to trust each other's intentions and abilities to solve problems for the good of the project. Team leaders committed to building trust between all team members, especially those installing the pieces of the building. This became a mission from the beginning to the end of the project.

15.4.2 Organization, Communication, and New Practices

Execute Work

Step 1. Communicate with and Listen to Those Producing Value Team leaders wanted new people and especially craft workers to realize that they would be provided with everything needed to produce quality work safely. That message was sent in a tangible way when the driveway and parking areas were paved as soon as possible. The very first structure erected was a 90 × 150 foot tensioned membrane tent large enough to hold 100 people seated at tables. Besides providing shade and protection from rain and the high winds that plague the area, the tent became a place for different trades to meet and get to know each other.

To improve working conditions during interior construction, trash chutes were installed and walk-behind powered floor vacuums were made available once interior construction had begun to make it easier to keep work areas clean.

After the electrical project manager's survey of electricians showed that they weren't aware of the project team's Lean approach, the project superintendent, Tom McCready, and project manager, Brent Nikolin, started inviting trade crews into the Big Room for lunch each Friday to discuss working conditions and listen to ideas for improvements.

Step 2. Make Accurate Information Available The team installed a wireless network as soon as the elevated decks were finished and modified rolling job boxes to serve as communication kiosks on each floor where crews were working. The boxes were outfitted with a computer and printer. Anyone who wanted was given access to use the kiosks and was able to review and print views and sheets of the latest virtual models and drawings. They could check and adjust the work plan, schedule inspections, and note damaged or incorrect work using cloud-based software. Supervisors, foremen, and crew leads could also see the status of requests for information (RFIs) and change orders, and read design bulletins without walking back to the job office and asking a project engineer for help. All of this information was available on iPads, which all of the partner companies and many of the fixed-price subcontractors issued to their foremen.

Step 3. Provide the Right Equipment and Tools The TVH team created a consolidated rental program, which allowed any supervisor or foreman to rent scissor lifts and forklifts from the inventory stored on site by the equipment rental company they had selected. They could place the order using a special iPad app, walk into the project office to pick up the key, and start using the equipment right away.

The project team agreed to pay for modification of the nut drivers the plumbers used to make it easier for them to change sizes. Time was saved for all trades by providing crews with right-angle drill motors whenever they worked in tight spaces. Cordless drills were provided to eliminate a major tripping hazard and the time lost stringing power cords.

The project superintendent designed and built plan tables for the scissor lifts so installers wouldn't have to hang drawings over the safety rails. The drywall contractor IPD partner, mounted chop saws on carts.

Step 4. Reduce Inventory and Pull Materials   The IPD partners' general foremen agreed that only small batches of materials could be moved into the building and had to be placed on rolling carts, with the exception of metal studs and wallboard. They also made the 30/30 rule, that all of the materials, tools, and information a crew was going to need for a specific task were to be located within 30 feet or 30 seconds of their activity. Fixed-price contractors were so impressed that they elected to fall in line with the program.

Step 5. Redesign Operations   The drywall contractor agreed to frame garage doors in the patient rooms to allow for the mechanical, electrical, and plumbing (MEP) trades to access larger areas and install pipe racks and duct work in the longest lengths possible. During design, the drywall contractor convinced skeptical team members that modeling every framing member would be worthwhile. Detailed modeling of the plumbing and drywall systems enabled the drywall contractor to agree early in production scheduling that the cast iron pipes for the drain, waste, and vent system could go before wall framing. This resulted in significant plumbing labor savings with little impact to drywall installation costs because both trade foremen could see where the pipes and metal studs were to be installed. The net result was reduced project cost.

All of the metal deck inserts and penetrations for the fire sprinkler trade partner were laid out by other team members due to the lack of experience using inserts. Fire sprinklers are typically a deferred approval, and they usually miss the opportunity to use inserts and are forced to drill the deck and use traditional anchorage. Time was saved and quality increased by having a single crew lay out all of the metal deck inserts and penetrations for MEP systems using a surveying total station reading building information modeling (BIM) data. The location of medical equipment backing in walls was modeled, which enabled the drywall team to make story poles for different room types so that carpenters would not have to find the dimensions and measure in each case.

Early in construction, the drywall contractor proposed prefabricating the ground-level exterior wall panels. Representing the wall framing with a BIM began immediately after the IPD team agreed, followed by coordination with structural steel and MEP. At that point, the drywall team broke the elevations up into panels of similar size to minimize how many different templates they would have to build. Once they had all the panels broken up into similar-sized pieces, the drywall modeler/detailers created spool sheets with all the callouts and detailed information regarding dimensions, stud size/gauge, dimensions for openings, and dimensions for specific stud layout for coordination issues. The drywall team then reviewed the spool sheets with the field to make sure that they had everything covered prior to construction to avoid any delays. Once the design was done and the material was ordered, the constructors marked out a lay-down area for incoming material and final products to be stored prior to erection, and began to build their framing jig table.

The success of the panel prefabrication gave the drywall team the confidence to prefabricate the complex cupola roof structure five stories above the ground. Because the roof trusses were prefabricated by a truss company from plans approved by the Office of Statewide Health Planning and

Development (OSHPD), the only additional design that had to be done was to make sure that the roof was reinforced for hoisting. After the staging area was cleared and the trusses were delivered, the drywall team began construction of the roof per the documents as if they were building it up on the roof level. The only difference was that it was much safer and the materials did not have to hoisted up five stories, which allowed the work to go much faster. Once completed, the drywall framing engineers came to the site and worked with the installers to find the most cost-effective design using materials that were on site to "sturdy up" the structure to allow it to come off the ground. A second part of the reinforcement design was the drywall team sitting with the riggers and engineers to make sure that their design incorporated "pick point" locations where the rigging could attach. Once this was completed and inspected, the plywood sheeting was added and the structure was ready to lift. The cupola roof was then installed exactly as planned, quickly and safely.

Step 6. Maintain Workflow   The team agreed that each trade would respect the production scheduling flow plan. Except for an occasional lapse, everyone kept their promise not to race ahead and to stay out of spaces designated for another trade. This was difficult and required discipline not to do what was best for themselves at the moment. The foremen soon realized that they could count on the time and space they needed to be productive without hurting other trades. The team members also agreed to leave their scissor lifts in the area that they just completed for the next trade so they were not constantly driving lifts past each other. This reduced waiting time at the man lift. Everyone was better off.

Production Planning

Step 1. Check Daily on Completions and Reasons for Failure   Each afternoon the level 1 and tower superintendents met with the general foremen and foremen in a stand-up check-in meeting lasting 10 to 30 minutes. The foremen would report on whether they had completed tasks they had promised. They would also describe how their crews were doing on work-in-progress. Foremen also reported on material deliveries, inspections they had scheduled and damaged or defective work. The inspectors-of-record (IORs), hired separately by the owner, also attended these meetings.

For example, the last planners used the daily check-in to coordinate metal decking above under-slab work. Because of an environmental issue that delayed the start of foundations, the structural steel was ready to be erected prior to completion of under-slab utilities. The delivery team decided to erect the steel and install metal decking ahead of the underground utilities. This created a potentially unsafe condition when welding studs to the deck and steel over the people below putting in piping and conduit. During this time, the foremen used the check-ins to communicate very explicitly the progress and locations of the welding overhead in the next 24 hours. In this way, they were able to coordinate the movement of materials and crews below. This allowed the last planners to make minor tweaks to the work plan to keep everyone safe and allow the crews below to take advantage of any increases in productivity of deck installation.

When their crews hadn't succeeded, the foremen were asked for a new completion date and to choose from a predefined list of reasons for failure to complete. Having this list allowed team leaders to see patterns indicating persistent issues. Completions, failure reasons, and adjustment were entered into the production planning software as foremen responded. Their Last Planner® System software, in turn, calculated plan percent complete (PPC), the percentage of committed tasks that had been completed. If the foremen had promised to complete 10 tasks and completed only 7 in a given week, the team's PPC score would be 70 percent. Everyone could see how reliable they and the team had been at doing what they said they would.

Because the software was cloud based, the updated production schedule was instantly available on kiosk computers and iPads.

Step 2. Check Daily for Constraints   After the foremen reported in, attention shifted to problems inhibiting workflow. Were there potential safety hazards? Did the foremen have a full crew? Did they have the right equipment for their crews to work safely and efficiently? Did they have all the information they needed, and was it clear and complete? Were materials being delivered? Was another trade's material or equipment in the way? Would prerequisite work be completed in time and correctly so that their crews could start?

Whoever was best able to solve a problem was asked to commit to take action. Problems that could not be resolved within the group of last planners were brought to project managers or project engineers in the Big Room. Solutions were communicated to affected trades by radio or in person as soon as they were decided.

One example was the discussion between the ductwork foreman, Ken, and fire protection foreman, Greg. Ken mentioned that Greg's sprinkler pipe arm-overs were in the way of his duct installation and it would hurt his productivity. Greg understood the issue as soon as Ken brought it up. He replied that it would be a very easy fix to pivot those arm-overs up out of the way, and he said he would have his journeyman do exactly that before 9 A.M. the next day.

Production Scheduling

Step 1. Workflow Planning   Five months after the IPD team formed and still very early in design, project superintendent Tom McCready asked the partner companies to name the general foremen (GFs) for the project and to promise that they would work all the way through completion. He also requested GFs' participation in a two-hour scheduling meeting each week.

At the first meeting, Tom introduced Chris O'Dwyer, the scheduling engineer, and asked the GFs to work with him to develop their own schedule. Chris suggested that they use scheduling software intended for planning the flow of work by location. The GF scheduling team met each week for fourteen months, with Chris as the facilitator, asking questions and building the production schedule that created a predictable workflow, acceptable to the major trade partners.

Initially, the GF schedule mirrored the master schedule. As they continued to meet each week, these men, one level above the last planners, began to think harder about what their work crews could do if they had clear space to work and everything they needed. The GFs also learned more about what they would be building as design developed and BIM progressed. The GFs designated areas on each floor based on the quantities of material that had to be installed. This allowed them to plan prefabrication and size their crews.

As the GFs worked together, they saw that they could trust each other to abide by the agreements they were making. Because of that, the GFs realized they didn't need to add a time buffer for each activity. Ten-day durations shrunk to seven days in each of the four areas on the four floors of the bed tower. Because of the increasing trust, the GFs became willing to share space with the following trade toward the end of their designated time. The overall duration shrank further still, ultimately by six months overall. The trade partners owned the construction schedule. Because of their location-based approach, the GFs had created the flow plan similar to the one created by the Palo Alto Medical Foundation (PAMF) Sunnyvale Center project team, shown in Figure 15.2, by clearly defining areas, the direction of work, and agreeing that no trade could advance its material, equipment, or people into the space designated for another trade without prior agreement.

Step 2. Pull Planning as Customers and Suppliers The Lean coaches had helped everyone see themselves as a customer needing work completed satisfactorily by someone working ahead of them, and also as a supplier providing what someone else required. The GFs and foremen came to understand that they needed to be clear about their conditions of satisfaction. Every team member was taught that the biggest waste was not delivering exactly what their customers needed when they needed it. CoP facilitators taught and mentored the skills of negotiating and making reliable promises.

The GFs and foremen saw that pull planning was the way to plan the hand-offs between crews so that suppliers produced what customers needed. They realized that pulling was the alternative to command and control by the general contractor. The IPD partner GFs and foremen planned each large phase of work for particular production targets by pulling back from one or more milestones in the master schedule. These planning sessions typically lasted about three hours in the beginning and progressively less as the team became better at the process. They learned to describe what their crews needed in the top half "I Get" space, and what their crews would deliver in the bottom "I Give" space on special sticky notes and put them in the right order on a whiteboard. Figure 15.3 shows a DPR "I Get | I Give" sticky note and a foreman pull-planning session.

Every Tuesday, the GFs and foremen met before the Big Room meeting to pull new tasks into week 6 in their lookahead plan window. They would also better define and break into smaller chunks the work closer to the present as they could see conditions on the job, including confirmed material deliveries and labor availability. Once this group of last planners was satisfied, each foreman entered and edited their own tasks into the cloud-based production planning application, which made their plan visible to everyone.

Step 3. Look Ahead to Identify and Remove Constraints Later on that same Tuesday morning, the last planners would reconvene in the center of the large Big Room meeting area, surrounded by IPD project managers, including the owner, to go through their lookahead plan. First, they would confirm completions and failures during the previous week and report what they expected to accomplish for the current week. This led to a discussion of problems jeopardizing tasks in their six-week planning window. The design and construction managers surrounding the last planners heard their concerns as they were expressed. There was no time lost in communicating them.

The general contracting team in the trailer realized that their job was to remove as many constraints as early as possible to enable predicable workflow. To that end, they tracked tasks anticipated (TA) and tasks made ready (TMR) and reported on these in the monthly update. TA was calculated based on how many of the tasks were identified in the lookahead schedule one or two weeks ahead of when last planners created the weekly work plan showing the tasks they committed to complete in the coming week. If only five tasks were identified one or two weeks before and nine were put into the weekly work plan, the TA score would be 55.6 percent (5 divided by 9). The other four were not anticipated and only added as the weekly work plan was being created. The TA metric indicated how well the

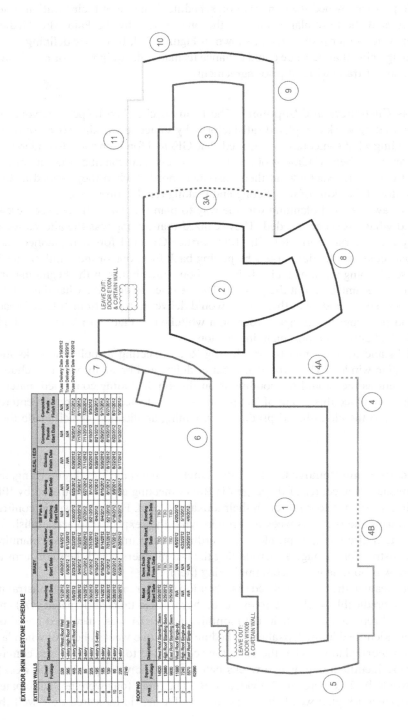

PAMF SUNNYVALE – EXTERIOR SKIN LOGISTICS & MILESTONE SCHEDULE
Rev. 5 - 5.8.2012

FIGURE 15.2 PAMF Sunnyvale Center exterior flow plan. Courtesy of Sutter Health; courtesy of DPR Construction.

FIGURE 15.3   I Get | I Give sticky note for pull planning. © DPR Construction.

team was was able to anticipate tasks during lookahead planning. TMR was the number of tasks in the lookahead schedule that were made ready, that is, all constraints removed so last planners could commit to complete them and be put into the weekly work plan. If only four of the nine tasks originally identified in the lookahead schedule were ready for commitments, the TMR score would be 44.4 percent (4 divided by 9). The TMR metric indicated how well the team was doing at making work ready to be performed (Hamzeh, Ballard, & Tommelein, 2008). TMR allowed the general contractor team to see how their performance impacted trade crews. Figure 15.4 illustrates why the calculation of Last Planner® System metrics for TA, PPC, and TMR together, rather than only PPC, is necessary.

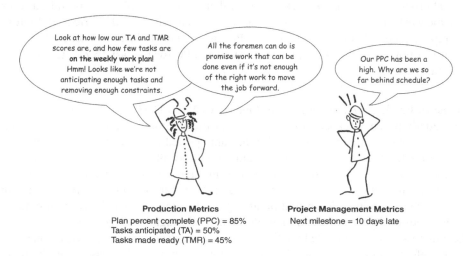

FIGURE 15.4   Using Last Planner® System metrics together. Courtesy of CDReed.

Experienced Last Planner practitioners know that PPC can be high if the last planners are able to commit to completing very few tasks and get them done as promised. These can be secondary tasks, meaning they don't contribute to getting high-priority work done. PPC is high, so managers think things are going well and report that the project is on schedule. Low TA and TMR scores would indicate that not enough tasks are being anticipated by the last planners and not enough tasks are being made ready by the design and construction project and field managers (project superintendents and trade project managers and general superintendents). This results in two dysfunctions. First, the project falls into crisis management because materials, equipment, and labor cannot be brought to the jobsite in time if work is planned just one week ahead. Second, last planners are unable to commit to doing priority work because it is not ready, and no one can predict when it will be. A high-performing team must score well on all three metrics in order to be assured that they are progressing enough work, in the right sequence, and with the right priorities to achieve the project schedule. The TVH team used all three Last Planner® System metrics along with productivity metrics to see the big picture.

Step 4. Commit to Deliver   After this discussion, the GFs and foremen would commit to the tasks their crews would complete the coming week. The meeting would adjourn for lunch, followed by a "gemba walk" where managers walked out and through the building with the last planners to see conditions for themselves.

Continuous Improvement

The TVH leaders realized that they could continually improve quality and performance by observing and measuring the work and the planning process for the work to understand the current state as the first step to improving it.

Step 1. Measure Plan Percent Complete   Both in the weekly Big Room meeting and the daily check-in, TVH production managers tracked the number of tasks completed as promised compared to the total for the week. Their production planning software calculated and charted PPC automatically, as foremen reported completions. The team saw PPC as the best indicator of schedule progress, and as an indicator of how well they were planning. Average PPC for the last 13 months of construction was 83 percent.

Step 2. Measure Productivity   Beginning in March 2012, the TVH team began to ask IPD partners to report productivity. They realized that they should use the metrics each trade saw as important and compare them with manpower counts for the same period to understand batch and crew sizes. Comparing output for trades working in the same area also allowed managers and supervisors to understand the impact of changing sequences and asking trades to work together in the same space.

Analyzing productivity enabled superintendents and general foremen to make adjustments. They saw and addressed a drop in underground plumbing productivity. They reduced plumbing batch size to get plumbers out of the way of electricians doing in-wall rough-ins. Tracking productivity and discussing the numbers as a team enabled the GFs to see that the decision to install cast-iron drain, waste, and vent lines before framing had paid off. The result was a dramatic increase in productivity for the plumbers without increasing labor costs for framing. The general foremen discovered the optimal crew size for installing cast-iron piping. Tracking productivity enabled the plumbing GF to convince the IPD managers to authorize his crew bracing the med-gas copper lines to work overtime. The numbers had shown that they were far more productive when they had stayed late and were able to do this work without other trades around. Productivity tracking and analysis also proved what the superintendents and GFs suspected, that productivity decreases dramatically, dipping below estimates, during the end of the year holidays.

DPR Construction's Southern California Drywall Group has continued to do First Run Studies, and extended the use of productivity dashboards in combination with the Last Planner® System. Just as they learned to do along with other IPD partners on the Temecula Valley Hospital, the Drywall Group managers produce a monthly report such as the one shown in Figure 15.5.

Step 3. First Run Studies The TVH team adopted the PDSA improvement cycle, putting it on the masthead of their monthly update. The CoP talked about "Ohno circles," the practice Taiichi Ohno had of asking people to observe and note what was happening around them as they stood in the circle he would draw on the shop floor. Video studies were a logical next step. Production managers realized they could use a movie camera as eyes for many people. They began filming and redesigning operations with work crews to help improve productivity. As soon as the elevated decks were placed, they shot video of the drywall crew framing the exterior wall. The resistance to "Big Brother," which some supervisors feared, didn't materialize because the workers knew "this job was different." The drywall supervisor and the crew watched the video together and saw how they could accomplish more with the same effort. The difference was dramatic. Exterior wall framing productivity increased 67 percent and rework decreased 100 percent on the third level after the changes made because of the second level First Run Study. The production team ended up doing 100 First Run Studies in all.

Step 4. Ask Five Whys to Understand Root Causes The TVH team learned to do root cause analysis by asking "why?" five times for what they saw as serious or persistent problems (Ohno, 1988). An example was discovering that the root cause behind having to reinstall humidifier piping was that the fire-proofing had been sprayed on too thick in some places. This, coupled with construction tolerances, resulted in what seemed at first to be piping installed at the wrong elevation. The TVH team learned that the prerequisite for asking five whys is getting the people involved in the same phone call or in the room together. Solving the humidifier piping bust required bringing the HVAC piping general foreman and the fire-proofing foreman together. Another "five whys" uncovered the root cause of doors laid out incorrectly. The root cause was that the 3-D model had not been updated to reflect changes made to the door schedule. The door manufacturer did not know about the BIM, which was used to lay out the door openings. In this case, it was the carpenter doing the layout, BIM detailers, and the door manufacturer who had to be brought together.

Step 5. Swarm to Improve Without any expertise in collaboration technology or cognitive science, the TVH team adopted an approach exemplifying "swarming" for making improvements in productivity and attacking problems. These sessions were less formal, often coming on the spur of the moment. Swarming was possible because the right people were available. Besides having data in the form of metrics, they had strong intuitions from years of experience. They also knew that there would be no

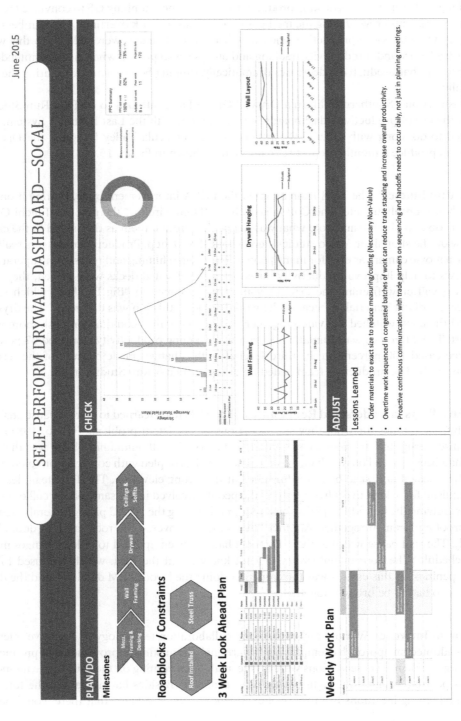

FIGURE 15.5 DPR Construction Southern California Self-Perform Drywall Group production planning and production dashboard. © DPR Construction, by DPR SPW Drywall Group–SoCal.

blaming, only a genuine effort to make something better, which would benefit them all as business partners.

Master Milestones

Step 1. Leverage Team Experience The very first request from Bill Seed after the mechanical, electrical, plumbing, and fire (MEPF) protection trade partners had joined the team was for them to estimate the cost of the project as they understood it at that time, with almost no design documents. Project superintendent Tom McCready knew that he had to provide the estimators with his best guess at how much time would be needed for construction. He had never built Lean before, and didn't want to allow too much time, which would drive up general conditions cost, nor did he want to be too optimistic. Tom decided that he would plan for all the work to happen as it should, as if there were none of the problems he had dealt with his entire career. How long would it take if everyone had what they needed and could work in the right sequence with no design or physical obstacles in their way? Then he added back a time buffer for surprises.

Step 2. Milestones Are the Deliverable Next, Tom solicited everyone's feedback. On this project he had an entire team of experts available. Tom organized his plan by phases and listed the processes for constructing and assembling the major elements and various subsystems based on his experience building many other hospitals. Tom discussed fabrication and procurement lead times, activity durations, and sequencing with the IPD trade contractor managers. He tested the logic using the Critical Path Method software he had mastered. Unlike on all of his previous hospital projects, he had learned that he would not be required to continue developing the plan to the level of describing the operations within each process so that it could be used as the primary tool for controlling the schedule for the project. No outside consultant would be reviewing the schedule each month to compare progress to something he'd planned many months earlier. Tom and the team had learned that they could control outcomes far more effectively by applying the Last Planner® System to plan and manage production (Ballard & Howell, 2003).

Tom, in particular, realized that the Last Planner® System made a lot more sense, especially the part about engaging the foremen to plan production to meet the master schedule milestones progressively, as they got close enough in time to see conditions on the ground. In fact, Tom had decided that his next move would be to challenge the IPD trade partner general foremen to create a better plan for the project than he had. He also knew that he would report progress at the phase and system level each month. The end of each summary bar in the Gantt chart was equivalent to a project milestone (Figure 15.5). He was satisfied with his effort and the fact that everyone on the team agreed that it was a reasonable schedule. Figure 15.5 shows a similar milestone schedule, which DPR superintendent Brian O'Kelly developed with IPD partners for the PAMF Sunnyvale Center project, which became an exhibit in the integrated form of agreement (IFOA) and the contract schedule.

Step 3. Challenge the Builders to Plan Flow At Tom McCready's request, the IPD trade partner companies committed individual general foremen for the project and agreed that they would be available a minimum of two hours each week to plan for Lean workflow. In fact, the GFs did meet each week for 14 months and built their own schedule based on crew flow. By the time foundation work began, the original 27-month master schedule had become 21 months as Tom incorporated the GFs' sequences and durations. Of no small consequence was the significant savings in general conditions because of the half-year schedule reduction.

Step 4. Communicate Master Milestones Every month, the TVH project management team reported progress against the master schedule baseline activities and milestones in the "Temecula Valley Hospital Monthly Update." The update looked and read like a newsletter. Anyone who read it could understand schedule and budget status and see the improvements and innovations the team had implemented.

15.5 INTERCONNECTIONS

The IPD contract changes the game from a zero sum choice between schedule and productivity to mutual gain where productivity delivers savings, and failure to collaborate results in higher costs and potential lost profit for all of the IPD partners. Traditionally, specialty contractors whose profits depend on the productivity of labor had no choice but to try to create the best workflow they could for their own crews. Because they were acting on their own, this often came at the expense of other trades and the project schedule. The IPD framework provides all partners a financial incentive to communicate effectively to resolve issues. There is no longer a reason for the general contractor to push schedule at the expense of trade contractor labor productivity.

Design documents, quality and performance specifications, budget, schedule, and fabrication and installation instructions can be integrated and shared among partners so that team members do not need to re-create them and run the risk of working with out-of-date or wrong information. Foremen from different trades can access the same set of documents and 3-D models in the field their project managers are seeing in the job and home office. Fabricators can see Last Planner commitments, deliveries, roadblocks, and inspections for the next week in real time.

The very best way of planning to eliminate waste while installing systems right the first time is by builders constructing virtually before they go into the field to do the physical work. Not only can they verify that components will fit, but builders can test their installation methods and sequence by simulating them. They no longer have to rely solely on judgment based on past experience.

Co-locating so that the team really works together gives leaders of the multidisciplinary/cross-functional teams the opportunity to coordinate their teams' work so that everyone is aligned. Co-location, combined with effective planning and coordination of information flow, dramatically reduces waiting and errors that come when people make assumptions when their questions are not answered in time. Involving constructors and operators in design much earlier than ever before challenges a team to think about how the design will be built and operated. These stakeholders inevitably raise concerns and ask questions that should be addressed early rather than during construction and at turnover.

It's impossible to establish and manage Lean production without following Deming's PDSA method for continuous improvement. This requires setting targets and measuring performance. PPC, TA, and TMR measure how well a team is planning, making work ready, and executing it. The quantity of material installed by a crew within a period of time indicates how productive they have been.

12-09-2010 | **PAMF 301 Old San Francisco Road Sunnyvale IFOA Schedule** | **Page 1 of 1**

Activity ID	Activity Name	OD	Start	Finish
PAMF Sunnyvale		788d	9-24-10	11-8-13
A1000	60 Day appeal period	50d	9-24-10*	12-6-10
Pre-Construction		259d	10-14-10	10-25-11
Design		136d	11-29-10	6-14-11
A1050	Shoring design	10d	11-29-10	12-10-10
A1480	Building Package Design	130d	12-7-10	6-14-11
Permitting		168d	10-14-10	6-16-11
Estimating and Buyout		159d	11-29-10	7-18-11
Materials Procurement		216d	12-16-10	10-25-11
Construction		752d	11-15-10	11-8-13
A1080	Remove materials from building at 301	8d	11-15-10*	11-24-10
A1710	Start Abatement	0d	12-7-10*	12-7-10
A1600	Start Demo and mass excavation	0d	1-3-11*	1-3-11
Site prep, excavation and shoring		187d	12-7-10	9-2-11
Structure		263d	3-25-11	4-9-12
Buildout		375d	12-2-11	5-29-13
Skin		154d	4-10-12	11-14-12
Sitework		44d	3-28-13	5-29-13
401 Parking structure and site		132d	11-16-12	5-29-13
Startup & Commissioning		66d	2-26-13	5-29-13
Occupancy		150d	4-10-13	11-8-13
A1730	IT networking/telecom	15d	4-10-13	4-30-13
A1720	Rad Onc and Diagnostic Imaging Equipmen Installationt	43d	5-1-13	7-1-13
A1520	Substantial Completion	0d	5-30-13	5-30-13
A1640	Punchlist	20d	5-30-13	6-26-13
A1630	Final Completion	0d	6-27-13	6-27-13
A1500	Furniture installation	39d	6-17-13	8-9-13
A1760	Clinical Equipment installation	68d	7-29-13	10-31-13
A1750	IT Desktop//telecom	15d	8-12-13	8-30-13
A1770	Bio med Check	10d	10-28-13	11-8-13

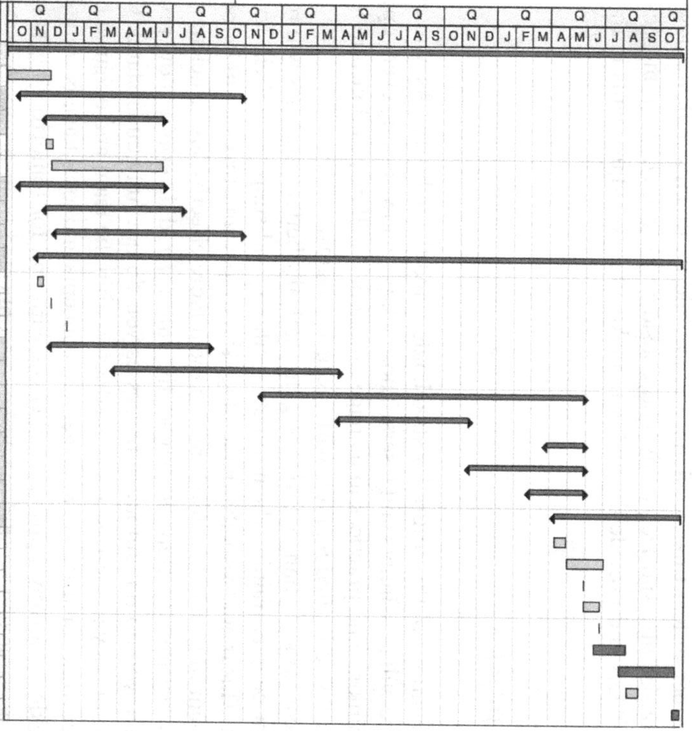

FIGURE 15.6 PAMF Sunnyvale Center milestone schedule. Courtesy of Sutter Health; courtesy of DPR Construction.

The number of inspections passed the first time, along with the amount of rework, are indicators of quality. The number of lost time accidents and first aid treatments indicate how safely crews are working. Taken consistently and frequently, and considered together, these metrics are the basis of the Study step in PDSA and provide the ability to make effective adjustments in the Act/Adjust step of that cycle.

15.6 REFLECTION

What are the implications of this future for stakeholders and the industry? As with any economic and social change, there will be winners and losers. Those who are now seeking every opportunity to learn and apply Lean thinking and practices to manage production will be positioned to lead in this future; those who don't will find limited markets with meager margins.

People are the agents of change. Producing dramatically better outcomes will require investment in finding the right people, training them in new methods, and supporting their development as production managers. This is not to say that technology is unimportant. Investment in the right technology can be an investment in people because it makes them more effective and efficient. But you can't just buy results. You must first develop the people, the culture, and determine how processes should be optimized. In Lean, you do not automate processes you do not fully understand. Once you do, it is possible to use technology to increase connectedness within the team and enhance collaboration.

Many of the current construction management concepts and tools have become obsolete; they are insufficient to tackle tomorrow's project challenges with and for tomorrow's generation (Koskela & Howell, 2002). The same can be said for design disconnected from production. People will need to learn and practice new methods and skills. The primary role of integrated team leaders will be to create and maintain alignment, not to think for the rest of the team. Superintendents will be rewarded not for being crisis managers, but for how much they engaged their trade partners in creating and executing a successful work plan to deliver a flawless product safely, on budget, and on schedule. Individual craftsmen and women will need to be as proficient in mapping and designing work processes as they are in executing work. Everyone will have access to the latest information, work plans, and videos showing best methods through information appliances like tablets, glasses, and so on.

The smartest teams will deliver the best results, not the teams who waste time and effort every step of the way. It will be clear to all that the workers making the product off site and on site hold the keys to value. Project managers and superintendents will see success in making the crews as safe and productive as possible while installing their work flawlessly.

15.7 SUMMARY

In a traditional project, the schedule is imposed by the contractor with limited feedback from trades and production planning, if it exists, is done by trades independently. When integrating a project, both production scheduling and production planning are joint responsibilities of the entire team. Milestones become team deliverables, and work is undertaken only if it is pulled from a milestone.

The key steps for managing production that supports the delivery of value are:

1. Execute work to produce value, that is, work that is needed by downstream customers.
2. The people (usually foremen) responsible for executing work jointly and continuously plan the work using the Last Planner® System.
3. The production management team must arrange and pace work to meet the master schedule.
4. Once production is under way, analyze work processes and outcomes through PDSA cycles and team problem solving to understand root causes and improvements.
5. Compare progress against project milestones, which become deliverables for the project management and production teams.

REFERENCES

Ballard, G. (1994, April 22–24). *The Last Planner.* Monterey, CA: Northern California Construction Institute. Retrieved from http://www.leanconstruction.org.

Ballard, G. (2000). *The Last Planner System of production control.* PhD thesis submitted to the Faculty of Engineering, University of Birmingham, Birmingham, UK.

Ballard, G., & Howell, G. A. (2003, August). An update on Last Planner. *Proceedings of the 11th Annual Conference of the International Group for Lean Construction,* Blacksburg, VA.

Hamzeh, F., Ballard, G., & Tommelein, I. (2008). Improving construction work flow: The connective role of lookahead planning. *Proceedings of the 16th Annual Conference of the International Group for Lean Construction* (pp. 635–646). Manchester, UK.

Howell, G. (2011). *A conceptual history of LCI—plus some thoughts on the rationale and history of "Lean Construction."* Lean Construction History. Retrieved December 31, 2013, from http://www.leanconstruction.org/about-us/history/.

Koskela, L., & Howell, G. (2002). The underlying theory of project management is obsolete. *Proceedings of the PMI Research Conference 2002* (pp. 293–301). Project Management Institute, Newtown Square, PA.

Latzko, W. J., & Saunders, D. M. (1995). *Four days with Dr. Deming.* New York, NY: Addison-Wesley.

Liker, J. K. (2004). *The Toyota way.* New York, NY: McGraw-Hill.

Ohno, T. (1988). *Toyota Production System beyond-large scale production.* New York, NY: Productivity Press, 1988.

CHAPTER 16

Avoiding the Pitfalls of Traditional Contracts

"Never interrupt your enemy when he is making a mistake."

—Napoleon Bonaparte

Structure matters. This is an essential premise of the Simple Framework. How you structure organizations, how you structure information, and how you structure work significantly affect outcome. The integrated project delivery (IPD) agreement, in the Simple Framework, is a supportive structure that allows and encourages effective implementation of each Simple Framework element. The IPD agreement, as described in the next chapter, is fundamentally different from traditional design and construction contracts.

But many organizations struggle with changing their procurement systems. Adopting full IPD requires rethinking their business and contract models—in addition to behavioral change. Some shy from confronting the changes required by rationalizing that they can get "enough" behavioral change under traditional contracting approaches. Others rationalize the status quo by blaming resistance from their procurement, legal, or executive groups. While considerable improvements to behavior and process can be made under traditional contracting structures, holding on to traditional forms creates an inconsistency between structure and behavior that inherently limits performance. And even if you obtain the desired behaviors, the dissonance between contract requirements and the achieved behaviors *increases* the parties' legal risks.

Another argument is that the IPD structure can be mimicked using traditional contracts that are interlinked through an addendum or rider. Although this approach, and the approach of custom interlocked contracts, is theoretically possible, in practice it is very difficult to rationalize the different terms among the different contracts. Inconsistency among the contracts is almost guaranteed. Making the addendum or rider approaches work requires exceptional effort and legal acumen. In practice, the single IPD contract approach is simpler, clearer, and more effective.

The next chapter discusses the structure of a full IPD agreement, including alternative business models and contract forms. Before exploring how an IPD contract should be created, it is useful to examine at least a few reasons why the contractual status quo is not appropriate for an IPD project. It takes effort to create an IPD agreement, but the advantages are well worth the effort.

16.1   TRADITIONAL CONTRACTS CREATE AN INHERENTLY ANTAGONISTIC ENVIRONMENT

Lump-sum and guaranteed maximum price (GMP) contracting create conflicting interests between owner, designer, and builder. When bidding a lump-sum job, the contractor must interpret the contract documents in the least expensive way possible. If work is omitted from the plans, it must be omitted from the bid. Once the project has begun, the contractor is encouraged, through the change order process, to discover and document each design error or omission. Knowing that the contractor may bid and manage the project aggressively, the designer will attempt to defensively overdesign and transfer design risk to the contractor through specifications requiring extensive verification, coordination, and contract interpretation clauses. None of this is in the owner's best interest.

The situation under GMP contracting is quite similar. For example, under a GMP with preconstruction services and shared savings (a modern GMP approach), the parties can be driven into contention. Because the contractor cannot know whether costs will remain under the GMP, it must issue a claim notice because it may need to include this claim to support a change order increasing the GMP. Often, this will lead the contractor to contend that the construction documents are incomplete or flawed. In response, the owner will contend that the change is within the original scope or was implied, and the designer will respond defensively to allegations of a flawed design. The upshot is that fear of exceeding the GMP drives antagonistic behavior among all the parties.

Consider also the approach taken in most contracts to change issues. If for example, a differing site condition occurs, what do traditional contracts require? Almost uniformly, they instruct the contractor to provide immediate notice of the situation and warn that failure to do so results in a loss of any claim for additional cost or time. A high-performance approach, in contrast, would require (either contractually or through the business model) that the parties focus on the solution to the problem rather than focus on claims procedures.

Simply stated, traditional contracting is focused on assigning responsibility rather than facilitating solutions.

16.2   TRADITIONAL CONTRACTS ARE BASED ON A PIECEWORK BUSINESS MODEL

Profit, in most design and construction contracts, is based on the number of units sold. For example, if a percentage of a designer's billing rate is profit, the designer makes more profit if he or she can sell more hours. This is even more true in construction because the builder makes a percentage markup on labor, materials, and equipment. As before, the more units sold, the higher the profit. Thus, inefficiency is incentivized, because this produces the highest profit.

This holds true even in lump-sum bidding. The designer has no incentive to reduce design effort in coordination with the contractor (who may not have been selected until the design is effectively completed). The contractor has no incentive to suggest a more efficient design solution. Moreover, the designers and contractors create their lump-sum bids based on projected units with a markup for profit. The suggestion that a competitive bid will "squeeze" out the inefficiency overlooks that a competitive bid is not the lowest price at which work can be performed. It is the highest bid that did not lose the job. And most construction markets have limited bidders, and thus resemble an oligopoly more than an "economically efficient" market.

Efficiency requires minimizing the amount of labor, material, and equipment, which is difficult to achieve if improving profit is based on increasing the units sold.

16.3   TRADITIONAL CONTRACTS RIGIDLY DIVIDE WORK BASED ON TRADITIONAL ROLES

Traditional contracts carefully segregate participants' roles and responsibilities. This has created a self-reinforcing dynamic because case law, insurance coverage, and professional regulations assume contractual segregation and circularly create a reason for contracts to segregate roles, which reflects back into insurance, professional regulation, and liability decisions. Two examples from the General Conditions for Construction (AIA A201) issued by the American Institute of Architects reflect this traditional segregation.[1]

> The Contractor shall be solely responsible for, and have control over, construction means, methods, techniques, sequences and procedures and for coordinating all portions of the Work under the Contract, unless the Contract Documents give other specific instructions concerning these matters (A201, §3.3.1).
>
> The Architect will not be responsible for the Contractor's failure to perform the Work in accordance with the requirements of the Contract Documents. The Architect will not have control over or charge of and will not be responsible for acts or omissions of the Contractor, Subcontractors, or their agents or employees, or any other persons or entities performing portions of the Work (AIA A201, §4.23).

Similarly, the contractor is not responsible for design. Thus, because neither architect nor contractor is responsible for overall project performance, there is no incentive— indeed there is a liability disincentive—to looking over the other's shoulder to make suggestions or spot errors. IPD reduces these disincentives by encouraging communication and collaboration, tying compensation to project outcome, and waiving liability among the parties.

How I Came to IPD

In the early 2000s, I began to investigate a new technology—object-oriented computer-aided design (CAD)—which was a precursor to building information modeling (BIM) and then became involved with the National Building Information Modeling Standard. It was an exciting time with promises of continual reuse of a common data library to accomplish multiple uses. Design, estimating, logistics, construction simulation, functional simulation, and building automation systems would dip into this single data stream using different tools to display views of the data appropriate to the user's needs. Beautiful as it was, there were a few problems with this utopian vision. The data standards necessary for full interoperability did not exist. The software of the time hinted at these possibilities, but could not deliver. But, most importantly, the structure of the design and construction industry did not support creating and using information in this integrated way.

If data are going to be used for costing, scheduling, or construction simulation, the right data, assembled in the right way, need to be in the data library. The users—trade and general contractors—need to specify what information they will need and how it needs to be incorporated into the models to allow extraction. Facility managers and operators need to be similarly involved in the creation of this data library. But in the standard design and construction model of the time, the design would be completed before the trades were even identified, making it impossible to have the data users (trades and contractors) participate in the data design and development. In fact, because of a sharp divide between design and construction, designers were (and still are) discouraged from considering constructability and contractors were discouraged from being involved in design. The upshot is that we had an exciting technology whose full promise depended on behaviors we did not allow. Worse yet, I had helped build, or support, the existing project delivery models.

This led to two converging realizations. The technology didn't care about how projects had been delivered in the past. It was going to move forward, and yelling "stop" wasn't going to work (and wouldn't be any fun). Instead, we needed to develop business, operational, and contract models that allowed BIM to flourish. The other realization was that after more than two decades as a construction lawyer, my daily work was spent solving the exact same type of problems that I had been struggling with throughout my entire career. Despite lots of hard work from lots of smart people, we weren't improving. It was 2006 and time to hit the reset button.

And that is how I came to integrated project delivery.

Howard Ashcraft

16.4   TRADITIONAL CONTRACTS CONSTRAIN COMMUNICATION TO SPECIFIC AND INEFFICIENT PATHS

Again, AIA A201 provides a convenient example:

> Except as otherwise provided in the Contract Documents or when direct communications have been specially authorized, the Owner and Contractor shall endeavor to communicate with each other through the Architect about matters arising out of or relating to the Contract. Communications by and with the Architect's consultants shall be through the Architect. Communications by and with Subcontractors and material suppliers shall be through the Contractor. Communications by and with separate contractors shall be through the Owner. (AIA A201, 4.2.4)

Traditional construction contracts also limit participants' reliance on information developed or provided by others. For example, standard language in the Engineers Joint Contract Documents Committee (EJCDC) takes a very conservative approach toward electronic information prohibiting reliance on the exchanged electronic information.[2]

More recent contract documents from the AIA[3] and ConsensusDOCS[4] have improved communication flows by permitting reliance under specified circumstances, although they have not eliminated concern that expanded reliance might result in expanded liability. Thus, they create pathways and standards but, by themselves, do not eliminate disincentives to communication.

IPD strives to shorten communication paths to encourage rapid information exchange, reduce information loss or misunderstanding, and stimulate innovation and creativity. This requires communication flows unburdened by the limited and highly structured paths of traditional contract documents and liability concerns.

16.5   TRADITIONAL CONTRACTS REWARD INDIVIDUAL, NOT GROUP, PERFORMANCE

Traditional construction contracts optimize individual performance. For example, a lump-sum plumbing subcontractor can increase individual productivity (and hence profitability) if its installation precedes the dry mechanical or life safety subcontractors—even though the dry mechanical should have first access to space because of duct size constraints.

Similarly, a contractor has little financial interest in pointing out design errors during preconstruction or bidding if they would support change orders after award. And designers have little incentive to consider the constructability of their designs because doing so could increase their liability exposure and they do not share in any productivity increases. Thus, the compensation terms of traditional contracting breed competition rather than cooperation.

The fracturing of a project into individual, competing contracts creates another source of antagonism and inefficiency. If each contract is an island, it can complain of interference from any of the other components. A classic example of this issue arose in a university laboratory that had significant dimensional conflicts in the interstitial space above the corridors. Each trade complained that the designer left insufficient space to install the necessary mechanical and electrical systems and complained of interference (lack of access to the worksite and to the interstitial space) from the other trades. The designer claimed that the space could have been coordinated, had the construction manager done a competent job. The construction manager blamed both the designers and its trades. By focusing on individual performance, rather than optimizing the project, the parties created (or at least severely exacerbated) a problem that would not have existed if they were jointly responsible for the whole.

And as noted earlier, project inefficiency may actually increase the parties' profit because profit is earned on the amount of labor, material, and equipment. The owner ultimately bears the cost of the inefficiency created by individual incentivization.

16.6   COLLABORATION WITHOUT AN IPD AGREEMENT CAN INCREASE RISK

Some projects have been successfully delivered by "pretending" to be in an IPD contract even though they were not. These projects are cited as evidence that contractual change is not necessary. But this logic has three major flaws.

The first flaw is that the absence of a catastrophe is used to show that the project was successful. Almost any project will go well if nothing goes wrong, but the test is whether the project can survive adverse conditions. In looking at these projects, you need to ask whether they were tested such that the parties had an incentive to resort to insular, selfish behavior that was greater than the incentives of maintaining relationships and future work. In these situations, a properly crafted IPD agreement will hold the team together, whereas a traditional contract will allow—and even encourage—the team to fall apart. As noted in a recent study, "The IPD teams exhibited resilience in recovering from cost or schedule impacts through collaboration that broke the silos typically found in traditional delivery" (Cheng, et al., 2016, p. 37).

The second flaw is that no one asks whether the project would have been more successful under an IPD agreement. In other words, would IPD have made a good project better? Data from a 2016 study found that the ratio of best to typical projects was dramatically better in IPD projects than the more traditional projects they studied (Mace, et al., 2016).

The final flaw is the assumption that everything will be OK, or at least won't be worse, if the project falls apart. But there is a serious problem if behaviors don't match the contract language. For example, if a project team is working on issues collaboratively, they are likely not giving contractually required notices. Yet if they haven't, they will have waived their rights under the agreement—which in a conflict they may need. And if the parties process changes without the notice or other procedural requirements, then the court may find that these requirements have been waived by conduct. If the contractually required submittal and request for information processes are not followed, then responsibility for decisions may be askew. And although the contracts may carefully divide the responsibilities of owner, designer, and builder, if these parties work as an intertwined team, perhaps in a collaboratively developed and managed building information model, they may develop responsibilities to each other and to third parties despite any contract language to the contrary. Contractors may find they have uninsured design liability and designers may be liable for means and methods. Owners may not be able to rely on their designers and contractors because of their involvement in the process.

The IPD agreement is as much a safety net as it is a catalyst for change. By addressing the real needs of a collaborative team, it creates a safer sandbox for the team's efforts. Without it, the team is engaged in a high-wire act.

16.7   AND IF TRADITIONAL CONTRACTING IS SO SUCCESSFUL, HOW DO WE EXPLAIN THE OUTCOMES?

Some of the most carefully crafted construction agreements are those used for public contracting. These agreements are written to positively transfer risk, to enforce a fixed price (or at least a guaranteed maximum price), and are tight and prescriptive. If this is a winning strategy, why don't they work better?

And, more importantly, what should our contracts look like in a high-performance IPD project? That is the question addressed in the next chapter.

16.8 SUMMARY

If traditional project delivery approaches and contracts are supposed to provide cost and schedule certainty, quality, all without waste, then how do we explain the cost and schedule overruns, litigation and claims, and the estimates of waste in traditional projects?

In reality, traditional contracts are designed to maintain distance between project members and insulate them from each other. Profitability is determined individually and is unrelated to project outcome. Rather than share risks equitably and have a governance system to avoid and mitigate project risks, traditional contracts work to shift responsibility—often to the party with the least bargaining power, even if that party has little or no control over the risk.

The upshot is that traditional contract approaches impede integrating the project and achieving the highly valuable/high-performing building. Teams seeking higher performance should examine their business and contract structures to assure that they haven't created obstacles to performance.

NOTES

1. The use of an AIA example is not a criticism of the AIA contract documents. A201 is being used as an example because it attempts to accurately reflect existing industry norms. AIA has recently issued a series of construction documents that are intended for IPD projects and are less structured and more collaborative (www.aia.org/ipd).
2. EJCDC Document C-700, §3.06. Electronic Data:
    A. Unless otherwise stated in the Supplementary Conditions, copies of data furnished by Owner or Engineer to Contractor or Contractor to Owner or Engineer that may be relied upon are limited to the printed copies (also known as hard copies). Files in electronic media format of text, data, graphics, or other types are furnished only for the convenience of the receiving party. Any conclusion or information obtained or derived from such electronic files will be at the user's sole risk. If there is a discrepancy between the electronic files and the hard copies, the hard copies govern.
    B. Because data stored in electronic media format can deteriorate or be modified inadvertently or otherwise without authorization of the data's creator, the party receiving electronic files agrees that it will perform acceptance tests or procedures within 60 days, after which the receiving party shall be deemed to have accepted the data thus transferred. Any errors detected within the 60-day acceptance period will be corrected by the transferring party.
    C. When transferring documents in electronic media format, the transferring party makes no representations as to long term compatibility, usability, or readability of documents resulting from the use of software application packages, operating systems, or computer hardware differing from those used by the data's creator.
3. AIA Documents E-201 Digital Data Protocol Exhibit, and E-202 Building Information Modeling Protocol Exhibit.
4. ConsensusDOCS Documents 200.1 Electronic Data Transmission Protocol, and 200.2 Building Information Modeling Addendum.

REFERENCES

Cheng, R., Allison, M., Sturts-Dossick, C., Monson, C., Staub-French, S., Poirier, E. (2016) *Motivation and Means: How and Why IPD and Lean Lead to Success*. University of Minnesota, Retrieved from http://arch.design.umn.edu/directory/chengr/.

Mace, B., Laquidara-Carr, D., Jones, S., (2016), *Benchmarking Owner Satisfaction and Project Performance*, Retrieved from http://www.leanconstruction.org/learning/.

CHAPTER 17

Contracting for Project Integration

"It is not the beauty of a building you should look at; it's the construction of the foundation that will stand the test of time.

—David Allen Coe

17.1 INTRODUCTION

The integrated project delivery (IPD) contract defines the relationships among the project participants and the processes that guide their actions. It embodies the project goals and creates consequences for success or failure tied to their achievement. It puts control in the hands of the project participants and makes them responsible for total project outcome, not just their individual performance. Correctly designed, it stimulates behaviors that increase creativity, improve productivity, reduce waste, and lead to better outcomes, whether measured by value, aesthetics, functionality, or sustainability.

The IPD contract is the result of an intentional process, purpose-built for the project, and adaptable to changing conditions.

> [P]roject delivery systems must be adapted to their contexts, and should be viewed as products of design. This design cannot be entirely completed at the start of a project, but rather must occur throughout project execution, responsive to emerging phenomena. Further, although the industry is strongly urged to create conditions under which the ideal project delivery system can be more completely realized, designing to context can be used to get the best outcomes from a given set of project circumstances—even when less than ideal (Ballard, Kim, Azari, & Kyuncho, 2011).

Like a constitution, the IPD contract establishes roles, responsibilities, structures, and processes, but does not dictate how the goals will be achieved. The parties, through joint collaboration, extend the principles in the IPD contract by developing collaborative work plans, co-location plans, building

365

information modeling (BIM) execution plans, and similar detailed elaborations to the contract structure. The IPD contract sets the stage for the performance that will follow. As discussed later, the IPD contract differs significantly from traditional design and construction contracting that tends to undermine, rather than reinforce, collaboration. Traditional contracting methodologies, and the difficulties they pose, are discussed in Chapter 16.

> It is possible to act collaboratively without an IPD contract, but doing so is less likely to achieve high performance levels. In order to facilitate information exchange and crossing of boundaries, IPD contracts have joint liability waivers or liability limitations. These are often absent from a noncontractual IPD approach. If the project never hits a snag, it may turn out very well. But if serious issues arise, the parties will be deeply exposed. Experience with partnering has shown that high aspirations, without contractual reinforcement, is not reliable.

17.2   IS THE IPD CONTRACT REALLY NECESSARY?

Researchers at the University of Minnesota, University of British Columbia, University of Washington and Scan Consulting, led by Renée Cheng, have recently completed an in-depth study of 10 successful Lean IPD projects (Cheng, et. al. 2016). The researchers were intent on determining the reasons for IPD success, having concluded, based on their own research and research by others, that Lean IPD produced superior results in delivering projects at or under market cost and schedule and the projects they studied were consistent with this finding. Although the study was limited in the number of IPD projects examined, it provides the most detailed examination of IPD projects, to date.

Several of their conclusions confirm commonly held beliefs. The quality of team members, coaching, and having a "champion," the depth in which Lean processes were used, the extent of initial planning, and the effective use of co-location all positively affected outcomes. The researchers concluded that IPD as a structure, and Lean as an execution process, reinforced each other.

Other conclusions, particularly concerning the value of the IPD contract, have clarified issues where opinions have varied. The researchers found that the process of developing and understanding the contract was valuable for establishing a foundational team culture, including appreciation of the differences among partners' business practices. Teams that collectively invested time in developing the contract believed it established a strong foundation of trust and respect and a deeper understanding of the business needs and practices of their partners. While many teams downplayed the role of the contract in the successful execution of the projects, the research team concluded that the IPD agreements bonded the team as a unit, thereby increasing their resilience in the face of challenges and protecting the teams from entering into a cycle of blame and defensiveness.

The research team also found that collaboration beyond the typical was most often seen by parties who participated in the incentive pool, sometimes in great contrast to those that were not participating. although there are examples when the collaborative culture extended outside the incentive pool.

The upshot of their conclusions regarding IPD contracting is that the process of developing the IPD agreement is both a training and alignment tool, that the concept of shared risk/reward and liability waivers enables deeper collaboration, and that the contract increases team resilience when faced with difficulties. Thus, the IPD contract is part of the full integration formula, but without the other processes and behaviors, as described in their research and throughout this book, not a guarantee of success.

17.3 DEAL FIRST, CONTRACT SECOND

Construction contracting is often a conflict among the parties' goals, the business deal they craft, and the contract they sign. IPD contracting attempts to harmonize these forces into a coherent whole. Because the contract defines the structure of the organization, and structure influences behavior and outcome, it is important to align the structure to the goals and business deal.

Every contract is a reaction to a specific project or an assumed project type. Using a form contract, or the contract from a prior project, imposes the structure from a previous project on the new project. This can result in a contract that is always pulling the parties away from the project objectives, because it was designed for a different project or project type.

A better approach is to develop the key goals and business terms before considering contract language. Once the business terms and key processes are settled, the contract can be drafted to document these agreements and reinforce the desired behaviors. Deal first, contract second leads to better and more usable contracts.

17.4 THE IPD CONTRACTING MINDSET

IPD is designed to encourage behaviors that lead to exceptional project performance and value. A properly drafted IPD contract reinforces these goals by:

- Removing impediments to, and stimulate, communication, collaboration, and creativity;
- Aligning participants to well understood and agreed objectives; and
- Encouraging and rewarding behavior that increases project value.

These attributes must be built into the fabric of the IPD contract. In practical terms, this means that no element of the contract should be inconsistent with the drivers of IPD, and that all elements should be consistent with IPD's values. Contracts built on these premises are fundamentally different from traditional construction contracts.

The transformation from traditional to IPD agreements requires a mental shift regarding how contracts are developed. Traditional contracts are designed to be prescriptive. The drafter attempts to envision all the possible scenarios and craft language that tells the parties what they must and must not do in each scenario.

The traditional mindset reflects two serious fallacies. It assumes that it is possible to actually foresee the many possible futures with sufficient precision to identify and handle all of the important scenarios. It also assumes that the team, armed with actual facts and years of experience and training, can't develop a better solution than the drafter might imagine. Neither is likely to be true. In practice, traditional contracting overemphasizes what could go wrong and underemphasizes what is required for success.

In contrast, IPD agreements are flexible and empowering. They fundamentally assume that a properly configured and incentivized team can best determine how to achieve project goals. Trying to predict what a creative team will develop and telling them what they must and must not do is futile and counterproductive. Instead, the IPD agreement focuses on collaborative project structures, enhancing communication, and providing opportunities and incentives for creativity.[1]

The transformation from traditional to IPD agreements also requires shifting negotiation from a zero-sum approach to a principled[2] and synergistic effort. This requires understanding that a sound structure that improves project outcomes may require understanding the deal from the perspectives of all of the parties, rather than from a single viewpoint. For example, understanding the financial and accounting issues of parties may result in a milestone distribution structure to provide owner security while accommodating the participants' legitimate cash flow concerns. Or an institutional owner may favor maximizing the value received from a fixed budget, whereas a speculative developer may prefer minimizing the cost of a defined scope. Unless you understand the parties' concerns, you will not be able to skillfully address them.

The business deal must also be fundamentally fair. If the opportunities for gain and the potential for loss are not equitably distributed, the parties will not be motivated to collaborate.[3] Resentment, not cooperation, will result. Financial targets and profit at risk should be carefully set to challenge the team members, but not discourage them.

In summary, the IPD contracting mindset focuses on how to create a strong, adaptive structure that appropriately considers the interests of all participants to improve the outcomes for all. For many, this is a departure from prior practice and their hard won experience may actually hinder their negotiation. It takes effort to suspend your existing mindset while you learn another. Once that is achieved, the new perspective, melded with prior experience, creates greater competence. Because of the difficulty in shifting mindsets, teams embarking on their first IPD projects should consider using a facilitator with IPD experience, to help them jointly craft a solid project structure.

Real-Life Example: Negotiating Is the First Collaborative Act

The owner issued an RFP to an architect and contractor team for a new IPD project. Several highly qualified teams responded, were interviewed, and ranked. Selection was difficult because the teams were ranked closely, but the owner chose the top team. Everyone was excited about working with this team because each firm had impressive individual credentials.

The project team conducted an IPD contract workshop and began to negotiate the business terms for the agreement. The proposed agreement had been forwarded with the request for proposal with a notation that significant changes would not be considered. Almost immediately owners and counsel could see a disconnect among the individual firms, and between the owner and the winning team. They didn't seem to be able to see each other's perspectives or how the IPD process differed from traditional project delivery. As negotiation progressed the gaps continued to widen, until it became clear that the "top" team had too much difficulty working with the owner and with each other. As a result, they turned to the second place team.

The difference was remarkable. Not only did this team understand the IPD approach, they seemed to blend together seamlessly. Contract negotiation flew by as the new team brainstormed better structures—better for the project, not just for themselves alone. Project facilitators started noticing the contractor addressing design issues and the architect focusing on the contractor's problems. And during the project itself, the contractor and architect "danced" together; never changing what they were, but interacting in new ways. When the project was finished, the contractor's project manager and the architect's project engineer both stated that they had never worked so hard on a project and couldn't wait to do it again. It was a great collaborative performance, and looking back, team leadership agreed that contract negotiation was their first collaborative act.

17.5 A NEW BUSINESS MODEL

An IPD project should be self-centering. If problems arise, the agreed business model should force the parties to solve the problem—not assign blame. Moreover, the business model should align the parties to the common goals and remove any incentive to forsake project outcomes in favor of individual gain. Although the business models do, and should, vary among projects, there are four characteristics seen in well–designed models. These are:

- Fixed profit
- Guaranteed variable costs without a cap
- Profit based on project outcome
- Limited change orders

17.5.1 Fixed Profit

Profit in traditional projects is related to the amount of work done. Fees are based on the cost of the work and lower tier work is marked-up by the prime. For designers, profit is embedded in each hour billed. Profits grow as work increases, which incentivizes inefficiency. In an IPD project, the parties agree to a fixed prospective profit that is not linked to the actual labor, materials, or project cost.

A fixed profit creates an incentive to reduce the variable costs to increase each party's margin. Because variable costs (labor, material, and equipment) account for most of a project's cost, reducing these costs directly benefits the owner. Moreover, because adjustments in scope do not affect the fixed profit, work may be easily transferred between parties. For example, if one party can efficiently install all the hangers needed for electrical, mechanical, plumbing, and fire protection, then the work can be shifted to that party. No one needs to fight for scope in order to maintain their profit. This allows the project team to look at what resources the project needs, remove duplications, and improve efficiency.

17.5.2 Variable Costs without a Cap

A lump-sum, GMP, or not to exceed (NTE) contract transfers—at least on paper—the risk of a project overrun to the contracting parties. If this really worked, then you would never see change orders or litigation on these types of projects. What actually occurs is that the contracting parties insert contingencies into their prices to protect against a cost overrun and then use change order and claims provisions to escape the constraints of the lump-sum, NTE, or GMP. Worse yet, this padding is inserted into each subtier contract because the fixed price cap is imposed on these parties, too. This results in multiple (and in aggregate, excessive) contingencies. Moreover, it locks the contingency resource into little packets that can only be used for a specific party, rather than be available to the project, as a whole. And in addition, it creates the possibility that the owner will pay for this risk three times through initial excessive pricing, followed by change orders during the project, and finally litigation expenses to resolve claims.

In the IPD business model, the owner agrees to pay for the variable costs (not the profit) without any cap. This removes much of the overrun risk and, thus, there is no need for the excessive contingencies carried in most projects. Moreover, because IPD projects generally use a design-to-cost-target approach, design contingencies are not required, either. The result is that the owner only pays for what the project actually costs, not for the parties' excessive perception of their risk. Although the absence of a contractual cost cap may seem bold, it is balanced by strict limitations on change orders and the ability to use party profit as a buffer against overruns. And as mentioned above, lump-sum and GMP do not prevent cost overruns and claims.

17.5.3   Profit Based on Project Outcome

The fixed profit is contingent on project outcome. If the goals are not met, project profit is reduced or even eliminated. If the project performs better than the goals, the project profit is increased. Each party shares in the increase or decrease based on their percentage of the project profit, which should be 100 percent at risk. This increases the buffer against overruns and maintains the division between profit and variable costs. Alignment among the participants is strengthened because individual profit can only be preserved, or increased, by improving overall project performance with respect to the agreed goals.

Project outcome is often expressed in cost or schedule and the following section on contract structure discusses possible variations in risk/reward plans. But in addition to these financial models, compensation can be tied to nonfinancial goals such as design and construction quality, sustainability goals, or project functionality. The essential point is that the model that is used should be designed to align and support the owner and team goals.

In a recent project being executed for a Native American organization, community values and respect for the local environment were judged more important than the common goals of meeting cost and schedule. Team compensation, therefore, was tied to achieving these distinctly nonfinancial goals.

17.5.4   Limited Change Orders

Many IPD projects have a zero-change-order goal, and most have far fewer change orders than conventional projects. As issues arise, the business model combined with joint project management leads to a rebalancing of budget in response to change rather than issuance of change orders. Moreover, properly crafted, the IPD agreement should limit change orders to conditions outside the control of the IPD team. For example, design errors and omissions, a fertile ground for contractor claims in traditional projects, are a team risk that needs to be managed by the team together, not be an opportunity for change orders. Similarly, contractor delay is not a reason for designers receiving augmented construction administration fees. The limits on change orders force the team to plan thoroughly, coordinate closely, and react swiftly when problems arise.

These four business terms create a new balance among the parties and, combined with the new IPD contract structure, create a durable, high-performance contracting platform.

17.6   A NEW CONTRACT STRUCTURE

"We must look into the underlying structures which shape individual actions and create conditions where types of events become likely."

—Peter Senge

Achieving specific and better outcomes is the purpose of the IPD contract. The process of defining a structure starts with determining the outcomes to be achieved, deciding which behaviors and processes are necessary to achieve the outcomes, factoring limitations that are imposed on the contract, and then designing a structure. In this process, structure is the servant of outcome.

In the influence diagram in Figure 17.1, the project outcome requires some blend of the behaviors shown in the ovals. These behaviors are encouraged and shaped by the five key structural elements shown in the rectangles. In a full IPD contract, all five structural elements are present and harmonized to the project objectives. The importance of these elements is described below.

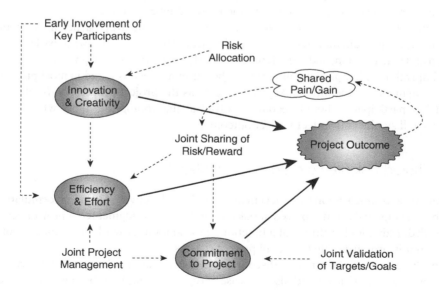

FIGURE 17.1   IPD elements and outcomes. © Howard Ashcraft.

17.6.1 Early Involvement of Key Participants

Early involvement of key participants—those who have the greatest influence on project outcome—is a critical factor for IPD success. A project participant deeply influences project success if it can impart knowledge that improves the effectiveness or constructability of design or if its interactions with other organizations enhance project productivity. Identification of key participants is specific to a given project, but—in addition to the owner, designer, and builder— key participants generally include the mechanical, electrical, and plumbing (MEP) designers and contractors because their knowledge strongly affects design and these parties must cooperate closely for the project to proceed smoothly. Depending on the project, steel erectors, framers, curtain wall contractors, major equipment vendors, and others may similarly be key participants.

The key participants' diverse viewpoints improve project performance in many ways. Studies of creativity in commercial contexts note that teams with diverse backgrounds are more creative.

> ... [O]ne common way managers kill creativity is by assembling homogeneous teams. The lure to do so is great. Homogeneous teams often reach "solutions" more quickly and with less friction along the way. These teams often report high morale, too. But homogeneous teams do little to enhance expertise and creative thinking. Everyone comes to the table with a similar mindset. They leave with the same (Amabile, 1998).

The broad experience of a diverse team also benefits target value design. Designers provided with information concerning effectiveness and constructability of a variety of concepts can more accurately choose systems and layouts that efficiently achieve the project goals. Moreover, key specialty contractors can provide pricing information that is current and accurate, leading to better price control and fewer surprises. Finally, when parties are engaged in developing the project design, they develop a commitment to the overall project, not just to their individual component.

The timing of key participant involvement is also important. Key participants should become engaged when their participation will benefit the project. This is almost always earlier than in traditional design and construction, and the reference to "early" is meant to highlight this change in practice. It does not imply that all key participants come on board simultaneously, and in most projects, the core team will be augmented by additional key participants as the project progresses. In other words, the rule is that key participants should become involved at the *appropriate* time, which is when their contributions will significantly affect project outcome.

17.6.2 Shared Risk/Reward Based on Project Outcome

IPD agreements tie compensation to achievement of project objectives. Although formulations vary, all or part of the participants' profit is placed at risk and profit may be augmented if project performance is met or exceeded. Individual profit is not a function of the amount of work performed, or of individual productivity, but is proportionate to overall project success.

Tying profit to project performance discourages selfish action that is inconsistent with optimizing project outcome. Shared risk/reward also increases project commitment because the parties perceive that they are rowing the same boat and they benefit by providing suggestions or assistance to other team members. Thus, parties become engaged optimizing the whole project, not just a single system or element.

Shared risk/reward also serves to align the parties to the project objective. If compensation is based on achieving that objective, it behooves each party to understand precisely what the objective is and how it is best achieved.

Shared reward not only makes risk more tolerable, it provides a rational basis to prefer IPD projects. Thus, a workshop of design and construction managers concluded:

> The group felt that structuring participants' compensation to be raised or lowered according to performance against predetermined targets is the most important and effective driver—it provides a monetary reason to collaborate (American Institute of Architects, California Council, 2009).

Shared risk/reward is a balance of four components: (1) reduced liability (which is discussed separately below), (2) limited options for change orders, (3) owner guarantee of direct costs, and (4) profit tied to project outcome.

In an IPD project, change orders should be limited to those issues that the team cannot control, such as *force majeure* events, or owner's elective changes in scope. Issues such as the quality of the construction documents or the productivity of field crews are team risks that are not grounds for change orders. Typical grounds for change orders are:

- Owner's elective changes in scope.
- Owner's suspension or termination for convenience.
- Changes in laws and regulations after project commencement.
- Differing site conditions and *force majeure*.
- Reconciliation of allowances.

No other changes should be permitted. This is not as draconian as it might seem because the owner has guaranteed the direct costs without profit. Thus, if an event occurs, such as omission in the design documents, the owner will pay the cost of the omitted item, but will not pay any additional profit, and the increased costs may reduce the team's profit distribution.

The risk/reward system should reflect the project objectives. Virtually all systems increase or decrease profit based on project outcome, but the outcome metrics can vary from pure cost. They can be based on value, life cycle costs, sustainability, functionality, or even aesthetics. Moreover, there are many different risk/reward models that address cost certainty, cash flow, and other real project concerns. The key to a successful system is to determine what outcomes are most important, and to align the risk/reward metrics to their achievement.

At its most basic, the risk/reward model is based on paying the direct costs incurred (costs without profit) and a fixed profit. The amount of fixed profit is adjusted based on comparing the total project direct costs with a previously agreed target. If actual costs exceed the target, the profit is reduced by the same amount until it is exhausted. If actual project costs are less than the target, a portion of the savings is added to the profit. In either event, the available profit is distributed to the team in proportion to their respective interests. This approach is graphically shown in the Figure 17.2.

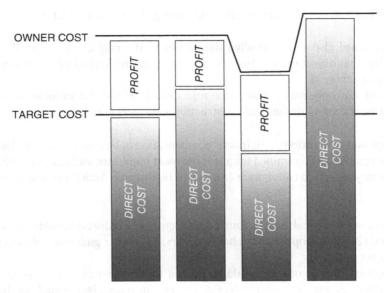

FIGURE 17.2   Basic compensation model. © Howard Ashcraft.

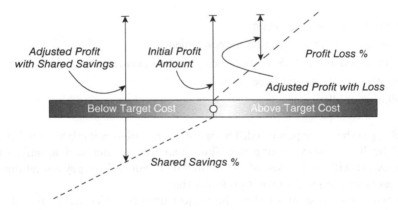

FIGURE 17.3   Profit at risk with shared savings. © Howard Ashcraft.

The increase or reduction in profit depends on the final actual cost and the shared savings and profit loss percentages. A common approach is to share 50 percent of the savings with the team and to reduce team profit, dollar for dollar until the profit pool is exhausted. Figure 17.3 graphically represents this approach. To more finely tune this approach to achieving outcomes other than target cost, an assessment of specific—and often nonmonetary factors—can be used to adjust the shared savings percentage from a minimal percentage to a large portion of the shared savings.

Under any model, behavior and outcome are affected by when the targets are set. If set too early, the targets will likely contain excessive contingencies to buffer uncertainty. If set too late, opportunities for innovation and target value design will largely be lost. There is no clear-cut rule that applies to all projects. Simple projects that have cost data from comparable projects can set their targets quite early. Complex and rapidly evolving projects must set their targets later.

There are several strategies for coping with rapidly evolving and complex projects. One option is to segregate the project into packages with individual targets that are aggregated into a single project target on a rolling basis. Another option is to use allowances to segregate the unknowable portions of the project from the portions that are better understood.

Still another option is to set the targets early, but use a neutral zone to reduce risk to the owner and to the project team. In Figure 17.4, below, the team's profit is at risk if they deliver the project above the at-risk threshold. If the project is delivered below the shared savings threshold, the team shares in any savings below that level. If the project is delivered in the neutral zone, the initial profit is neither increased nor decreased. The use of the neutral zone reduces the team's risk because there is a buffer above the shared savings threshold and reduces the owner's risk because the owner retains any savings in the neutral zone. Because it does not pay any "bonus" until the team is below the shared savings threshold, it is protected against the risk of an excessively high target cost.

A neutral zone may be appropriate if uncertainties preclude an early target setting. But the zone should be kept as narrow as possible because the incentives and risks are diluted as the zone expands.

Although cost is almost always a factor, *value* may be more important to an owner. Many institutions prefer getting the best project their budget will allow rather than the least expensive project that meets their base requirements. Because they often keep their facilities for decades (or more), quality and functionality are paramount concerns.

Scope expansion within a budget is often difficult in conventional projects. Even if savings are projected, the parties (bound to a fixed price or guaranteed maximum) want to hold on to savings to protect against the unexpected. When they are finally comfortable with releasing savings to the owner, the opportunity for reinvestment may have passed.

The value cost model shown in Figure 17.5 incentivizes innovation during the planning (design/preconstruction) period and provides the team with potential profit augmentation if they can increase project scope. Once the scope is fixed, however, the project substitutes execution efficiency for design innovation. In this way, the value cost model allows the owner to reinvest potential savings and still achieve construction efficiency by reducing design variation during construction to allow for smooth execution.

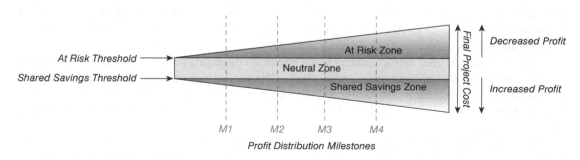

FIGURE 17.4   Risk/Reward with a neutral zone. © Howard Ashcraft.

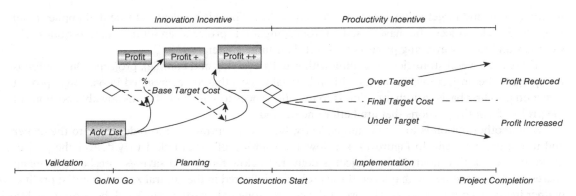

FIGURE 17.5   Value cost model. © Howard Ashcraft.

These are only a few of the risk/reward models that are possible, but they demonstrate how the models can be tuned to favor specific outcomes and manage specific risks. There is an art to developing an appropriate risk/reward model, determining when and how to set targets, assuring transparency, and defining appropriate metrics. Unless the contracting parties have significant IPD experience, they would be well served to seek assistance in developing an appropriate risk/reward model.

17.6.3   Joint Project Control

Joint project control requires real communication among the parties. To achieve consensus, the parties must clearly explain the issues from their perspectives and listen to the perspectives of others. The increased understanding provides a clearer and jointly held understanding of the issues. Miscommunication, although certainly possible, is less likely.

Joint project control also reinforces the communal nature of the undertaking. It is not "their project"; it is "our project." In addition, joint project control balances the interests of the parties and provides a check against favoring the interests of one party over the other. It also reflects a fundamental fairness. In IPD, parties are accepting risk based on project outcome and should certainly have a voice in decisions that affect those risks.

Joint project control also affects the perception of risk, as well as risk itself. Risk perception research indicates that perils a party cannot control are feared more than those they can (Slovic, 2000). As noted below, fear chills creativity and results in defensive behavior. It also results in excessive risk hedges through explicit or implicit contingencies. Thus, joint management serves to reduce defensive behavior and avoids unnecessary contingency expense.

In an IPD project, joint project control is accomplished through a project management team comprising at least the owner, contractor, and designer. The project management team is authorized to manage the project to achieve the jointly agreed objectives. Thus, each member of the project management team must have the authority to bind its respective entity, and each party must be able to rely on the agreements of the others.

Jointly reached decisions are generally final and not subject to later review. They are made by parties closest to the data and are tested by the parties' different perspectives and interests. Keeping control close to project operations speeds decision making and project schedule. Moreover, senior management "second-guessing" is wasteful because work done in reliance on the decision must be abandoned. And, more importantly, "second-guessing" can paralyze project management, which may feel that there is little point in making decisions that may later be reversed.

Joint project control is a significant paradigm shift for many owners. Traditionally, the owner's project representative functioned as the owner's "eyes and ears," but did not actively participate in the development of design or construction solutions. Instead, the contractor or designer proposed options and solutions that were approved or disapproved by the owner's senior management after being communicated by the project representative.

The IPD owner, in contrast, is actively involved in the development and analysis of options and solutions. This level of owner involvement and control is, in fact, one of the major advantages of IPD for owners. In no other project delivery method does the owner have such a strong role in fashioning the project to meet its needs. But this strength implies responsibility to commit sufficient capable resources authorized to make reliable decisions. This change in practice can be particularly difficult for owners who have traditionally vested their project representatives with little authority.

Joint project control is not leaderless. For IPD to succeed, the key participants must be able to freely exchange information and comment on each other's work. Ideas should be respectfully discussed and differing opinions must be voiced. But in the end, decisions need to be made and documented so that the project moves forward. This requires leadership.

Leadership in an IPD project may fall on one person or party or may be distributed among the parties based on the level of project development or topic issue. The owner will naturally tend to lead during the early portions of the project with a transition to the designers as needs and values are transformed into conceptual plans. The builders will often become the leaders as physical construction commences. In addition, leadership may be distributed based on systems, such as mechanical or structural systems, with systems leaders reporting to the collaborative management team and the overall team leader. The structure of leadership will vary between projects, but the need for effective leadership will not.

IPD leadership requires mentorship, facilitation, and accountability. It is not a command-and-control relationship, because the leader must encourage innovation from the bottom up. Moreover, the IPD leader needs to know when discussion needs to be transformed into action and how to forge consensus from differing opinions. But IPD is based on supporting behaviors that will achieve superior results. And if behavior does not meet IPD expectations, whether due to lack of teamwork, failure to meet agreed obligations, lack of transparency, or discord within the team, action must be taken to correct the behavior or remove the noncompliant party from the team.

The owner's role in IPD differs significantly from the owner's role in other project delivery methods. As noted above, the owner in an IPD project must be actively involved during every stage of the process, not just as a reviewer or approver, but as a contributing member of the design and construction team. It thus follows that the owner has a special leadership role as well. The owner must continuously communicate his or her needs and vision while recognizing the legitimate interests of the other parties. By working with and for the other parties, the owner can expect them to work for the project to achieve the owner's vision.

Although all current IPD agreements have some form of joint project control, the detailed decision process and ultimate authority of the participants varies significantly. Variation is inevitable given the needs of specific projects and participants. But joint project control is designed to provide parties at risk with some control over the risks they have undertaken and to increase parties' commitment to the project as a whole. Thus, skewing control in favor of one party or the other may undermine the behaviors IPD seeks to create.

17.6.4   Reduced Liability Exposure

The primary reasons for limiting liability are to increase communication,[4] foster creativity, and reduce excessive contingencies.

> Information sharing and collaboration support all three components of creativity. Take expertise. The more often people exchange ideas and data the more knowledge they will have. The same dynamic can be said for creative thinking. In fact, one way to enhance the creative thinking of employees is to expose them to various approaches to problem solving.
> With the exception of hardened misanthropes, information sharing and collaboration heighten peoples' enjoyment of work and thus their intrinsic motivation (Amabile, 1998).

Sometimes, freely exchanging information can lead to greater liability. For example, many states[5] permit actions for negligent misrepresentation under the guidelines of section 552 of the Restatement of Torts, Second.[6] Under that standard, a person providing errant information is liable for the damage caused to anyone whose reliance was intended. This has led to bottling up information to limit liability expansion, although this diminishes creativity and performance. Liability waivers support creativity by removing this concern.

In addition, liability waivers serve to generally reduce fear of failure. In a creative project, there must always be a safety net below people who make suggestions. A climate of fear is not conducive to creativity and undermines intrinsic motivation (Amabile, 1998).

Liability exposure also directly raises project costs through increased contingency allocations. A rational negotiator assesses the risks his or her organization faces, attempts to quantify the risk, and includes an allowance in the project cost. This rational action is repeated by each participating organization with the result that the summed risk allowances exceed the actual contingency required for the project. Moreover, the division of project contingency into many smaller allocations impairs effective contingency management.

Liability concerns create hidden costs caused by defensive design and reluctance to consider using new materials and techniques. Old practices may be costly and inefficient, but they are comfortable and appear safe.

Liability waivers also reduce litigation costs, and can be justified on this ground alone, but as noted above, the primary reasons for liability waivers are to increase communication and creativity, and to limit unnecessary contingencies.

17.6.5   Jointly Developed and Validated Targets

Jointly developed targets are the parties' first collaborative act. They document the parties' agreement regarding objectives and confirm that they are achievable. In addition, the targets serve as metrics for compensation adjustment and as goals for target value design. Because they are jointly developed, each party owns the objectives and is committed to their achievement.

Jointly developed and validated targets are the mission statement of the IPD project.

17.6.6   Multiparty or Polyparty Contract?

The five structural elements can be arranged in several configurations. It is possible to address all elements in multiple, interlocking agreements, but this is complicated, requires additional drafting time, and runs the risk of inconsistency. The more common approaches are to use a single contract between owner, prime designer and the contractor (multiparty contract) or to have a single contract that all of the risk/reward parties execute (polyparty contract).

In the multiparty contract, shown graphically in Figure 17.6, the parties with their profit at risk and eligible for incentives are within the risk reward group. The primary parties—owner, designer, and builder—execute the prime agreement and the risk/reward consultants and risk/reward trade contractors execute specialized subagreements with the designer and builder, respectively. The key incentive, control, dispute, and liability allocation provisions flow down through these contractual arrangements. There will usually be some consultants and subcontractors who are outside of the risk/reward group, and they are contracted traditionally.

The multiparty contract approach transfers management of subcontractor relationships to the contractor and management of consultant relationships to the principal designer. This mirrors traditional management arrangements and uses the skills of the prime designer and the contractor to manage these relationships. This allows the owner to concentrate on the project at a higher level and reduces its administrative burden. Many teams favor this approach for their initial foray into IPD because it

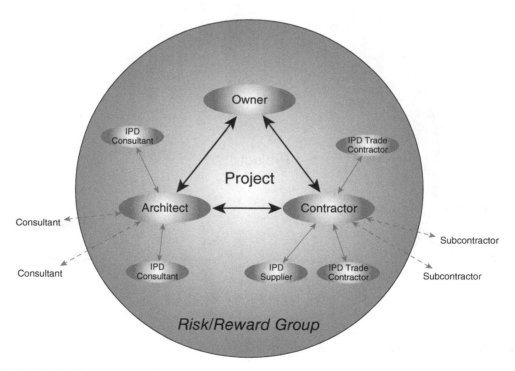

FIGURE 17.6   Multiparty contract. © Howard Ashcraft.

is relatively familiar. In some instances, an additional party is added to the contracting group where the party has a large percentage of the project or is especially significant. In those instances, the added party is usually aligned with the principal designer or with the contractor—and must jointly make decisions with its aligned party—in order to preserve the balance among the owner, designer, and contractor.

In the polyparty contract, depicted in Figure 17.7, all parties that are within the risk/reward group sign a single IPD agreement. This can be done initially, or additional parties can be added sequentially by using joining agreements. The parties outside of the risk/reward group sign traditional consulting agreements with the prime designer or contractor, as appropriate.

The polyparty contract is very transparent. The owner has direct access to all of the key participants (and they have access to the owner), and all stand on equal contractual footing. This can lead to a higher level of communication and commitment from trade contractors and consultants and ensures that they have a voice in project management. It is administratively more complicated than the multiparty agreement because the owner must manage many relationships simultaneously, including administrative issues, such as invoicing. The hierarchical flatness of a polyparty contract improves communication and commitment, but also creates leadership challenges that must be met by the owner, in the first instance, and by other participants, thereafter.

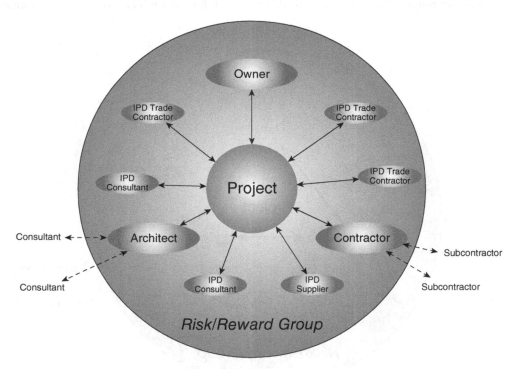

FIGURE 17.7   Polyparty contract. © Howard Ashcraft.

17.7 NEGOTIATING THE IPD CONTRACT

IPD projects, at least at the current state of development, must be crafted to match the specific project and project team. In current practice, the IPD contract is created through facilitated contract negotiation. Unlike traditional contract negotiation, the IPD agreement is developed through simultaneous negotiation of multiple parties with a high degree of transparency. The parties need to be able to openly discuss their goals and concerns and to trust the information being provided by their prospective teammates. Because the IPD agreement is based on fundamentals that are radically different from traditional construction contracts, the contract negotiation workshop should be led by, or should include, an attorney or attorneys experienced in IPD agreements. In some projects, the IPD team uses a project counsel—an attorney jointly retained by all parties to facilitate contract negotiations and develop the IPD agreement—to lead the contract negotiation workshop.[7]

17.7.1 The Contract Negotiation Workshop

The contract negotiation workshop will usually have two phases: negotiation of the prime IPD agreement and negotiation of joining agreements or IPD subcontracts and consulting agreements.

Negotiating an IPD Agreement (Ashcraft, 2010) contains a thorough discussion of principles, issues, and options affecting IPD agreements. Parties that have not negotiated an IPD agreement should review this or another similar work before entering into IPD agreements. This is more important than studying individual contract examples because you need to understand the reasons for contract provisions before you can safely accept or modify them.

One of the first issues to address is who needs to participate in the negotiation. Every entity that will sign the agreement should participate, but in some instances significant at-risk trade contractors and consultants may participate, as they may need to agree on business and legal terms that will flow through to them from the prime agreement.

The negotiation workshop[8] should focus on issues that affect the behaviors desired. The significant issues list and their relative priority vary between projects. However, there is a series of questions that can focus discussion and highlight the most important issues:

- Which parties are within the shared risk/reward group?
- Will the contract be multiparty (owner, architect, contractor) or polyparty (all at-risk parties sign a single agreement)?
- What is the size of the at-risk pool, and how is it funded?
- Is the at-risk pool distributed at project end, or at milestones? Is any amount withheld?
- How will the project be governed?
- Who has the authority and responsibility to make decisions?
- How are the targets defined and measured? When are they set?
- Are contingencies explicit or implicit? If explicit, how can they be used and when must they be used?

- What is the role of allowances?
- How should escalation be factored into the targets?
- What are the consequences of missing a target?
- How is non-collaborative behavior prevented or discouraged?
- What events justify changing a target?
- What liabilities are not waived?
- How are warranties handled and funded?
- How does insurance relate to the unwaived liabilities?

17.7.2   Subcontractors, Consultants, and Joining Agreements

Contractors and architects rarely perform their full contractual scope of work, delegating much of their scope to subcontractors and consultants. Architects may retain less than half of the total design fee and some contractors retain no self-performed work, at all. If IPD seeks to energize the people actually doing the work, it must clearly engage subcontractors and consultants. Moreover, if IPD is to provide the owner with a sufficient buffer against cost overruns, the subcontractors and consultants—or at least the key participants—must also share in the risk/reward structure.[9]

Because IPD is a collaborative, trust-based delivery method, the consultants and subcontractors chosen must embrace IPD and must be able to work cooperatively with the other parties. Thus, in most IPD structures, the subcontractors and consultants are jointly chosen by the owner, designer, and contractor team, or the team has interview and veto rights over the designer's and contractor's preferred consultant and subcontractor choices.[10]

There are two primary methods for incorporating the key consultants and subcontractors that did not sign the prime IPD agreement: subagreements and joining agreements.

In the subagreement approach, the key IPD elements flow through the prime agreement (designer or contractor) into the subagreement (consultant or subcontractor). These include key risk and reward terms as well as any liability limitations and waivers. The at-risk compensation of the subcontractor or consultant is a portion of the at-risk compensation of its respective prime. In almost all instances, the business structure of the subagreements mirrors the business structure of the IPD agreement, except that the subcontractors and consultants are less involved and have no or limited voting rights at the project management level.

In a joining agreement approach, the key subcontractors and consultants execute an agreement that amends the IPD agreement to add them as a party. The risk/reward provisions are amended with each added key subcontractor or consultant to reflect the amount of compensation the added party has placed at risk. If all parties are added to a single agreement, the IPD agreement must be able to adjust terms, such as project management processes, to accommodate the changed nature and number of participants.

The IPD team must decide how subcontractors and consultants are selected. In some instances, the designer and contractor may propose with their preferred subcontractors and consultants and be viewed as an existing team. In other instances, the team will need to develop a process for assuring that the selected subcontractors and consultants will work effectively with each other and with the overall IPD team.

There will always be some subcontractors that will be procured on a fixed price or time-and-materials basis. These parties are typically outside of the risk sharing and liability waiver provisions of the IPD agreement, but their contract values are included in the target cost as reimbursable costs to a risk/reward party.

17.8 IPD CONTRACT FORMS

A team planning an IPD project can choose to use a standard contract form or to use a custom agreement. Although there are no reliable statistics, it appears that as many or more IPD projects are done using custom agreements as are executed under a form agreement. Moreover, none of the form agreements are usable without modification, and in many instances, they should be substantially modified before use. Thus, one of the primary advantages of form agreements—reduced negotiation and drafting effort—may be illusory. And, in any event, it is better to conform the agreement to the business model and the procedures the team will actually use to manage the project than to force fit the project into an ill-fitting contract.

This section briefly describes the currently available form agreements and two influential private agreements that serve as models for custom agreements used across the United States and Canada. It provides an explanation of how each agreement addresses key IPD issues. This information should be used in conjunction with an attorney who specializes in IPD agreements and who can assist the team in deciding which approach best suits the team and the project, and what modifications will need to be made to use the contract for the specific project, or whether a custom agreement is preferable.

17.8.1 Association Agreements in Brief

1. ConsensusDOCS CD-300

    ConsensusDOCS, a joint effort of the Associated General Contractors and many other organizations, has issued a form multiparty IPD agreement. It was the first IPD form agreement and was significantly influenced by the integrated form of agreement used by Sutter Health for its IPD projects. One of the unusual features of the CD-300 is that it uses a checkbox approach to many key decisions such as liability limitation and payment and dispute resolution. It assumes a minimum core management team of owner, architect, and constructor and is neither clearly a multiparty or a polyparty agreement. It includes a mixture of Lean and traditional processes. The CD-300 was revised in 2016 to incorporate some features from other agreements and to remove some innovative features of the first version that proved difficult to use.

2. AIA C195 Series

    The AIA C-195 documents are the most unusual, because it uses a limited liability company (LLC) to integrate owner, architect, and contractor. This approach is more complex than the multiparty or polyparty agreements, but may be preferable in limited circumstances. It requires significantly more legal and accounting involvement, because it relies on a separate legal entity. It also has a different approach to project governance (reflective of the LLC structure) and treats compensation differently from the other form contracts.

3. AIA C191 Multiparty Agreement

The AIA C191 agreement is a multiparty agreement that was designed to follow the basic principles set forth in the AIA/AIACC IPD Guide. One of the unusual features of the C191, which is also in the C195, is the option for an "achievement bonus" that is separate from, and unaffected by, the performance of the project against target cost. The C191 is a multiparty agreement (owner, architect, and contractor) and integrates into the AIA contract document set.

4. AIA A195/B195/A295 Interim "IPD" Agreements

Although the AIA lists this set of documents within its IPD portfolio, they are fairly close to traditional documents. There is an owner-architect agreement (B195), an owner-contractor agreement (A195) and a joint set of general conditions shared by both agreements (A295). When first issued by the AIA in 2008, they were viewed as "transitional" documents, rather than an actual IPD document set.

17.8.2 Private Agreements

Many, and possibly the majority, of IPD agreements are private, proprietary forms. They are customized to specific project and client needs and are normally used only on the client's projects. But these documents have been circulated within the industry and have strongly influenced the development of IPD agreements. The two most influential are the Sutter Health IFOA and the Hanson Bridgett LLP IPD agreements.

1. Sutter Health Integrated Form of Agreement (IFOA)

Sutter Health was a pioneer in using IPD and Lean project delivery. Will Lichtig, and other attorneys at the McDonough Holland & Allen[11] law firm developed this agreement. The 2008 version of this document was circulated at various conferences and has become the model for other IPD projects, particularly in health care. The Sutter Health IFOA influenced the development of the initial CD-300. Although primarily a multiparty agreement, Sutter Medical Center Castro Valley was delivered as an 11-party (polyparty) IFOA. There are also various other versions of Sutter IFOAs that reflect negotiation and adaptation for specific projects and the IFOA form is also being updated by Sutter Health. In 2015, the Sutter IFOA was rewritten by the Hanson Bridgett law firm. However, it is expected that the original IFOA will still remain influential because of its wide distribution.

2. Hanson Bridgett IPD Agreements

Hanson Bridgett, a San Francisco law firm, developed a suite of IPD documents in multiparty and polyparty versions. In various editions, these documents have been used across the United States and Canada. An early version of the multiparty form influenced development of the AIA C-191. It uses a two-level management structure, strong liability limitation, limited change orders, and custom risk/reward/incentive plans. It is written in a plain, business English format.

17.9   A PARALLEL PATH: THE U.K. EXPERIENCE

IPD owes much of its initial framework to the Project Alliancing approach developed initially in the United Kingdom for oil exploration in the North Sea. U.K. practitioners also recognized that design and construction was fraught with inefficiency and waste due to fragmentation and antagonism within the industry. When evaluating IPD, it is instructive to reflect on their findings and their adoption of relational contracts. While not identical to IPD, their solutions reflect—within a U.K. commercial and legal setting—similar concerns and responses.

17.9.1   Recognizing the Problem

In 1994, Sir Michael Latham chaired a task force that issued the influential Constructing the Team: Final Report of the Government Industry Review of Procurement and Contractual Arrangements in the U.K. Construction Industry (Latham, 1994). The report recognized that the U.K. construction industry was capable, but was seriously lagging other industries. The Latham task force proposed 30 recommendations to achieve a 30 percent improvement in performance. The Latham Report recommendations were widely hailed and resulted in significant new commissions and endeavors, but also suffered from slow and incomplete adoption (Cahill & Puybaraud, 2008). But the gauntlet had been thrown.

The Latham Report was shortly followed by a report issued under the chairmanship of Sir John Egan, Rethinking Construction (Egan, 1998), that defined the existing state of U.K. construction and proposed strategies for improvement. Despite a high level of individual competence, it found that the construction industry was unprofitable, invested little in research and development and regularly disappointed its clients. It commented on surveys that found that over a third of all clients were dissatisfied with their contractors' and consultants' performance and that they believed that significant value improvement and cost reduction could be gained by integrating design and construction. It summarized studies from the United States, Scandinavia, and the United Kingdom that suggested that 30 percent of construction is rework, labor is only 40 to 60 percent efficient, accidents absorb 3 to 6 percent of construction costs, and at least 10 percent of all materials are wasted. The upshot was that the existing industry was fragmented, adversarial, and inefficient.

The Egan task force recommended a radical change in project philosophy, structure, and execution. Although some of the recommendations, much like the recommendations in the Latham report are applicable only to U.K. practice, many of the recommendations are almost identical to the principles and practices of IPD. A few excerpts from the Egan Report demonstrate the parallel thinking:

> 34. If we are to extend throughout the construction industry the improvements in performance that are already being achieved by the best, we must begin by defining the integrated project process. It is a process that utilizes the full construction team, bringing the skills of all the participants to bear on delivering value to the client. It is a process that is explicit and transparent, and therefore easily understood by the participants and their clients.

35. The rationale behind the development of an integrated process is that the efficiency of project delivery is presently constrained by the largely separated processes through which they are generally planned, designed, and constructed. These processes reflect the fragmented structure of the industry and sustain a contractual and confrontational culture.

50. We are impressed by the dramatic success being achieved by leading companies that are implementing the principles of "lean thinking" and we believe that the concept holds much promise for construction as well. Indeed, we have found that lean thinking is already beginning to be applied with success by some construction companies in the USA. We recommend that the UK construction industry should also adopt lean thinking as a means of sustaining performance improvement.

58. As we have already emphasized, in our experience too much time and effort is spent in construction on site, trying to make designs work in practice. The Task Force believes that this is indicative of a fundamental malaise in the industry—the separation of design from the rest of the project process. Too many buildings perform poorly in terms of flexibility of use, operating and maintenance costs, and sustainability. In our view there has to be a significant re-balancing of the typical project so that all these issues are given much more prominence in the design and planning stage before anything happens on site. In other words, design needs to be properly integrated with construction and performance in use. Time spent in reconnaissance is not wasted.

59. There is a series of practical consequences that flow from this:
    ○ suppliers and subcontractors have to be fully involved in the design phase. In the manufacturing industry, the concept of "design for manufacture" is a vital part of delivering efficiency and quality, and construction needs to develop an equivalent concept of "design for construction";
    ○ the experience of completed projects must be fed into the next one. With some exceptions the industry has little expertise in this area. There are significant gains to be made from understanding client satisfaction and capturing technical information, such as the effectiveness of control systems or the durability of components;
    ○ quality must be fundamental to the design process. Defects and snagging need to be designed out on the computer before work starts on site. "Right first time" means designing buildings and their components so that they cannot be wrong;
    ○ designers should work in close collaboration with the other participants in the project process. They must understand more clearly how components are manufactured and assembled, and how their creative and analytical skills can be used to best effect in the process as a whole. There is no longer a place for a regime of design fees based on a percentage of the costs of a project, which offers little incentive to build efficiently;
    ○ design needs to encompass whole life costs, including costs of energy consumption and maintenance costs. Sustainability is equally important. Increasingly, clients take the view that construction should be designed and costed as a total package including costs in use and final decommissioning.
    ○ clients too must accept their responsibilities for effective design. Too often they are impatient to get their project on site the day after planning consent is obtained. The industry must help clients to understand the need for resources to be concentrated up-front on projects if greater efficiency and quality are to be delivered.

64. One area in which we know new technology to be a very useful tool is in the design of buildings and their components, and in the exchange of design information throughout the construction team. There are enormous benefits to be gained, in terms of eliminating waste and rework for example, from using modern CAD technology to prototype buildings and by rapidly exchanging information on design changes. Redesign should take place on the computer, not on the construction site.

These recommendations clearly fit well into the IPD structure.

17.9.2   Relational Contracts

The Latham Report recommended use of relational contracts, such as the NEC3.[12] Although the NEC3 is designed to increase project collaboration, it falls short of contractually integrating the project team into a virtual organization. Another agreement, the Be Collaborative Contract, was created by the Be (Building and Estates Forum), which joined with a series of other organizations to create Constructing Excellence.[13] The Be Collaborative agreement is primarily between the Purchaser and the Supplier, which may have design responsibilities and can be viewed as a guaranteed maximum price agreement with a subtarget cost. Although clearly heading toward an IPD approach, its flexibility can lead to a variety of project structures.

Dr. David Mosey, on behalf of the Association of Consulting Architects, developed a *partnered* contract[14] that in many ways resembles an IPD agreement (PPC2000). The term *partnering* in this contract differs from U.S. practice because collaboration under the PPC is a contractual obligation of the parties, not an unenforceable aspiration reflected in a project charter. Moreover, the PPC2000 is a multiparty agreement with the key parties all signatory to the same agreement. It favors early involvement of key participants and joint management through a core group. Risk and reward can be shared by the project team.

The PPC2000 contract naturally differs from U.S. practice in areas that are particular to U.K. design and construction. It also relies on use of a partnering facilitator who helps create the basic structure and contracts and then provides assistance to the team throughout the project. It allows, but does not prescribe, risk limitation. The basic business structure also differs in that there is a contract price rather than a price target, although shared savings are possible. Management is somewhat different from U.S. IPD practice because the management team (core group) does not include the owner, who has a separate client representative. The PPC2000 does not explicitly adopt lean practices (but does not prohibit them) and is silent regarding technology, such as BIM.

Although there are differences between the United Kingdom and the United States regarding collaborative project delivery, the similarity in reasons for change, the proposals to remedy deficiencies, and the structures proposed is far more significant than the differences.

17.10   INTERCONNECTIONS

The IPD agreement underlies all of the elements of the Simple Framework. Like a skeleton, it provides the structure that enables the muscles of the Simple Framework to work efficiently. It removes obstacles to collaboration and communication, it aligns the parties' commercial interests with the project goals, and creates an economic model that favors efficiency over quantity.

17.11   REFLECTIONS

The world has significant needs but limited resources. Paraphrasing a statement from the 1970s, if buildings are not part of the solution, they are part of the problem. While we may not be able to predict all of the future developments in project delivery, we are certain that successful methodologies will be highly collaborative and tightly integrated. Even now, the pressure for integration is building. Lean requires integration. Sustainability requires integration. BIM requires integration. The Simple Framework shows that integration is required from the highly valuable/high-performing building to the beginning of the design and construction process.

The road to fully integrated business and contract models will not be easy. Some will cling to traditional contracting practices. But the need for integration will eventually overcome resistance. And the successful contract models of the future, like the IPD agreement, will promote and enhance collaboration, not stifle it.

17.12   SUMMARY

The IPD agreement incorporates a business and contractual model that aligns the parties' commercial interests, removes barriers to communication and collaboration, enhances creativity and engagement, favors problem solution to risk transfer, and melds the key project participants into a virtual organization that is aligned to the project goals.

Key elements of the new business and contractual models are:

- Early involvement of key participants.
- Profit disassociated from units of cost.
- Costs without profit paid to completion.
- Profit adjusted (painshare/gainshare) based on agreed outcomes.
- Joint setting of targets.
- Joint project management.
- Reduced liability among team members.

NOTES

1. An enlightening discussion of a systems approach to contracting is contained in Larman (2010). He notes that, in a study by the International Association for Contract and Commercial Management, not one of the top 10 contractual terms focused on by corporate lawyers were likely to affect the day-to-day working of the project and that none of them involved the essential object of the contract—the very thing that the contract is about.
2. Principle-based negotiation is described in Fisher and Ury's *Getting to Yes* (Penguin, 1981).

3. Axelrod's *The Evolution of Cooperation* (Basic Books, 1985) discusses how, even in a simple game, cooperation is an optimal strategy. Moreover, it is even more effective if the parties can divine each other's strategy and thereby account for the decisions others will make. In an IPD context, this means that you need to understand the other person's interests, be transparent with your own intentions, and be trustworthy. As long as there are repeated interactions (which is guaranteed in construction), the optimal individual strategy is collaboration.

4. The liability concern, and its potential harm, was neatly summarized in the commentary "Intelligent Building Models and Downstream Use, Comments of the Technology in Architectural Practice Advisory Group" submitted for the 2007 revisions to AIA Documents B141 and A201, AIA 2005. "We fear there will be a tendency, driven by valid concerns about liability and insurability, to prevent such use of the architect's design data. We believe this is the wrong answer and would jeopardize the future of architectural practice as we know it. ... Obstacles to a free flow of data among the project participants should be overcome so that the architecture firm can deliver the full value of its work to the client and be rewarded commensurately."

5. See, for example, *Bily v. Arthur Young & Co.,* 3 Cal. 4th 370 (1992).

6. § 552, Information Negligently Supplied for the Guidance of Others

   (1) One who, in the course of his business, profession or employment, or in any other transaction in which he has a pecuniary interest, supplies false information for the guidance of others in their business transactions, is subject to liability for pecuniary loss caused to them by their justifiable reliance upon the information, if he fails to exercise reasonable care or competence in obtaining or communicating the information;

   (2) Except as stated in Subsection (3), the liability stated in Subsection (1) is limited to loss suffered;
      (a) by the person or one of a limited group of persons for whose benefit and guidance he intends to supply the information or knows that the recipient intends to supply it; and
      (b) through reliance upon it in a transaction that he intends the information to influence or knows that the recipient so intends or in a substantially similar transaction.

   (3) The liability of one who is under a public duty to give the information extends to loss suffered by any of the class of persons for whose benefit the duty is created, in any of the transactions in which it is intended to protect them.

7. Attorneys usually represent a single party and owe an obligation to act vigorously on behalf of that party, alone. Under the ethical rules applicable to attorneys, an attorney cannot represent multiple parties negotiating among themselves unless there is a written waive of the conflict of interests. The implications of joint representation should be explained before the parties decide to use joint counsel.

8. As noted previously, it is very helpful to have this workshop lead by a person who has significant experience in structuring IPD projects.

9. As a general rule, at least half of the anticipated construction cost should be within the risk/reward structure, and preferably more.

10. Another option is to have each new project participant interviewed by the entire team that precedes it. Although this may work on smaller projects, it becomes increasingly cumbersome as the number of project participants increases.

11. McDonough Holland & Allen ceased operation in 2010, its construction group merged with Hanson Bridgett LLP, and Mr. Lichtig subsequently joined Boldt Construction in an executive capacity.

12. http://www.neccontract.com/about/index.asp

13. http://www.constructingexcellence.org.uk/aboutus/

14. The PPC2000 contract is part of a set of contract documents that includes contracts for term agreements, specialty contractors, and an international version less tightly coupled to U.K. practices.

REFERENCES

Amabile, T. M. (1998, September–October). How to kill creativity. *Harvard Business Review, 76*(5), 76–87, 186.

American Institute of Architects, California Council. (2009). Experiences in Collaboration: On the path to IPD, 9. American Institute of Architects, AIA California Council.

Architects, Association of Consultant. (2003). *Guide to ACA project partnering contracts PPC2000 and SPC2000*. Tatsfield, Kent Association of Consultant Architects.

Ashcraft, H. (2010). *Negotiating an IPD agreement*. Retrieved from http://www.hansonbridgett.com/Publications/pdf/~/media/Files/Publications/NegotiatingIntegratedProjectDeliveryAgreement.pdf.

Cahill, D., & Puybaraud, M.-C. (2008). Constructing the team: The Latham Report (1994). In *Construction Reports 1944–98* (pp. 145–160). Blackwell Science Ltd.

Cheng, R., Allison, M., Sturts-Dossick, C, Monson, C, Staub-French, S., Poirier, E., *Motivation and Means: How and Why IPD and Lean Lead to Success,* (2016) University of Minnesota, Retrieved from http://arch.design.umn.edu/directory/chengr/.

Egan, J. (1998). *Rethinking construction*. London, UK: UK Construction Task Force to the Deputy Prime Minister, John Prescott.

Ballard, G. Kim, Y. W., Azari, R., & Kyuncho, S. (2011). Starting from scratch: A new project delivery paradigm. *Construction Industry Institute Research Summary 271* (11).

Larman, C., & Vodde, B. (2010). *Practices for scaling Lean & agile development: Large, multisite, and offshore product development with large-scale scrum*. New York: Addison-Wesley.

Latham, M. (1994). *Constructing the team*. London, UK.

Slovic, P. (2000). *The perception of risk*. Risk, Society, and Policy Series. Sterling, VA: Earthscan.

CHAPTER 18

Delivering the High-Performing Building as a Product

"The future has already arrived. It's just not evenly distributed yet."

—William Gibson

18.1 WHAT IS THE HIGH-PERFORMING BUILDING AS A PRODUCT?

The prior chapters have focused on how to integrate project delivery to create the high-performing building. Using the Simple Framework, we have laid out how the component pieces of design and construction can be brought together in a virtual organization. Although the end product may be a building, the project team is providing a valuable service to their client.

Designers provide services. They even refer to design documents as "instruments of service." The general contractor/construction manager predominantly provides management services. The traditional construction warranty only guarantees that the components are free from defect and meet the requirements of the contract documents. There is no warranty that the building is actually fit for its purpose or will do what is intended. Quite the reverse, under traditional construction law, the owner warrants to the contractor that the plans and specifications are sufficient.[1] The parties are required to perform in accordance with professional and industry standards, but they are not required to succeed.

An automobile is also constructed from components designed and manufactured by different organizations. But from the consumer's perspective, a car is not an assemblage of services. It is a product. The consumer expects that it will meet its requirements for mileage, maintainability, capacity, and style. All of its parts will work in harmony with each other. In fact, the consumer expects that the manufacturer will guarantee these qualities and will stand behind its product throughout a lengthy warranty period and perhaps beyond.

What would it mean for the owner if the high-performing building was a product? The owner would know that the building would perform "as advertised." The owner would expect that the building would not only meet specifications, it would meet their needs. From the owner's perspective, they purchased a whole building that would operate holistically. And if it does not, the owner expects the seller to make it right.

What would it mean for the project team if the high-performing building was a product? Most importantly, it would mean that the team was responsible for creating a product that meets the owner's needs—not try to meet—meets. It would also mean that the project—now product—team would have to execute the elements of the Simple Framework exquisitely. And it puts greater emphasis on digital prototyping, optimization, visualization, and simulation, because a product team guaranteeing a result must be highly certain that they have correctly identified the owner's needs and that the proposed building will satisfy them.

There are also opportunities for the project [product] team. Services are essentially cost-based. The owner pays the cost of the services and a reasonable profit. But the purchaser of a product pays based on the *value* of the product to him or her. Profit, in products, is the difference between cost and a sales price that is based on the value to the consumer. For a skilled and committed team, value is the best economic metric.

Example: Energy Retrofit

Energy retrofits are an example of product pricing. The provider designs and builds a solution, guarantees energy savings, and receives a portion of the savings. Profit is the difference between the value (energy savings) and the cost of realization.

Note that we are not proposing manufactured buildings, or mass-produced housing, such as companies have been doing since the 1940s and 1950s. Instead, we are proposing a high-performing building as a customized product that meets a particular owner's specific needs. In manufacturing terms, this is "mass customization." Each product is individual, but it is still a whole product, not an assemblage of services.

We recognize that the high-performing building as a product is not yet here. But this chapter describes ways that these types of buildings could be delivered as products for individual owners with unique requirements. In this way they would be similar to the complex products sold by aircraft manufacturers and ship builders rather than automobiles. And, as our real-life examples demonstrate, there are leaders who are currently exploring portions of the product space, and beginning to bring the high-performing building as a product from concept to reality.

18.2   WHAT DOES SUCCESS LOOK LIKE?

Building owners get exactly what they want for the price they agreed to pay on the date agreed, completely ready to use. The quality and scope of the finished product are as promised by the company or companies at the time of sale. No person is injured doing this work and the environment is not harmed in any way. The project organization will warrant operational performance and fitness for use based on criteria described in the contract.

The project delivery team is entirely responsible for the means and methods and is completely transparent. The owner or tenant's capital facilities staff can focus on stakeholder concerns and building performance throughout its life cycle. In many situations, the project delivery team will be a virtual enterprise made up of companies that come together right after the real estate or capital facilities director signals interest in construction of a new facility, equivalent to an IPD team. In other cases, the delivery organization will be a legal joint venture comprised of design and build companies with possible inclusion of facilities management, operations, and financing. In either case, most team members will have worked together on many projects and know each other well. They will understand how to leverage all elements of the Simple Framework and have refined their work processes outside of and between projects. Project managers from different disciplines who "possess group facilitation skills, organizational management skills, people assessment and change management skills, along with the tactical skills" will lead and manage these projects (Seed, 2014).

The project delivery team will explore and determine value for stakeholders and the project using a structured process such as Value in Design (VALiD), described in the Chapter 11. The stakeholders and team will use methods like those pioneered by John Haymaker to make stakeholder values visible to each other so they can be negotiated into a coherent set of project goals. Next, the owner and project team will focus considerable attention on performance to achieve triple top and bottom line sustainability as well as a highly useful and operable building. Their task will be to translate stakeholder goals into objectives with criteria that can be known, measured, and evaluated consistently. The project team will already be masters of integrating process knowledge, integrating information, employing simulation, and collaborating because these are the essential competencies for knowing what they should and can deliver.

Team members will be circular life cycle thinkers, always intent on using only healthy and sustainable materials that can be reused at the end of their product life (McDonough & Braungart, 2010). Above all, team members will practice the "Predict/Test/Adjust" cycle, striving to predict performance and engineer ever-shorter feedback cycles to see how they compare against the metrics they've established (Spear, 2010). They will see problems as opportunities to learn and improve. The team's goal will be to get smarter together and better every day. They will do this by creating a learning culture where failure is appreciated and experimentation is encouraged. Leaders and both design and build team members will understand that both are necessary for creativity and innovation (Catmull & Wallace, 2014). Early on, during value definition and design, the team will plan for iteration to allow for creative interaction.

Product Development

New products and services are generated, not by asking the consumer, but by knowledge, imagination, innovation, risk, trial, and error on the part of the producer, backed by enough capital to develop the product or service and to stay in business during the lean months of introduction.

W. Edwards Demming

The delivery team will develop, simulate, and analyze solutions for sustainability, usability, operability, and buildability. The team will propose sets of design solutions that meet stakeholder goals.

Delivery team members and stakeholders will work through the solution sets together. Both groups will learn and understand needs, desires, constraints, and possibilities better. Solutions will be refined. Ultimately, a single set with a range of alternatives will be chosen. At the end of the Conceptualization and Criteria Design stages (Eckblad et al., 2007), the team will develop its offer, which will focus on how the facility will perform to meet the owner's objectives.

Every team member will understand that trust is the prerequisite for true collaboration. They will know how to make reliable promises and protect the time to make good on them. Team members will be transparent about their failures and learning. They will set aside time for reflection, comparing their performance against their goals, as members of the California Pacific Medical Center (CPMC) Van Ness and Geary Campus team did (Lostuvali, Alves, & Modrich, 2012). The benchmark they chose, Toyota's principles, and system characteristics for product development, seem very appropriate for delivering the building as a product (Morgan & Liker, 2006). A willingness to look at problems openly and learn will be an essential part of challenging and fulfilling work, as described in an inspiring look into the future in the paper "Tomorrow's Workday, Spontaneous, Creative, and Reliable" (Fischer, 2007).

Members of the integrated product team will be Lean thinkers intent on making information and materials flow between individuals, workgroups, and construction crews. They will also be experts in virtual design and construction (VDC). They will model and simulate the performance of everything, including their own organization and work processes. Team members will not tolerate multiple different versions of the same information and establish single sources of truth (SSoT) for design requirements, and all performance data including first and life-cycle cost, quality, and schedule. The project will showcase live integrated dashboards that connect the different construction and planning processes (bidding, procurement, work packages, lookahead scenarios, field productivity, etc.) allowing the product team to exercise better control over the supply chain and cash flow.

The owner will accept or reject the fixed price, which may include the possibility of the owner deciding to expand the scope by including options described and priced previously. Funding these efforts will vary. In some cases, delivery organizations will find ways to pay for the design and even construction. In other cases, likely the majority at first, the owner will fund at stage gates based on the delivery enterprise's demonstration that the building will meet performance requirements. Owners and tenants will be motivated to change their own practices because this will allow them to gain much greater value, no matter how they calculate that.

Financial success will be based entirely on a product organization's ability to deliver value. Profit velocity will figure heavily into the equation because the enterprises that use these approaches will be able to deliver products faster than their competitors. Customers and suppliers, meaning the product delivery teams/enterprises, will understand and accept their liabilities. Top managers providing integrated high-performing buildings as products will know their responsibilities, which will be spelled out in a new kind of contract for product delivery.

Product versus Service Liability

Services, such as design and construction, are legally different than products, such as cars or mass-produced housing. In general, a service provider is responsible if it acted negligently or if it breached its contractual agreement with the owner. Products, in contrast, are warranted (impliedly or expressly) and the product provider may be responsible for losses even if it is not negligent, especially if the losses include personal injury or property damage. Moreover, the product provider may be responsible to everyone in the distribution stream, including subsequent purchasers or even third parties who are injured by the product. Consumer protection regulation is more likely to apply to products than services.

Mass-produced construction (e.g., tract housing in large subdivisions) is more product than service and has long been treated as a product by courts. (See *Kriegler v. Eichler Homes, Inc.* (1969) 269 Cal.App.2d 224.) In general, the more repetitive the product, the less likely it will be considered a service. An individual, fully customized building, however, falls into the gray area between clearly service and clearly product. But the more automated and repetitive the process, the more likely that the deliverable will be classified a product.

The expansive liability attached to products requires careful consideration of warranty terms and procurement of appropriate insurance. In addition, custom manufacturing may raise professional licensing issues. If a design is generated by computer algorithms, for example, who is the engineer for professional licensing purposes? And does the manufacturer need to be a licensed general contractor?

Although the legal issues related to the transition from services may feel strange to design and construction professionals, they have been successfully addressed in other industries. They should not impede the transition to product-based delivery. Rather, they are items that should be considered—and resolved—during the transition process.

Howard Ashcraft

18.3   HOW CAN THIS BE DONE?

If product profitability is the difference between value and cost, it is critical to know what will maximize the value to the customer and to be confident we can achieve that value, or more, without exceeding a maximum cost. Thus, we need to accurately assess and quantify value to the customer and then validate that it can be achieved. This need for certainty operates at an immediate scale, and a project scale. At the immediate scale, we need to be certain that we have adequately planned everything necessary for physical realization. For example, have we considered everything we need to know before beginning to drill piers? At a project scale, have we engaged in sufficient planning and digital simulation to commit our resources to building the product? The certainty requirement also leads not only to validating the product; we need to validate the processes we use to develop it. Creating a product is difficult enough without having to develop new organizations, processes, and information flows. The processes of the Simple Framework need to be in place before product development, rather than being developed in tandem. Finally, the certainty requirement also leads to considering the entire supply chain and how durable relationships can enable repeatedly delivering the high-performance building as a product.

It is a tall order, but let's start with the first step: clear instructions and directives.

18.3.1   Clear Instructions and Directives

The product value must be clearly defined. The project team must assure that they truly understand what the owner wants and needs. This may also be an opportunity for the team to create demand by discovering value that customers have not seen. As previously noted by Deming and as seen with products like the iPhone, innovation can create new opportunities, markets, and demand. Next, product value must be defined so clearly that the designers can instruct those who will be procuring the materials and components on what is needed, as is done in Lean manufacturing today. The engineers must hear the voices of those who will be making, assembling, installing, and commissioning the building and its systems. Although those executing work must be engaged in creating the directives, it is the responsibility of project management to communicate clarity at every step. Moreover, project—or rather product—management must determine when direction and planning is sufficient to proceed.

We agree completely with the assertion made by Digby Christian, Director of Integrated Lean Project Delivery, Sutter Health, that the lowest-risk strategy for project delivery is one where there are no questions about engineering, fabrication, or installation before a purchase or production order is placed, as shown in Figure 18.1 (Digby Christian, "Where Ambitious Owners with Ambitious Projects Need Ambitious GC's to Focus," unpublished presentation to DPR Quarterly Meeting, April 25, 2012). To the left of the point of release, the team is engaged in the many planning processes that must precede physical realization. Before the point of release, iterative design, set-based design, rapid prototyping, and experimenting can be valuable. But the variability that may be valuable to the left of the point of release is very damaging when it occurs to the right and undermines certainty of outcome. Thus, project management needs to determine the sequence of decisions and what tasks must be complete before authorizing procurement and fabrication. There should be no variability after the point of release.

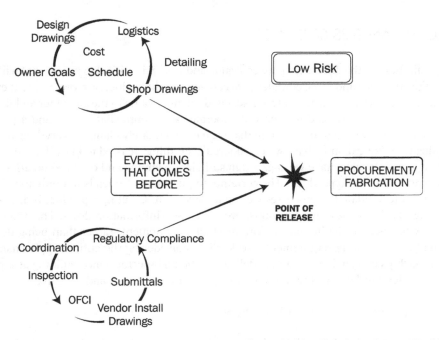

FIGURE 18.1   Digby Christian's "Point of Release" strategy. Courtesy of Digby Christian and Sutter Health. Illustration provided by DPR Construction, Lyzz Schwegler.

18.3.2   Prediction and Verification

The point of release focuses on decisions related to individual components or systems. But at some point, the project team must commit to creating the product. Success in product development will hinge on the ability to predict outcomes with much greater confidence than current practice. High competence in visualization and simulation will be required; this will be the price of entry into this new game. "Digital prototyping," which has been used for some time in manufacturing to simplify and test assembly, will also be a critical technology.

The authors of the paper "Four-Phase Project Delivery and the Pathway to Perfection" propose using a systems engineering V-Model, adapted by Zigmund Rubel of Aditazz, for design and construction in combination with Lean construction theory to compare current practice against an ideal, perfect state (Christian, et al., 2014). The "Rubel diagram for the building industry" in Figure 18.2 shows the steps for "building the right building" and "building the building right" in the shape of a "V."

The "System 'V' diagram of Four-Phase Project Delivery," which is an overlay (shown in Figure 18.3), includes phases for "Value Definition," and "Value Capture" leading to a vertical line marking a "Boundary of Realization" between the "Representation" and "Realization" phases. This will be the point where project delivery organizations make their offers to owners and budgets are set.

The "Looking forward" and "Comparing back" arrows across the top, and description labels noting that "Analysis," "Modeling," and "Simulation" are the focus of Representation, present a useful high-level model of building product development. Feedback and analysis all along the way are critical to making highly reliable predictions and verifying that a project team is delivering the right building as a product. The purpose of the V-Model is to show the link between definition and development on the left side and verification and testing on the right. The idea is that managers of complex systems cannot do one without the other. This comparison is the same as comparing the function and behavior in the POP framework described in earlier chapters in the book.

The "Boundary of Realization" is a critical project juncture. The investment to the left of the boundary pales in comparison to the investment to the right. It is similar to a "Point of Release" for

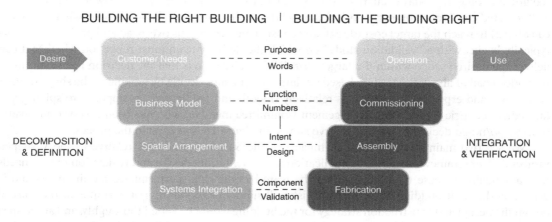

FIGURE 18.2   Zigmund Rubel's System "V" cycle for the building industry. © Aditazz.

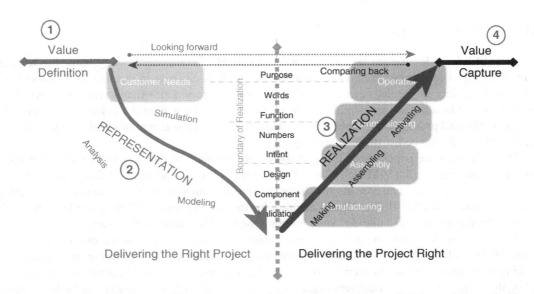

FIGURE 18.3   System "V" diagram of four-phase project delivery. Courtesy of Digby Christian and Sutter Health.

the project. It is the point where the team has made an irrevocable financial commitment. Confidence, once again, becomes a critical factor.

But it would be a mistake to equate confidence with perfect knowledge. Validation, in Figure 18.4, below, is the process of developing a reasoned, and reasonable, belief that the outcome can be achieved, even if there are details still unknown. Confidence is represented by the gap between perfect knowledge (certainty) and what we actually know (validation). Using tools, such as simulation and visualization, we can reduce the gap between validation and certainty to create adequate confidence to proceed. This point of sufficient certainty, which will be project and team specific, is the "Boundary of Realization."

In fact, design-build and IPD teams applying the target value design approach developed by Glenn Ballard are doing this today (Ballard & Morris, 2010). Every team of which we know has had to reduce an "expected cost" (current market price) that is often well above the "allowable cost (what the owner can afford) to reach the target cost (design and construction costs plus overhead and profit). The effort typically begins with a "validation study" on which the delivery team must report on whether it can deliver the desired scope within the target cost. What we've seen is that the decision to commit must be made ahead of absolute certainty based on data. Smart teams decide by looking at the shape of their cost curve, and especially whether it is relatively smooth and predictable, as opposed to spiking and unpredictable. Eric Lamb, DPR Management Committee member, who has been involved in several of these go/no-go decisions, drew the curve shown in Figure 18.5 to explain the process.

We see two main applications and impacts of the Simple Framework relative to the V-shaped project delivery framework. First, its application in the definition, design and realization phases needs to dramatically increase the certainty that the team has the right performance requirements and is designing the right building, otherwise the risk will simply be too large. For example, at the start of design the design and construction strategy for the building would be vetted thoroughly, and at the start of construction, there should be few remaining questions about the building design. Since the team has

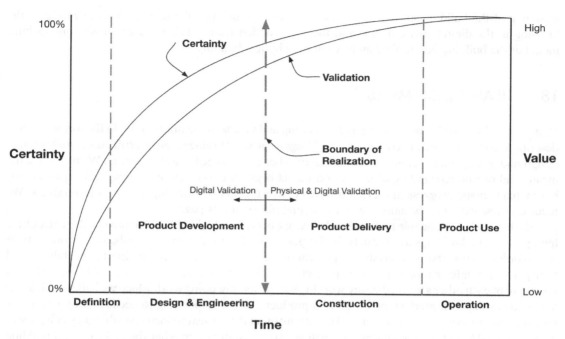

FIGURE 18.4 Confidence, validation and the Boundary of Realization.

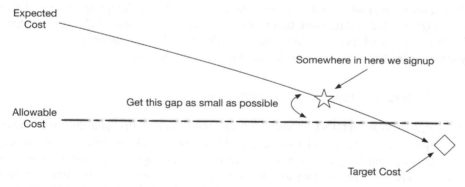

FIGURE 18.5 Eric Lamb's "Closing the Gap Curve." © DPR Construction, by Eric Lamb.

access to all the expertise and information needed and can validate a design internally (through expert feedback, and simulation and visualization) and externally (through benchmarking with other similar structures), the validation of the design follows the definition and design of the building very closely, dramatically reducing the risk for all team members and the owner. Second, the Simple Framework continues to be applied during construction for the remaining design decisions, but also to design the supply chains and project schedules so that the designed building is built as designed and can be in productive use quickly. The team will also stay ahead of the physical construction with the digital modeling of the building so that the operations team can practice the hand-over and start-up of the building in the digital environment, and the construction team can hand over an accurate as-built model of the building before the handover of the physical building.

18.4   REAL-LIFE EXAMPLES

Until recent times, we've not had the understanding and means to use information in the way we've just described so that we could even consider buildings as products rather than as artifacts created through design and management services and craft labor. That has changed (Fischer, 2013). We now have the means and our understanding of how to do this is beginning to catch up. People and organizations have already made progress in moving towards delivering high-performing buildings as products. We focus on these examples because they are essential pieces of this puzzle.

We have asked six people to describe their work because it highlights what must be done to deliver high-performing buildings as products. In the pages that follow, they will describe their work in their own words. Each is trying to produce significantly better outcomes by improving the usefulness and transparency of information early in the project life cycle, where leverage is greatest. Although they all are aware of each other's work, they are working independently, doing work which we believe will allow teams to develop and produce buildings as a product. All are responding to needs and opportunities they see and are working within financial constraints that drive them to focus on what they believe will be most valuable to their customers. You will see strong attention to what the end user of a building really needs and the corresponding performance targets. You will also see a strong understanding of the processes needed to achieve the required outcomes, and you will notice that the six are thinking differently and inventing tools to automate mundane tasks so that an integrated project delivery team can focus on higher value-adding work that moves it closer to delivering much greater value.

Table 18.1 shows each person's focus, their starting premise, the theory on which they operate, how they apply their ideas in the real world and what they are learning.

18.4.1   A Different Perspective on Architecture

Markku Allison, AIA, is an architect-turned-consultant who works with organizations and teams, especially those employing integrated, collaborative, and Lean models. He has over 25 years of experience both as an award-winning designer and as a thought-leader on design and construction industry transformation issues. He spent seven years as a Resource Architect on staff at the American Institute of Architects focused on industry transformation issues before founding Scan Consulting in 2012.

Markku described this work and his view of the future of architecture as follows.

Creating Value through Design

We've designed and delivered buildings in essentially the same way for hundreds of years, with very unreliable outcomes—many projects go well, but many do not. Highly collaborative and integrated methodologies enabled by technology have emerged in the last ten years or so that have great promise for improved outcomes, but require a significant cultural shift on the part of all of the players.

TABLE 18.1 Different Approaches to Product Development and Delivery

#	Innovator	Focus	Theory	Application	Learning
1	Markku Allison/SCAN	Designing valuable solutions	Rich information informs better decisions.	Create and explain examples in word and deed.	Changing thinking is difficult but possible through changing work practices.
2	Michael Bade/UCSF	Holistic performance-based design and construction	UCSF will receive significantly greater value by clearly articulating performance it needs rather than prescribing design requirements.	Develop a database of performance requirements that can be use selectively across projects.	This can work! It takes proactive thought by the owner before prior to the project.
3	Stewart Carroll / Beck Technology	Integrated information tools	Teams with rich, visual integrated information will provide greater value.	Develop software for integrated information. Facilitate use of tools in teams.	People must trust information. They must be involved in creating it.
4	Andrew Arnold/DPR Consulting	Validated information for facility operations	Building owners can harvest valuable information for operations created during design and construction.	Understand what information is useful to operators and how they will use it. Test building information against criteria digitally at the time of production.	Useful information can be validated digitally if operations and use criteria can be determined. This work can be done in phases.
5	Forest Flager/CIFE, Stanford	Multidisciplinary design optimization (MDO)	Designs can be improved over thousands of iterations if objectives can be made clear enough so that criteria can be determined for testing.	Understand how well defined and precise criteria must be to leverage computational resources on projects.	MDO is viable. More applied research is necessary to understand range of application and best use.
6	Zig Rubel/ Aditazz	Componentized design and construction	A significant percentage of building design can be based on rules, opening the door for automation. Certain types of buildings, such as hospitals, are better candidates. Building systems can be made up of automated components designed how microchips have been designed for many years.	Domain experts develop rule sets that can be embedded in computer programs that optimize component design for performance, fabrication, and installation.	Component design can be automated. Questions remain about scalability, i.e., at what point are there enough components to make this viable economically.

What my work has really been about for the last ten years, both during my time as Resource Architect at the American Institute of Architects and in my own consulting practice since 2012, is exactly that cultural shift. My efforts have largely focused on helping industry stakeholders, organizations, and project teams—and architects in particular—better understand how design delivers value across the life cycle of an enterprise, to learn to behave differently in collaborative environments, and to embrace new ways of thinking and acting around project delivery.

My work is based on belief in a fundamental premise that higher value outcomes can be achieved in complex problem solving if tackled in a collaborative fashion. Collaborative models are not the norm for most designers and constructors, and so a big part of my work has been focused on trying to help clarify their value proposition for industry stakeholder partners in general and architects in particular.

As an architect, my particular personal interest is in the design side. Why is it beneficial for architects to function and behave in a more collaborative fashion? I've mostly been helping teams in early stages of projects clarify their value propositions to define more clearly the goals and objectives and outcomes for not just the project, but all its individual stakeholders, to identify the values that they collectively wish to embrace as part of those adventures. To help frame the entire design conversation within a much larger time view of the project.

Rather than having a conversation focus on just the design phase, all of these conversations need to be rooted in an understanding that the building is really a tool for an owner's enterprise to create value with. The conversations need to occur from well before the building begins to well after the building is complete, and the conversation about values and goals and objectives and alignment needs to stretch across those same time boundaries. That's really the crux of my work.

Some fear an exposure of the architect to more risk with their expanded role. This fear is situated in a practice where a firewall has been instituted between the perils of designing and the perils associated with executing those designs. In reality, the more significant risk to the architectural profession is that architects will experience an increasingly smaller role. Morphosis never faced these risks because our work has always been about the complex interactions that produce something; never about the thing itself. We started with small-scale projects, making them and constructing them; we have always been builders. Thirty years ago, we were working at a level of craft where we began collaborating with the subcontractor from the preschematic phase. With BIM, we have been able to return to this initial level of engagement, working with the manufacturer to strategize fabrication during the first phase of drawings.

The practice of architecture is an integrative act. Information sourced from many specialists and forces emanating from many directions are synthesized by architects into a single work. The sharing of information between architect, engineer, fabricator and contractor erases the divisions between design and realization. The 3-D environment facilitates this integration and coordination across specific areas of responsibility. This environment enhances the spirit of cooperation, resulting in improved two-way communication. Designers are better able to incorporate contractors' input into their work, and contractors can contribute to the design process while gaining a better sense of design intent so they can reinforce its integrity as they execute the built work.

Thom Mayne, from Shift 2d to 3d, in *Digital Workflows in Architecture* (Marble, 2013)

Industry Response/Evidence

Everything that I have learned in my experience about Lean and IPD in the last 10 years has been over-whelmingly positive in terms of response from all industry stakeholders. That doesn't mean that all the problems are solved. There are still challenges that teams have. In any changing world, there will be naysayers. People who are just resistant to trying things in a new way, but overall, in my experience anyway, the response is quite positive.

However, a challenge that we face as an industry in trying to promote new models of practice or new models of design and delivery of construction projects is that everybody wants evidence. We want a proven track record. We want confidence. Many people are reluctant to embrace new models until they have affirmation that it will be a successful path. In response to this need, I've been working with Renée Cheng at the University of Minnesota on a couple of research projects. There hasn't been a great deal of research in the integrated and Lean project delivery worlds quantifying the benefits that accrue to owners and teams who tackle things in these new ways. Renée and I have been working on two projects designed to help build the body of evidence that supports collaborative models. (Published subsequent to this interview, see below.)

The first project was a simple survey. After collecting some demographic information about the stakeholder who was answering the survey, we asked essentially three questions:

1. How did this project perform compared to traditionally delivered projects that you've worked on?
2. How did your expectations compare from the beginning of the project to the end of the project?
3. What is the likelihood that you would use IPD again?

The results were profoundly positive. We had expected positive but not overwhelmingly positive answers in every category: schedule, budget, quality, changes, morale, and overall value delivered. Respondents from across the stakeholder spectrum overwhelmingly said that IPD projects were performing significantly better, that they would be significantly likely to do it again, and significantly likely to recommend it to others (Cheng et al., 2015).

The second project we are working on is a case studies effort. It's a deep dive into 10 Lean and IPD projects looking at what is working and what is not working in order to help teams that are starting down those roads identify tools or processes that might better help them connect collaborative approaches to improved business case outcomes for the owners (Cheng et al., 2016).

The Future: Building as Product

The work I've been doing has allowed me to spend quite a bit of time thinking about what the future might look like. What will architects do? At one level, we will continue to do that which we have always done well. We've always excelled at expression and exploration, and those outputs won't change. It's more the input side of the design equation that will dramatically change how architects work.

Personally, I'm not a believer in design by committee. However, that design can be far richer and more robust with more sources of input than it has traditionally considered. The real difference for how architects will work is that the number of inputs into the design process will increase dramatically. If, along the way, those input sources also happen to have ideas or design suggestions that are beneficial to the overall project, that's fantastic. But architects are still going to be designers. It's the multiplicity, volume, quantity, and quality of input information into the design process that will increase dramatically in these highly collaborative models. In the future, an owner enterprise will engage a project team—including architects—much, much earlier in the enterprise life cycle. Here's a scenario of what it might look like:

When an owner first sees that business case metrics are suggesting that a building might be a solution to better their enterprise outcomes, they'll form a cross-disciplinary team to start to monitor and analyze those metrics and test different scenarios. Architects will be a key player on those teams. They might be assisting in terms of definition of values, identification of scenarios or outcomes or purpose, and will help flesh out a stronger internal pro forma for the enterprise case for the development of a new capital project if indeed that ends up being the direction that things go.

Suppose all indicators suggest that the value proposition says, "Yes, a facility or some kind of a capital project is the direction that enterprise needs to go." The architect will move into a second phase of work within this cross-disciplinary team where we'll be involved in sort of a validation study and performance design phase. This part will really be about creating a yes or no gate for the enterprise owner to decide to go forward or not.

The architect will, as part of this team, be gathering and analyzing massive amounts of data and information. With their stakeholder partners they'll be identifying scope and performance objectives, looking at feasibility, assigning current market values to work. They'll be framing questions and gathering necessary information to answer those questions. They might be making some sketches or models or designs, but really, they're going to be answering questions and providing data within the context of a pro forma business case.

We don't need designs to make design decisions. At this phase of our projects, architects will be largely working with information and descriptions of performance rather than actual building designs. It's going to require us (and our partners!) to be more adventurous about what we traditionally think of as deliverables. It will be a key component of our future. This step isn't really about schematic design. This is a detailed analysis and information-driven response to a value proposition that's rooted firmly in an owner's business case.

At the end of that phase, assuming the project goes forward, then we start to get into territory that will be much more familiar to architects today, where this collaborative team will start to work together to really define what is it that's going to be built. One distinction from today's more typical practices: the future will be filled with very robust simulations. The architect and this collaborative team will be working with large amounts of data and information gathered from all of the stakeholders together fed into the design team's work to inform simulations along a whole range of different axes. One cluster might be around systems. One cluster might be around envelope. One cluster might be around performance. One cluster might be around aesthetics.

The team will be modeling, simulating, and designing in response to the criteria that we've set up as part of that original value proposition and business case. The team collectively might be starting to think about opportunities for prefabrication or automation during construction, modularization, working to come to an overall building design, but basing that design on a detailed description, a detailed estimate, a detailed performance specification, and business case versus having the design come first and then pulling other stuff out the back end and hoping for alignment. This design phase is the team's collective opportunity to drive value in and maximize the enterprise outcomes along the dimensions established in the original enterprise value proposition.

The next phase is about implementation. A distinction here in what architects will do as compared to what they do currently is a clear recognition that implementation begins well before construction.

The idea of implementation documents will become very common. That means a lot of information traditionally relegated to stakeholders further down the line, like shop drawings or very detailed decision making, will be pulled way forward and integrated into these documents.

In this phase, the architect's work will be different from what they might be used to historically because there will be many other contributors to the documentation about what this thing exactly looks like and how it is going to be built. The architect will continue to be the champions for expression and experience and holistic thinking about tying all of this stuff together. But we'll be working closely with our stakeholder partners to really maximize the value that they can bring to the table in terms of detailed instructions and thinking about assemblies, whether it's modularization or prefabrication or whatever.

Since it's simply not practical to create a 1:1 real-time simulation of a construction project, there will always be information gaps in implementation documents that will pop up during the construction phase. Which means there will always be detailed decisions that will be required to be made during construction and the architect will, of course, play a continued role as part of that multidisciplinary team to keep an eye on those detailed decisions and help contribute their particular expertise as it might be appropriate in that particular arena.

That's my rough idea about how things might look for us in the future. In the end it is going to be an initial step that's about an enterprise business case; a second step that's about performance and validation; a third step which is going to be about simulation and alignment and definition; and then a final step which is about how and implementation. And then repeat.

On Compensation

The future income streams for architects and other design professionals will be different than it is today. Today, architects and consulting engineers mostly exchange time for money. If this plays out optimally, integrated teams will be compensated for delivery of value. Architects will be part of those integrated teams. It's going to be more about a conversation about what is the overall reasonable compensation for a team to deliver this value-creating product for an owner.

If we think of the building as a tool that creates value for an owner's enterprise, we're toolmakers; the owner can commission a tool and we can deliver that tool. The tools that we make happen to make are much more significant than, for instance, a crescent wrench or something like that, because they exist in human culture, and so have additional dimensions of cultural and aesthetic importance. But in the end, it's still a tool that allows an owner's enterprise to create and deliver value.

The owner commissions a building and assembles a team; the architect is part of that team. How compensation filters down to the individual stakeholders within that team, will be an ongoing process of experimentation. If I were an owner, I would be interested in developing compensation models that had components that were based on proof that I have gotten what I asked for, along multiple dimensions of the business case—in some cases, long after the fact of the completion of the building.

We are going to be far more intimate partners with an owner's enterprise than simply associated with the delivery of a building. We will be woven into the fabric of the enterprise decision-making and value-creation process. We will be tasked with creation of a tool to maximize owner enterprise value and will be compensated accordingly. Architects will be part of these conversations, and there will be a ton of different compensation models that will emerge and get tested. Architects, as well as all of the stakeholders in this arena, should be bold experimenters in compensation but focus it ultimately on creation of value tied back to the owner's business case.

A Necessary Shift

What it will take for owners, architects, and other stakeholders to have a shift in perspective to where the building is thought of and delivered as a product is almost certainly going to rest on a shift in the perspective of time. A shift where we collectively—architects, owners, constructors, the whole spectrum of industry stakeholders—begin to look at the building in the context of an enterprise value-creation stream and not just as a design and construction project.

Right now, the industry has largely confined its value proposition between the beginning of design and the end of construction. The building becomes the end-all, be-all because that's our entire focus. But if we start to really see how the building supports and fits into an enterprise picture, the idea of the building as a product will become a far easier step. Robust simulations will be a critical supporting component here.

If we zoom out 30,000 miles and we see the facility as a point in time along, say, a 50-year life cycle of an enterprise, it becomes much easier to see the building as a tool. I want to emphasize again that doesn't mean the building doesn't have all of the cultural, aesthetic, or experiential significance it's always had to designers and our clients, but rather that it enables us to shift our perspective to create a clearer understanding of connection to the enterprise business case, and to use that as a driver for decision making.

This is a hugely positive shift! If we architects along with our partners embrace it fully, it will allow us to focus more of our energies on the creation of value in every dimension we care to consider (Markku Allison, personal communication, January 6, 2016).

18.4.2   Holistic Performance-Based Design & Construction

Michael Bade is associate vice chancellor for Capital Programs and Campus Architect at the University of California San Francisco (UCSF). He is responsible for the design and construction of classrooms, offices and laboratories, and community facilities such as housing supporting UCSF's mission at all UCSF sites, including the Parnassus and Mission Bay campuses. Since assuming these responsibilities in 2008, Michael has led his organization's move from prescriptive to performance specifications and design.

Michael described this work as follows:

We own a very large and complicated physical plant. I think we have more than 10 million square feet of space, and, in the way that the university classifies such things, about 70 percent of it is deemed complex space. It's laboratories and hospitals, and things like that. Those buildings are very expensive, and they consume loads of energy, and they contain critical operations, patients, and experiments. The performance of the buildings is very, very important to us. We came to the conclusion that using traditional methods where the owner tries to target building performance through standard specifications and the like just weren't working. It wasn't resulting in buildings that have the performance attributes that we wanted.

In thinking about this, we realized that we needed to take a really different approach than the traditional approach. We needed to objectify our performance objectives into top line project objectives. We needed to achieve that through a holistic design process, rather than a reductive design process. By that, I mean if we're going to target a high level of energy performance, we target a high level of energy performance. If we were going to target maintainability and durability, we need to do that straight up, as a front-end objective. We need teams designing our projects to employ a scoring system that would allow us collaboratively to determine what we're getting for our money, and to measure how what we're getting for our money contributes to our goals. The only way to do that is working from the big picture down, rather than from the nuts and bolts up.

We decided to change our attack with Mission Hall (shown in Figure 18.6) and we created our first stab at performance specs. Being an office building, the complexity of it was limited compared to our other buildings. What we were really looking for was a proof of concept that we could procure, design, and construct a building using a comprehensive performance specification and a rich set of programming information. The answer turned out to be yes, and it had some very interesting effects. Sam Nunes of WRNS Studio said to me one day over lunch, "This process which unfolded in the context of a design-build contract puts the designer back in the driver seat in a really positive way."

Mission Hall

- Design/build 265 KGSF office building at Mission Bay
- Three-level performance specification
- Fixed cost of $93.8 million (including furniture and IT | Fixed Program)
- Lean construction required
- Best Value Contractor Award
- Project awarded—August 2012
- Complete design—9 months
- Temporary certificate of occupancy & substantial completion—9/5/2014

FIGURE 18.6   UCSF Mission Hall.© Mark Citret Photographs.

The builder is usually in the position of always telling the designer it costs too much. If the builder has performance objectives that they understand, that they and the designer are partnering to try and achieve, then that just changes the dynamic. We really like that. We regard Mission Hall as a successful proof of concept. The building that we got has exemplary structural performance, especially for an office building. It's basically a Level 2 Essential Services Facility structure, which is important to us because we value being able to operate soon after a major seismic event.

It has energy performance that is in excess of the university's policy. Our policy is to be 20 percent better than California Building Code (Title 24). The building was planned to be roughly 25 percent better than Title 24 as modeled during design. We have not been in it for two years, so I look at that two-year-period as being something that gives us the requisite time frame to begin to ascertain energy performance. It's been a really good energy performance overall. Facilities feel that the building is, in the main, quite maintainable, and very durable, and investment was well placed. We've got Huntair air-handling equipment for instance in a building that cost less than $300 a square foot for the building structure.

Usually, we only get Huntair air-handling equipment on a premium laboratory building. Because the project was able to have a holistic approach, basically, they distribute air under the floor in a raised floor situation, so there's no complex sheet metal ductwork. The savings in ducting cost went into buying premium air handlers. It's exactly the kind of trade-off we're trying in engender.

We stopped trying to prescribe for the team how to achieve our goals, and usually what happens when you're trying to do that, your goals are fuzzy or not expressed clearly. Instead, we try to express our goals clearly, and we created a performance specification in as many areas as possible as tiered levels, so that at least there's a base level and one aspirational level higher than that, often two. That framework becomes then a framework for a complex nuanced calculation of value.

What have been the challenges to doing this? The first challenge is the people challenge of course. Everybody in the industry is used to doing things in a different way. It's just like moving to Lean construction (which is a required ingredient of the kind of process we are trying to achieve). In the construction industry, you've typically got a whole bunch of people who are experts at fighting with each other and avoiding responsibility, and you're trying to move to a situation where they're all pulling on the same rope in the same direction and accepting responsibility for top-level outcomes. Their self-interest and the owner's interest diverge if the process isn't correctly structured contractually, and so you have to find a framework within which you can get everyone's self-interest and the owner's interest to converge instead of diverge.

This wasn't just about a performance approach to design. That performance approach to design is supported by a commitment to Lean construction practice, which allows for the self-interest of the team, and the owner's interest to converge. Incentivizing achievement of high-performance in both the design and in project delivery is a high-level intellectual, and legal, and management challenge. Legal, because, in the public contract code arena, you have to be able to have your incentive programs pass legal view. Management, in that you have to have a clear vision of what you're driving the bus to, and stick with it. Human, because you have to use goodwill, and you have to break things down, so the people can understand them. You have to give them a chance to learn. You have to allow them to make mistakes, so that people can learn from their mistakes—learning cannot happen without mistakes.

This is fundamental, and my wife, who's an educator, constantly reminds me that there is no new information value in a mistake-free process. Toyota, I think, is emblematic of that. Those are the challenges: clarity of concept, both for the performance specification, for the contract, for the ways the contract is interpreted, for the vision that everyone shares. Clarity and shared vision are really important. Everyone commits to a learning process that allows mistakes to happen as long as they're learned from and not repeated. These are fundamental concepts that underpin a value-seeking process, which is what we are trying to engender.

Your business partners must be committed as deeply as you, the owner, are. We were really lucky to have, for Mission Hall, the team that we engaged to design to build the building was as committed to this as we were. They dedicated themselves from the first day that they got together to start to put their proposal together. They committed the leaders of their companies. I had Martin Sizemore's (general contractor, Rudolph & Sletten) personal commitment. We got one of their best superintendents. We got a terrific project manager who wasn't really up on Lean when she started, but she dedicated herself to learning in the most open-minded way. We had Southland for MEP, and Walters & Wolf for X-wall who have been through this with us before.

We had some strong backs pulling the ropes. WRNS Studio was all in. In fact, we all worked hard together, but because we had equal partners and people who were as committed to us, it wasn't as difficult as it might have been. We have had some other projects where we've used Lean ideas, and we've not had the commitment, and that's resulted in difficult situations. I think to go in a new direction, you need people who want to go there with you, and who are wholeheartedly committed to it, and we had that.

What did we do? I'm going to give you two answers. The first answer is going to be on Mission Hall, and the second answer is going to be on what we are doing now. I think that they need to be taken together. On Mission Hall, we hired Gensler Architects to help us develop the program for the building. We hired the consulting group at Gensler. The consulting group, at that point in time, was an interesting group of people. They were free to respond to the needs of the client, as opposed to doing things in any particular way.

We got a really strong partnership from John Duvivier, the leader of the consulting group, and he understood what we were trying to do, and he galvanized the effort within the firm to create the performance specification and the project program. They did a really good job of it for the first time out. It was clear and it answered just about every question that the designers later had about our intent. The standard answer for them was, "If you have a question, look in the book." Nine times out of 10 or better, they were able to find the answer to the question in the book, which I think is extraordinary. We did get questions, but these were really sophisticated questions.

One example is in the confluence between structural design and exterior envelope design and interior systems performance. Like I said before, we got a building structure that was roughly equivalent to an Essential Services Level 2. The commitment that the team had made was to an immediate occupancy in the maximal Hayward Fault Event as predicted by the USGS, United States Geological Survey. When we were going through structural peer review, the peer reviewers who work for UCSF and I asked about nonstructural performance, and we got into this big discussion. They said, "Well, you never said that we had to do nonstructural performance, too." I said, "Well, gee, I thought that immediate occupancy didn't just mean that the building didn't fall down. What if it rains? We have a raised floor. We can't have big areas of the raised floor caved in." They went back, and they looked at the envelope design, and all the anchorages were fine. They'd all been engineered at the same level as the building structure. All the structural work within the panelized exterior envelope, we had 12-foot high by 36-foot long exterior panels, 351 of them. That was all fine.

What hadn't been thought about was the width of the panel joints, in terms of the amount the building was going to flex. They actually had to widen some joints. Well, we had this conversation in a timely point in the design and construction cycle. They were able to make the joints wider. The cost impact was really minimal, and so they went ahead. Now on the raised floor, we did pushover tests in the field with the floor support system, and we found that the supports were stronger than had been identified by the manufacturer. They were steel columns on a base plate, and the base plate is epoxied to the concrete floor slab.

The adhesive has only been around for about 15 years, and nobody knows how long it will retain its properties, or whether it will brittle over time. We had this long debate over epoxy brittling, and there's literally no engineering information on this. We searched, and we did an A3 on it, and we looked at all those different ways that we could improve our comfort level. I agreed with the team that nobody in a design-build competition situation should be expected to understand the full implications of something as esoteric as epoxy brittling. I said, "Well, UCSF will pay for whatever enhancement we make over the epoxy," which as I mentioned earlier, had performed well. In our tests, we put 16 in and bent them until the base plates broke, and the epoxy didn't break, the base plates did, and the concrete wasn't torn up either.

We decided in an A3 process; we had four different alternatives. There was one alternative for about $280,000 where we bolted one base plate every 10 square feet as a sort of belt and suspenders approach to ensuring that the raised floor would perform, which we felt was a reasonable outcome on this. The performance approach, as you can see engendered this holistic approach to a very critical issue, which is how well will the assembly perform in an earthquake. That's an issue that's of great interest to us as an owner who builds buildings for our own account, and who will occupy them in perpetuity essentially, or until we tear them down. In fact, we are about to retrofit a building that we built 100 years ago. Long time means long time.

The dialogue that I just described is exactly the kind of dialogue that a performance-seeking owner who plans to own in perpetuity wants to have with their design and construction teams. We are such a performance-seeking owner. Not all owners are after what we're after. If you're a real estate developer and you're building to flip into a REIT (real estate investment trust), you sure aren't necessarily interested in what I'm talking about. This is a profoundly different way of thinking. The people who are involved in this have to be up to the task of developing and holding to an investment rationale that is different from the norm.

What next? Right now, we are confronted with a problem that a lot of people would like to have, but it's a hard problem to solve. We now have to do this at scale, and I don't mean just any scale. UCSF has a very large capital program. At this moment, we're starting four major projects simultaneously, two of which have some of the most sophisticated biological laboratory components, and there are just major challenges with schedule, cost, and ensuring the performance outcomes we seek. Rather than approaching this issue project by project, we have hired a single architectural firm to work with us to further develop the performance specification as a system that can be applied across multiple building types and occupancy classifications.

They are also going to develop the programs for all of these major building projects, to ensure that our programming standards are consistent. They are barred from going after the design contracts, but this is an important services contract that deserves its own recognition and reward. The firm is Perkins + Will. We're working and learning with their staff, and their staff and our staff are beginning to work this out. It's an exciting time right now because of that. We're putting the performance specifications both into a book, and into a relational database, out of which we will be able to extract the requisite sections that are appropriate for our building types on individual projects.

In the future with this database we'll be able to customize books of performance specs for individual projects. We're in the procurement process for the first building project and have at least three more following. We are also in the process of evaluating what we did with Gensler against the screen of the results, and looking at how we can improve it.

It's a very significant commitment to develop this material. In parallel with this, we are implementing a new business system. In that business system, we will characterize our construction cost using Uniformat, which is to say by building system. We will be asking our business partners to make their estimates and their pay applications in Uniformat, so that we can track the costs we incur back to building system and measure performance of systems against this cost. That's a fundamental system architecture decision that we've made, that is going to become foundational for everything that we do in terms of capital investment going forward. We're going to be coding our building information models in Uniformat, so that in a 5D design and construction process, we can also map costs for individual component line items back to the BIM.

I think that it's important to know that what we learned from the Mission Hall experiment in terms of performance-based design and construction was really, really positive. It resulted in a commitment to go in a new direction, and I think that some would wonder, "Is that new direction going to have a positive return on investment?" Certainly, there's no guarantee, but the initial pilot project certainly did. We are confident that in about five years' time, we're going to be in a very changed place that's much more intentional about performance objectives when we make capital investments.

We are working hard with our industry partners to develop these ideas. We have a high level of demand from industry to be involved with us. I think that people get excited about doing things in a new way. There's a real hunger in the construction industry for positive outcomes and to do things in new ways that can improve the human relationships between customers, designers, builders, and suppliers. We really support that, and enjoy it; we get a lot of value out of being a part of it.

We are using normal project funds to do this. In the context of a project budget, we're redistributing our investment in the development of owner-provided information across the project. We are setting a very high standard for the information that we give our design and construction team, because I felt after more than 30 years in the business, more than 20 of which are spent on the owner side, that the fundamental flaw in the process that really gets traced back to the owner is that owners too often give people poor quality information about what we want.

If we can solve that issue, then we free the people working to achieve our goals to work in a much more efficient way. We can plow the difference into improving the front end, and then it becomes a virtuous cycle where investment in the quality of information reaps great benefits down the road. We got a terrific building for a very competitive cost, and the design process was very direct and linear as opposed to wandering in circles in the wilderness.

The other benefit that this approach has for an institutional client is that it gives the institution something objective that all of the different voices in the institution have to honor. It's very hard for people in the institution to undermine the approach, because it's transparent. It was achieved in a dialog process within the institution, and then its approval means that, "Yeah. This is what we're doing." It provides ballast to the process that otherwise can get reversed by voices that are coming out of the woodwork (Michael Bade, personal communication, December 16, 2015).

18.4.3   Developing and Using Tools for Information/DESTINI

Stewart Carroll is chief operating Officer for Beck Technology, Ltd., which he has led since its inception in May 2000. Prior to that he was software engineer for PTC (formerly Parametric Technology Corporation) and Reflex Systems where he wrote C++ based code across the Reflex platform. Reflex was a 3-D building design software application developed in the mid 1980s and—along with its predecessors RUCAPS and Sonata—is now regarded as a forerunner to today's building information modeling applications.

Stewart explained Beck Technology and his work there as follows:

Peter Beck is the third-generation owner of H. C. Beck Construction. When he took over the company in the early 1990s, one of his first projects was a project up in Philadelphia that went amazingly poorly. We ended up in a litigious situation as a result of poorly executed, incomplete deliverables that resulted in some major safety and project delivery issues. There were some major misconceptions about what we were building versus what the owner thought they were getting, and it ended up in court. It was such a big court case it could have taken Beck out. Peter really had this epiphany. The epiphany was we were taking way too much risk too late in the game where we have little to no control over what is really being designed. We are trying to fix things to the best of our abilities, but we're delivering poor quality. We are over budget and over schedule, and it's just not working. Out of that came this belief that as an industry we need to start to collaborate a lot earlier than the construction phase. In order to collaborate, he recognized we needed a new set of tools. The new set of tools weren't simply aimed at documenting what the architect or contractor needed. They were tools designed to enable teams to make better, more informed decisions. Better outcomes were really the goal.

DESTINI stands for Design ESTImating INtegration Initiative. It began as an internal software development effort inside Beck and later led Beck to purchase the intellectual property of a database-driven parametric modeling engine, Reflex, from PTC in 2000. Our first iteration of DESTINI was an integrated product that enabled an architect developing a concept to sit down and answer a series of questions. Based on the answers to the questions, the computer program would assemble a rendition of the project based on the criteria it was given. It would run engineering calculations. It would tie the means and methods to a cost database where we were working with subcontractor prices for installation and materials. It would use the subcontractors' production rates and crews to generate a schedule. In a very short space of time, it would produce a set of documents from which a builder could build.

Our intent was to enable many different options to be studied, so collectively the team could come up with the highest value-add for the project. Out of that early technology came this proof of concept that it worked. However, the way that it worked was not scalable. It was offensive to architects who took our intent to be that you could design a building at the push of a button. They, and we, knew that design is more complicated than that.

In 2003 we took a step back and said, "How do we still deliver on the same idea of better decisions earlier in the process and cooperation between the architect, owner, and contractor?" The tool would be fundamentally different. It would be more interactive and flexible. The original technology was designed for office buildings. If you tried to use it on a data center, it didn't work. The original technology had a limited set of subcontractors, so how do you move this technology around the country and world? Out of that came a new technology roadmap for Beck Technology. In that roadmap we identified a number of products.

The first product we brought to market in 2006, DESTINI Profiler or DProfiler as we called it, was intended to enable a contractor and an architect to sit down, collaborate on what is going into the early conceptual phase of the project, and tie that concept to cost. Where there weren't quantities to extract, we would backfill with rules of thumb and historical numbers so we could come up with a complete estimate and schedule into which both the architect and contractor bought in.

The second product we developed was a product called DESTINI Optioneer. The intent behind DESTINI Optioneer was to take that concept on which the architect and contractor had collaborated and provide them a platform in which they could study many alternatives. Instead of studying one, two, or three alternatives, we wanted to provide the team with the ability to study the entire design space.

In order to do that, we created a cloud-based application into which you could load the base model created in Profiler. The user could then define the project objectives, such as reducing the cost, minimizing the schedule, maximizing the life cycle value or even maximizing the amount of glass on a given façade. The team could then define the variables within the design that could be changed and in what increments. For example, the building could be three to five floors but no more than five. It could be 250,000 to 300,000 square feet but no less and no more. It could be steel and concrete but not post-tensioned. Whatever the means and methods or restrictions the team wanted to impose, you could define.

Once you set this up, you could use the power of the cloud to run millions or even billions of alternatives within that design space. The tool would provide back to the team a number of options that would potentially be better based on those objectives: maximizing façade glass, minimizing cost and schedule, making it as environmentally friendly as possible, and so on. The team could then evaluate and bring the human element to the evaluation process.

The funding for Optioneer came through Beck and it is not a commercially available product yet. What we discovered was that it really didn't work well just at a push the button. It had the same flaws that we had with the original DESTINI application. If you just brought back the best solution to an architect and contractor, they wouldn't trust it. What we found with Optioneer was that it worked best when it was used within and by a team. As a company, Beck Technology is now offering Optioneering services. We work alongside an architect and contractor to understand their goals, the scope of the project, and the owner's desires. The technology promotes collaboration, rather than the most optimal solution by pushing a button.

The third product we developed was DESTINI Estimator. We developed it in partnership with Sundt Construction. The intent behind Estimator was that we have the most optimal plan, whether it was developed manually, with Profiler, or through Optioneer. When we start to execute on a project, the question is how do we keep the project on plan, scope, budget, and schedule? The intent behind Estimator was to be able to take the outputs from either Optioneer or Profiler as inputs and then provide a high level of management control by extracting quantities from the design as it was developed. You could update the means and methods because, maybe, you discovered that the original assumption put into Profiler of 30-foot deep foundations is invalid because when you got the geotech report you found out that you hit rock at 100 feet, and we need to change the depth of the piers. Things do change. The intent behind Estimator was to be able to update the estimate in a very quick period of time, provide the team with tools to communicate what changed against those targets, developed in either Optioneer or Profiler, and facilitate a conversation on what the team needs to do to get the project back to those targets. (Figure 18.7 shows where Profiler and Estimator are in the life cycle of a project.)

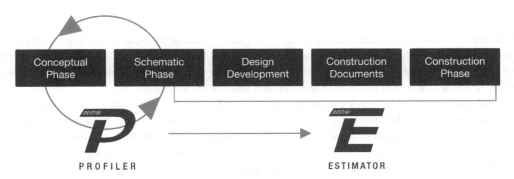

FIGURE 18.7   DESTINI in the project life cycle. Courtesy of Beck Technology.

The final piece to our technology vision, which is not being developed out yet, is another product called DESTINI Management Studio. The intent behind Management Studio is to take the successes that our customers have had across multiple projects of a similar type and to be able to mine that data so that again we can make better, more informed decisions throughout the life of the project. For example, we've successfully used Profiler, Optioneer, and Estimator on several office projects. Instead of the team coming up with rules of thumb, why not compare that rule of thumb against the data we have actually seen across three or four similar projects? The intent behind the Management Studio is to provide bench-marking information for any user-defined query based on our company's history to compare where we are with similar information across other projects. We have a working prototype of Management Studio, but it has not been developed out. Our intent is start developing that next year (2016).

By the end of next year, we hope to have all of those technologies available so we can really start to deliver on target value design. At the end of the day, that's what we believe is the best kind of delivery model. Set targets collaboratively. Use computers to test and come up with better solutions. Then, bring people together to use technologies to decide what needs to change based on where we currently are, to bring us back on target.

Peter Beck is a firm believer in the value of learning. He encourages us to try lots of things quickly in a limited way so that you can learn. The challenge that we've had along the way is that we are somewhat constrained by the financials associated with being a contractor. Margins are not particularly great. Several million dollars or more of investment annually before we went commercial is a big investment for a contractor. Once we went commercial, the challenge was how do you continue to push these tools forward while generating revenue.

In the early days when we just had Profiler, finding a use for it without having the complete package of tools was a challenge. What I mean by that is if you look at Profiler today in combination with Estimator, we have a tool you can sit down and come up with targets. We have another tool you can take those targets and manage throughout the life of the project. If you don't have that second tool and you can just come up with the targets, what's the value? What we found was that contractors saw some value, but it was limited. The value was a different way of communicating with the owner.

We have some early adopters like DPR, Sundt, Mortenson, some of the bigger guys that were moving more and more towards negotiated work and wanted to sit down at the table earlier. They needed a way to do that instead of reacting to what an architect had thrown over the wall at them. Beck and these

GCs wanted some way to be able to come to the table before there was a design solidified so that we could be part of the conversation, to provide more feedback and value.

What we found with the early days was that our industry is siloed. Contractors are not used to putting models together. When you look within the contractor, the estimating group above all others have been the most reluctant to change. Providing them with a tool that they could sit down with an owner and present a model was a challenge. We spent a lot of time with Profiler trying to simplify it and make it as easy to use as possible. We definitely stubbed our toe more than once. It was complicated. We discovered that different contractors, even though they may not be in the same market, view their data as proprietary. They're not going to use industry-wide data like RS Means. We needed to provide tools that they could integrate with their own data. We discovered that every contractor's data was organized differently. Some had it in Excel files. Some had it in estimating tools like Timberline. The way they classified the data was different. The way they quantified the data was different. We ran into a number of challenges with getting Profiler into the mainstream.

Over the course of several years, we realized a number of things. One, we really needed to target business development. Business development within the large GCs are typically the people trying to get in with the owner, and have a tool that estimators can use to support those business development efforts. We focused on making Profiler easier and a tool that you could do things in real time. Instead of the old school of taking what you were given and doing it behind the curtain, we put much more emphasis on progressive companies who were willing to sit down with an owner being able to use technology and work openly.

Out of that came some success stories, with Sundt being a great example. In 2011 they approached us and said: "We are getting more and more traction with Profiler, but the challenge now is we get an owner excited. We lock in a number, whether that's a contractual lock or just that we committed to an owner that this is the number that we think the project could come in. We're struggling to hold the project to that number."

We partnered together in developing our downstream estimating solution, Estimator. The intent, as I said earlier, was to provide not just a tool that you could calculate the cost of the project, but also a simple, easy to use management suite that would enable you to update the estimate, and visually be able to determine where the quantity came from, and understand what was driving cost. Our goal was, whether it was a 2-D deliverable or 3-D deliverable, to be able to click on an item within the estimate and see where that quantity came from. If I have two iterations of the estimate, I would be able to see within the context of the estimate what changed, and why. Curtain wall went up. Unit prices went down. Being able to see that in the model or set of PDFs, which that contractor and the team were working off of.

The other thing that came out of this management suite was the use of 3-D models becoming more and more the norm. Contractors are always pointing the finger at the architect, saying that they're not modeling for the purposes of cost or schedule. Inside our estimating solution we provided the ability that the contractor or estimator could draw over the top of the model, for example, if the wall needed an interior partition and it was modeled with the wrong object. Maybe it's a two-hour rated wall or should be, but it was modeled with a generic wall object in Revit®. It was modeled floor to deck with a little annotation that said, "Assume a 3-foot plenum." The estimator can actually draw over that wall, sketch the area they are going to use for the estimate, tag that area with a note saying: "On the next design iteration let the architect know they need to pull the wall down 3 feet so I can actually use the quantity from the model." They can also change the category of the wall from generic wall to two-hour rated wall. They could then run a report at the end of the design iteration that they can give to their design partner to correct or change the model for the purposes of estimating and scheduling.

Estimator is a relatively new tool. Sundt has been using it for about a year and a half. They are definitely getting some successes. They're starting to use more models. They're starting to integrate quantities into the estimating process instead of doing copy and paste from On-Screen Takeoff® into Excel, and then copying and pasting again into their estimating tool, which introduces the opportunity for errors. What we've seen so far is that estimators are using it. We think because it was designed from the ground up for estimators to collaborate with architects, and it's relatively easy to use. They're starting to use it as a collaboration tool and not just an estimating tool.

What are the lessons learned on Optioneer? We've done about half a dozen or so projects with Beck. Initially, we took what the design team and estimator came up with for their best concept. We would open up all the possible variables you could change within the design space. We would always look for the most cost effective solution and come back to the team thinking we saved them a ton of time and money. What we realized was it's not always about the lowest cost. The only way you can really use Optioneer is to integrate it into the team, and use it as a way for the team to test out certain assumptions or hunches.

Instead of opening up the entire design space, what we've seen work best is to run Optioneer several times and add more opportunities for change. It may be that in the first iteration you're only looking at shape and square footage. Maybe the team finds some designs that they like, and then maybe you start to layer on some what-ifs around means and methods. Finally, maybe, you start to look at the skin. It's not all or nothing. What we found to work best is do a little bit. Bring it back to the team. Let the team digest it and give you some input. Layer on another series of alternatives. Then, layer on another set of alternatives. Each of these series of alternatives can result in millions of possible permutations, but you're only throwing in little areas or additional scopes of the project as the team digests the last iteration.

Back in the early to mid 1990s when we started exploring these technologies right out of the gate, there were a couple of things we did that definitely support delivering the building as a product. What we ended up doing was creating this group called Beck Building Systems. It was intended to bring in certain vendors and subcontractors who we felt had knowledge. If we could capture that knowledge within the computer, we could put that in the hands of the design team. When we had come up with the concept, we could lock in the price along with performance criteria. We would have enough information because we in essence had already captured in the computer about how you would engineer, procure, and build it. The intent was that we could sit down with an owner. We could lock in a number. We would have a high degree of certainty that the number did capture the scope. Beck's team would deliver a superior quality product to the owner at a competitive price.

The goal was better decision making that led to better quality projects and more predictable outcomes for our customers and project teams. In our culture, collaboration is fun. Not having somebody at the table that needs to be at the table in order for a decision to be made is frustrating. It is a dream killer. Collaboratively we can come up with a solution. We can lock in a price. We can describe with enough detail that the owner knows they're getting a quality product. Then, we can actually deliver on that product, a high quality, and safer, more predictable one that's a better solution for everybody. That is still the vision. We've been doing technology. Technology is a means to an end. The end is a better quality, more predictable outcome for our customers.

Our mission statement at Beck is to revolutionize the industry and create a better future. To me that is what this whole journey has been about. It's not been a journey to develop tools. It's been a journey to deliver a product. That product will be better quality, best price, and schedule, and overall just a better way of delivering our products.

Our challenge at Beck Technology is developing tools. It feels like the "Field of Dreams" in some ways. You build it and they will come. We are trying to do our part in changing the industry. The industry is changing. We are seeing IPD. We are seeing more collaboration. We are seeing people trying to add more value earlier in the process. We're trying to be run as a business whilst going along this journey of build it and they will come. It's not easy. Not everyone in the industry is where we want them to be; yet we still want them as customers.

Along the way we have made many mistakes and it's taken a lot longer than I thought it would. I still think we have a long way to go, but I see light at the end of the tunnel. I think that leaders in the industry are starting to see the light. I think once you get some of these larger companies like DPR, Mortenson, Sundt, and Beck, I think we are going to hit a tipping point. At that point I think Beck Technology is going to have enormous value above and beyond what we have right now. I think we will then go into the next generation of research and development for our industry (Stewart Carroll, personal communication, November 16, 2015).

18.4.4   A Validated Database for Facility Operations

Andrew Arnold, PhD, is the director of DPR Consulting. Before joining DPR Construction in 2013, Andrew spent 13 years after earning a PhD in civil and environmental engineering from Stanford University leading work at companies focused on collecting and leveraging BIM data to improve the design and operation of buildings. At DPR, Andrew has led the development of VueOPS, a cloud-based searchable repository of construction project turnover documents and a service request ticketing system to help facility staff manage their new buildings. Andrew also leads efforts to assist owners in leveraging virtual design and construction (VDC) to develop facility operations databases as part of project delivery.

Andrew explained the work he and a DPR team are doing with a major international airport as follows:

We put together a team of young, BIM-savvy project managers and project engineers under the direction of two VDC wizards to help the owner specify a validated facility operations database as part of the project delivery process for the owner's in-house and large capital projects. This includes specification of FM data that should be included in BIM and which project team members author FM data at each project phase, organizational development for the owner to effectively manage the data, process engineering to tunnel through barriers to collecting and validating FM data through the project delivery process, and outreach to project teams to educate and assist them to incorporate the FM requirements into project deliverables. What I mean by "validated" is that the data quality is good because the owner and project team (architect, designers, general contractor, and trades) have tested it systematically and at the point of authorship for conformance to the owner's data standards, before the owner accepts models. We are helping the owner facility operations staff and stakeholder groups specify the results they seek technically, in a way that's measurable and supports better decision making about operations and maintenance of their buildings. This is low-level work that must be done to make VDC a reality. It's one of the gaps that have to be filled to reach the VDC vision of high-performing buildings.

If you're going to lay out a framework going from near future to far future, I would put first the ability for owners to get a lot better about specifying the results they seek out of a project at the database level. We're creating a performance-based BIM guide for an owner that I think will move the needle beyond what all the other institutions have done. Several leading universities have done BIM guides. None of them have specified the database that they want to flow out of a project at a level where the data can be validated programmatically.

It will be the first time an owner articulates a database schema and standard naming convention for the installed products and product performance attributes that an owner seeks to manage for facilities operation, as opposed to representing the information in a drawing,. You take that and merge it with trend data coming from building management systems and work order systems, and you finally get a picture of the original state of each product that was installed in your building, the current operating state, and the history of maintenance about it. That's a new ballgame for an owner.

The other goal is to reduce the time to zero for knowledge transfer from the design and construction team and the owner to zero so they're up and running on day one instead of taking years to figure out how to operate their building, and then years more to figure out how to deal with the problems in the building. Typically, when owners deal with temperature problems, they increase energy use in the building to solve the problem.

What we're doing for this owner is just getting them a database for the products and systems that are in their building. I consider this blocking and tackling at the data level; it's step one before moving on to bigger and better things. If you're going to try to match up, for example, the point data coming from the building management system, you have to have a database of the products and systems that are actually in the building to match that point data against.

We've created an info graphic showing how an owner can move in the direction of technically specifying the data that they want to flow out of a project and setting up an acceptance test framework for it. This is really just following what's been going on in the software development world for quite a long time, which is that when you write software you set up customer acceptance tests for the behavior of the system before you write the software.

Figure 18.8 is really saying what people are doing. It shows the artifacts we're creating to support their activities. On the left you've got project participants, architect engineers, contractors, and trades modeling.

The idea is that there's a defined technical database schema for the data that have been shared with the project participants ahead of time. Then we are using off-the-shelf tools to build an environment where a model author can post a model to a server and run tests on their model, acceptance tests. Those tests run either red or green. If green, you're good to go; the owner accepts, from a data perspective, your model for facilities maintenance (FM). Initially, we are focusing on a small subset of all the things you might want to test for. If you're red you've got a list of problems you've got to resolve before the owner accepts your data. Only after you get a test that run that's all green will it be considered a database that is shareable with the owner's enterprise systems. The diagram abstracts the flow of data to stakeholder applications where owner groups are using the data to manage and operate their building.

What we're finding as in all things related to VDC is that the technical side of this is not rocket science. It's basic relational database design. However, the process piece and the organization piece are critical. What we're achieving at the airport is to demonstrate how this can happen by having a team do

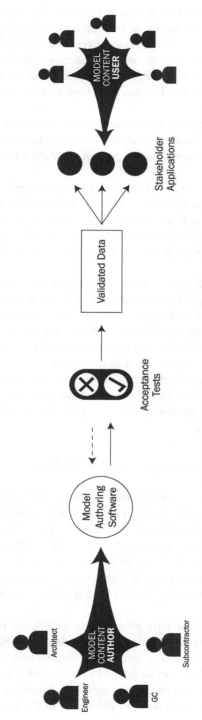

FIGURE 18.8 Data flow for a validated operations database. © DPR Construction, by Andrew Arnold and Lyzz Schwegler.

the technical work to demonstrate the data collection and validation workflows. And we're specifying the owner's requirements in a performance-based, not a prescriptive way, so that consultants can innovate how they implement the requirements. We also have team members who are working on several capital projects, so those project teams understand the requirements. We improve our understanding of the technical, organization, and process barriers and continue to tunnel through them. This level of engagement helps these project teams see that they don't have as much risk as they think and therefore the pricing shouldn't be outrageous.

The owner has to get behind putting the acceptance test framework in place. The owner has to require it and has to agree to pay companies to do this work if it costs money. The bet that the owner is making is they're going to get a validated facility operations database and decrease their cost of data collection after the project.

Presently, there's a lot of redundancy and some very suboptimal processes with a lot of waste in most facility operations. What we've been doing is really blocking and tackling work to line up the organization, the process, and the technical infrastructure, to allow an owner to receive what I call a facility operations database, a database of installed products that can flow within their work order system, their space management system and also interact with their building management system at the end of a project.

That is the foundation for putting together systems that do a better job of helping the owner manage the building over the life cycle because you need a rich database for the products and systems that are in your building. You need the building management system to give you the real-time feedback about how they're performing. Then you need the logic to start comparing how they're performing with their preventative maintenance plans to do better maintenance and extend the service life of equipment and also to optimize building performance. This is in some regard low-hanging fruit, but to move an owner to a point where they can say, "At the technical level, you've satisfied my requirements, or you haven't for the data you've provided" as part of the design and build, I think is a step forward for the industry. I don't know of any other owners that are doing it systematically (Andrew Arnold, personal communication, November 13, 2015).

18.4.5   Multidisciplinary Design Optimization (MDO)

Dr. Forest Flager is a research associate and lecturer at Stanford University's Center for Integrated Facility Engineering (CIFE). Dr. Flager's research explores how computing can be used to augment human capabilities for design exploration and decision making. Prior to his current role, Forest served as the product manager for a venture-backed technology start-up in San Francisco. Forest has practiced as a structural engineer for Arup in San Francisco and London, United Kingdom. He holds a BS and PhD in civil and environmental engineering from Stanford University, a master of design from Harvard University's Graduate School of Design, and a master's in structural engineering from MIT.

Dr. Flager and his colleagues at CIFE have pioneered the development of Multidisciplinary Design Optimization (MDO) methods for application in the AEC industry. MDO involves formalizing design coordination and iteration for teams working on complex engineering systems. Once the design process is formalized, computer algorithms can be applied to systematically generate and evaluate design alternatives with the goal of finding optimal or near optimal solutions given specified product performance goals and constraints. MDO methods were pioneered in the aerospace industry in the 1960s and have been subsequently applied in the automotive and electronics and architecture, engineering, and construction (AEC) industries to improve product quality and reduce time to market by enabling design and engineering teams to analyze orders of magnitude more alternatives in far less time than using conventional design methods.

Applications

Dr. Flager and his colleagues have successfully applied MDO methods to a variety of different types of problems in the AEC industry, which are summarized below.

Structural Design Determination of the layout and sizing of structural members for two large stadium roof structures, shown in Figure 18.9, in collaboration with the global engineering firm Arup: The application of MDO resulted in a 22 percent cost savings on average, saving over US$1 million on structural steel cost per project compared to designs developed in parallel using conventional methods. Table 18.2 compares the efficiency of the MDO method to conventional design practice in terms of the time required to set up each process (i.e., model development and software integration) as well as the time per design cycle (i.e., time required to generate and analyze a single design alternative). The MDO method generated over eight times the number of design alternatives as the conventional practice method in approximately 20 percent less time. For more information see Flager, Adya, Haymaker, & Fischer (2014) and Flager, Soremekun, et al. (2014).

Building Envelope Design Specification of glazing percentages and window and wall construction types for a large commercial office building in Atlanta, Georgia, in collaboration with Beck: Results show that the method identified design solutions that offered significant reductions in the building's life cycle cost and carbon footprint while requiring fewer man hours than conventional approaches (Flager, Basbagill, Lepech, & Fischer, 2012).

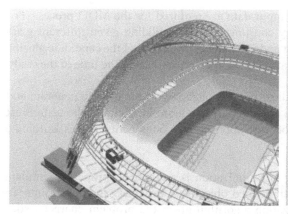
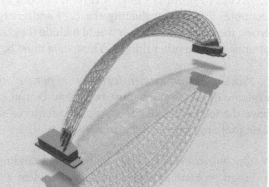

FIGURE 18.9 Building information model of the case study roof structure. Courtesy of Forest Flager, developed in collaboration with Arup, Manchester, UK.

TABLE 18.2 Comparison of Process Efficiency for Conventional and MDO Design Methods

Design Method	Setup Time (man-hours)	Design Cycle Time (avg.)	Number of Alternatives Evaluated	Total Design Time
Conventional Practice	60	4 hrs	39	216 hrs
MDO Method	140	5 min 45 sec	340	172 hrs 35 min

© Forest Flager.

Urban Energy Systems Determination of building mix and energy supply technology of urban district to maximize energy efficiency in collaboration with Disney Research China (Best, Flager, & Lepech, 2015).

Wind Farm Layout Layout of wind turbines on two large U.S. wind farms in collaboration with Mortenson Construction, reducing total project construction cost by 7 percent with equivalent or greater net energy production.

What's Required to Successfully Implement MDO in Practice? How Could This Technology Impact Practice in the Future?

Technical Prerequisites

1. Formulating the right problem Since MDO involves the automated generation and evaluation of product alternatives, one must be able to assess performance virtually (e.g., using analytic or simulation models). In addition, since a computer algorithm ranks the alternatives, all performance criteria must be measureable by a machine.

2. Access to product data A certain amount of input data are required for the MDO process. For example, if one is evaluating the cost and energy consumption of a building given different glass types, the input information would include the cost and mechanical properties of the candidate glazing options (among other things). These data must be available to the project team at the time of the study.

3. Process automation/integration expertise All software tools/components used to generate and evaluate design options must be able to be run in an automated fashion. In addition, the team must have the technical skills to "wrap" the required software applications/components, that is, automate data exchange between the analysis tools.

4. Optimization expertise Some understanding of optimization is required to properly formulate a design problem as an optimization problem with defined objectives, variables, and constraints. In addition, the selection of an appropriate optimization algorithm given the problem characteristics (e.g., single versus multiple objectives, discrete versus continuous variables) is important to ensure a successful MDO process.

Behavioral Prerequisites: Leadership, Organization, Process, and Culture

1. Design Space Thinking Most designers are used to sequential design iteration, that is, generating and evaluating a design option and then generating the next design option based on those learnings. MDO requires the project team to specify design variables that define a range of design options for exploration at the beginning of the process. Architects and engineers who don't have a background in parametric and/or generative design often struggle to understand how a given design parameterization might impact the range of possible design aesthetics and performance, but it just takes practice to develop these skills.

2. A Focus on Process In conventional design processes, project teams spend a relatively small amount of time planning the design process. By this, I mean specifying the design problem and the information exchanged for each design task and how the tasks and information will be shared amongst team members. Typically, this only makes up 5 to 10 percent of the total time spent by the design team. Most of the time is spent executing the design process (Flager & Haymaker, 2007). In an MDO process, these percentages are reversed: project teams spend the vast majority of the time formalizing the process and data requirements and integrating this process using software. Once the process is integrated, the computer executes the process automatically. This can be challenging to project teams that are accustomed or expected to immediately begin generating content (e.g., design options/analysis).

3. Holistic Perspective MDO studies are most effective when they are focused on improving the overall performance of the system, rather than optimizing individual subsystems. Design and construction specialists must be able to keep their disciplinary goals in perspective and have constructive discussions to identify the most important performance criteria for the project as a whole.

4. Sharing Culture MDO processes require project teams to share information both horizontally across design disciplines and vertically between designers, contractors, and suppliers. Strong leadership (and, ideally, the proper business incentives) are required to get project team members to buy into the concept that sharing information will result in a better outcome for all involved.

The Future for Automating Design through MDO

Role of the Designer The designers will spend more time doing more creative activities that require significant expertise and judgment, such as formulating the design problem, defining the range of options to explore, and selecting appropriate analytical representations. Machines will take care of the more routine tasks of modeling and generation and analysis of alternatives. This shift should help good designers and engineers to demonstrate their value in the marketplace more clearly.

Vertical Integration MDO is one approach within the data-driven design movement that seeks to use information to reduce uncertainty and improve decision making early in the design process. As practitioners continue to demonstrate the value of data driven approaches to facility owners and users, we will see pull-driven demand for tighter collaboration and increased sharing of information between designers, contractors, and the product supply chain.

Toward Mass Customization MDO processes enable custom design solutions to be generated automatically simply by changing the inputs to the process (e.g., climate, site conditions, design-construction team capabilities). These methods coupled with CNC and additive manufacturing methods enable customized solutions to be designed, fabricated, and constructed without the cost premium that has traditionally been associated with bespoke architecture (Forest Flager, personal communication, December 19, 2015).

18.4.6   Componentized Design & Construction/Aditazz

Zigmund (Zig) Rubel is a founder and chief of Building Sciences for Aditazz Inc., where he has worked since November 2009. Before joining Aditazz, Zig was a Principal-in-Charge on healthcare projects for Anshen + Allen Architects, where his most significant projects as the Project Director were Kaiser Permanente, Santa Clara Medical Center Replacement, Laguna Honda Hospital Replacement Project, San Francisco, and University California San Francisco, Mission Bay Hospitals.

Zig explained Aditazz and his work there as follows:

> The main driver of what we call BAAP (Building as a Product) is to find components within the building process and reduce their variability, but not have an impact on the design. The five components that Aditazz is focused on (shown in Figure 18.10) are the structural frame, the concrete slab, the interior and the exterior walls, with MEP devices and routing within the walls, and the MEP distribution. That constitutes, at least in a healthcare setting, between 60 percent and 80 percent of the cost of the work. That's what we're focusing on. That's really what the effort is about.
>
> The basic "big idea" for the company really started while Deepak Aatresh was an entrepreneur-in-residence (EIR), which is a program that some Silicon Valley venture capital firms use because they've recognized the real risk in a startup isn't necessarily the idea, but it's the person. They incubate these different entrepreneurs to work with them until they find that idea. Deepak was an EIR, and he was actually looking at the building space, and he saw a time lapsed video of the Sutter Health Eden Medical Center replacement hospital project in Castro Valley. That's when he realized that buildings are made very similar to computer chips; components are just stacked. The scales are just very different. The big idea was to leverage both the tools and the methods that have been used in the computer chip industry for the last three or four decades, and to apply them to buildings. At a very high level, that's the big idea.
>
> There are several different vectors as to why now? The first one is that the AEC industry is really starved for innovation. Many of the innovative pursuits that are pursued in the AEC industry are evolutionary and incremental, and not revolutionary. The opportunity that this presents is to ideally

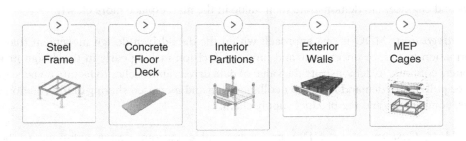

FIGURE 18.10   Aditazz "Building Kit-of-Parts." © Aditazz.

be a game-changer. Aditazz doesn't see itself as the only beneficiary to this process. We hope to find, I think the term is, co-travelers in this process who also believe that there's a better way to plan design and build facilities. That's the first thing.

The second thing is the reason why we focused on health care. Because it's rules based, and it's very complex, health care can really benefit from an algorithmic engine that leverages technology. As you know, health care in most of the world is broken. Our approach was to leverage the technology and really solve two problems, one being design and construction, and the other, creating more efficient and better operating health care facilities. The complexity in health care can also showcase our technology.

The third reason and what is allowing us to do this now is that our effort coincides with cheap computing power. For the past few decades, ever since the invention of the integrated circuit, the cost of compute cycles has gone down. What only was available to Fortune 50 companies having these huge server farms is now available to the masses. Aditazz is able to leverage a large amount of computer infrastructure on demand, whereas, beforehand, it was really only afforded to very few firms. We're really aligning what's being made available to everyone with some technology that was used by very few people to solve the problem that everyone is dealing with.

The fourth reason is that the tech industry is really looking for difficult problems to solve. They don't have many other interesting problems that they're addressing, and a lot of technologists are actually interested in this problem because it's difficult. If you look at venture capital investment in other industries, it's in the sciences, pharma, clean tech, and other types of industry. There's very little money that's being put into the AEC industry because it's just hard to grapple with and make it scale for a venture capital firm. Our investors saw an opportunity to fund the transfer of this technology to the AEC industry using the framework that Aditazz proposed and is pursuing.

There are two types of people at Aditazz. There are what I would call technologists; they are the differentiating people who are here to help us solve problems in an innovative way using technology. It's a combination of computer scientists, software engineers, and mathematicians. We have people from a manufacturing standpoint, people who worked at Intel on the operations side. People that really understand a really efficient industry in ways that people in the AEC industry don't understand. Deepak also needed someone like me, an experienced architect, to help him as a domain expert. In this case, we're people that really understand how to plan design and build buildings. The one thing that ties all of the domain experts together is our strong belief that there's a better way to do things.

When we started the company, our efforts were on the hardware and the software needed to transform the building industry. The hardware, which is what BAAP is all about, is a very capital-intensive problem to solve. Our focus has shifted to creating and providing software that can use the industry's available manufacturable hardware. This relates to the kind of people we need at the company.

One of the things that excited me about what Deepak conceived was the possibility that we could really provide value and not sell hours. I'd seen that construction services in that sense is similar to design where you have to justify your fees based on the number of hours, but ultimately you're not always providing value. If we could create automated tools so people could focus on the creative aspect and let the tools do the mundane aspect that would be a game-changer. And, in fact, all of our engagements have been what I would call lump sum. We don't charge by the hour. We just say we think this is going to cost say $200,000, or whatever the number is, and if we spend 200,000 hours on it, we're not going to change our price. We want to get paid by the value and not by billable hours.

We use Agile management to deliver our projects. Agile is in many respects similar to Lean, where you're really looking at the tasks as individuals. I would say the main difference between Lean and Agile is Lean is more focusing on the last responsible moment to make a decision, and Agile is more of a dynamic peeling of the onion, so to speak, to just focus all the different tasks in a non-sequential way. We are using similar tools and mindsets that the software developers are using on our projects to create more predictability and accountability as we're doing things.

One of my first responsibilities at Aditazz was to convince many of the tech people who were going to possibly join us that the AEC industry did not use the sophisticated tools that are in the chip design world, and that we're as archaic as we are. Many of the technologists just couldn't believe that the AEC industry was as backward as we are.

Morten Hansen (2009) in his book, *Collaboration,* talks about the barriers to collaboration. He identifies four barriers. I wanted to bring it up because we struggle with parts of it as well. The first one is the "not invented here" concept where people are unwilling to reach out to others. They sit at their desk and focus on what it is that they're focused on. The second barrier is the hoarding barrier, where people are unwilling to help one another. The third one is what he calls the "search barrier," where people are unable to find what they're looking for. The biggest one, which I think is really affecting our industry, is the transfer barrier, where people are not able to work with people that they do not know well.

Let me just focus on the last one. Our industry hires people that are credentialed. You want to hire someone that has 30 years of experience of doing nothing but the kind of thing that you want to hire them for. When you look at the star tech entrepreneurs, they're typically just beyond teenagers that are just really smart and came up with a really good idea. Typically, those two kinds of people don't interact well. We've had to figure out how to overcome that struggle because a lot of times our technologists would speak to our domain experts, and it's really about transferring that knowledge and overcoming that barrier of, "Oh, yeah. I hadn't really thought about it being done that way."

I think what our industry needs to get better at is marrying people who don't come from our industry because it feels like in all disruptive innovation that's out there, most of the disruptors are not from that particular industry. We have to embrace those people that have a really keen sense of what might be better, and figure out how to engage them in our mindset.

We're constantly evolving the how. This is a part of how all innovation works. Over the past few years, we looked at ourselves as a provider of services differentiated by technology, and those services could be design. They could be construction management, like with Dave Higgins, a former owner of HMH Builders—our construction domain expert. The point is that we want to learn by doing and prove our innovation in house before we open it to others. This sits especially well with some of our customers and partners who are risk averse and want to adapt a proven model.

Now, as some of our technology and methods have matured, we're opening our platform to partners outside our company to use. Starting next year, we will engage with partners where they can leverage our innovation. By releasing our platform in a controlled and calculated way, we hope to make a highly positive impact on those who engage with us. We expect both our customers and our collaborators to adopt the platform. The positive side effect of doing this is that we will begin to scale like a technology firm. The industry as whole will benefit since we are able to bring efficiency to both the design and the construction phases.

The other thing that Aditazz has done is invest time to create, protect, and share our intellectual property (IP). The model in the AEC industry for creating IP is via what is called trade secrets. Much of the technology industry operates with both trade secrets and patents. The real difference to point out is that a patent tells the world what an invention is and shares the knowledge for others to learn from, as opposed to trade secrets, which are differentiators that are not to be shared. The AEC industry does not have a two-tiered approach. Aditazz currently has over two-dozen patent applications, with two patents granted. All can learn from what is published.

We want to enable an owner to adopt our platform and ask their providers to use it. We are pursuing relationships with a couple of owners; one is in health care, the other is not. They're both in the Bay Area and are exploring how they could engage us more from a platform perspective and not providing services. For example, there's one particular client that has invested a tremendous amount of time in defining their standards. At the same time, when they hire architects and builders they have no way, without spending a lot of time, knowing that their service providers, being designers or builders, are complying with those standards.

We're investigating how we can provide some sort of, for lack of a better term, compliance checking to show that that particular owner's standards are met. Today, for example, you would have a client and they give you a book of standards. You say, "I designed it exactly the way you told me to," but the owner has no way of knowing that it was really the case. If you're really productizing a building, there's a higher degree of certainty that the design and the outcome of the building meet certain criteria. If our industry is going to change to a more productized method, we need more transparency in the industry to go from criteria requirements to outcome based on metrics. What's missing currently is a means for giving the owner some sort of real-time feedback that their requirements are being met. Aditazz hopes to be one such approach to do that. The first thing we do is sit with our customer to develop a brief, which we call "Building Facts" (shown in Figure 18.11).

Aditazz Customer Requirement

Building Facts
Business Size – 1 Location

Location – 123 Main Street, Springfield, State USA

Typology – Comprehensive Ambulatory Building

Managed Lives – 80,000 – 120,000

Key Drivers	Start Target	Annual Value		Physical	
Primary Care:	73%	70,000		Building	Type IA / Steel
Specialty Care:	69%	65,000		Massing	Contextually Responsible
Procedures:	66%	122,000		Facade	Articulate
Images:	72%	54,000		Structural	Best Performing
Scripts:	71%	113,000		Systems	Best Value
				Interior Construction	Fixed
Quality Targets		**%**		Interior Experience	Memorable
Patient Caregiver		98%		Material Durability	Everlasting
Care Coordination / Patient Safety		98%		Energy and Resource Efficiency	Optimized
Preventive Health		99%		Sustainable Materials	Least Costly
At Risk Population		99%		Integrated & Adaptive Technology	Conservative
Performance Indicators				Remodelability	Least current cost
Operations		7 Days / Extended		LEED	N / A
Financial		Upper 1/3 for Profit		Schedule:	
Staffing		Suburban FTE		Design: 87d Manufacture: 120d Commission 21d	
EUI		112		Permit: 20d Assembly: 43d Remodel 11Y	

FIGURE 18.11   Aditazz customer requirement/"Building Facts." © Aditazz.

I think the "how" really has to be tied to data. Coming back to an earlier topic of what were some of the earlier discussions when we started the company, it was all around how do you measure design? How do you know that one scheme is better than another scheme? The designer would say, "I would just know." How do you measure that?

If we're going to use tools, we have to educate the tools in an objective way to say the square scheme is better than the round scheme, or vice versa. Until we're able to use metrics and data to quantify our decision making, we're going to stay in the same rut for a long time. What we've been spending time on in Aditazz is coming up with a quantifiable way to measure our decision making. Health care is much easier to do it with because it's more rules based than other building types such as a museum or cultural center, which tend to be much more subjective. I would argue there's probably ways to measure decisions for those building types as well.

I think we need to realize that there will always be a mixture of technology and subjective thinking, a hybrid approach. What excited me about what Deepak was proposing is that if you just look at a hospital, half the cost of the hospital is in the walls and the ceilings. There's no artistic perception of the building systems inside those walls and ceilings. They just have to work. I think any building type is going to have that kind of mixture of where you're going to have someone who's going to want to create a delightful, creative response to a design challenge. Then there's a big part of it that just has to work. We think the "just have to work" can all be automated. There's no question. Do you really care about how the wire from the outlet gets back to the circuit breaker panel? No, you just need to know that it works, and that it's going to be provided.

The color of the wall and the shape of the wall, all of that will likely be subjective. I don't think a computer, at least right now, is going to be able to do that. Maybe eventually, when there's enough heuristics and objective thinking around it. If you look at all the problems that most architects and builders and engineers deal with, it is the mundane tasks that can all be automated. It's not the creative ones. Very few designers get sued for creating an ugly building. They do get sued for building systems not working.

For the last five years, we were pursuing the commercial applicability of our idea and how the market would respond. Initially, the interest has been around what we call our operational planning and validating that people are building the right building. We started this effort in the early days. We met a hospital executive, and she poignantly said, "I don't understand why my kid brother has the entire Internet playing 'World of Warcraft,' and I'm saving lives and managing a hospital with a pivot table." She was exactly right. There's very little technology that helps executives in the healthcare arena on how to operate their hospitals. One of our first pursuits was taking an operational simulator that is used in the chip design world, and validating how an emergency room operates. Many of our pursuits and interests are around the functioning of an emergency room. We are expanding that kind of analysis with other departments. The emergency room is a great place to start because so many hospitals make and lose money in their emergency rooms.

That has been a huge interest commercially from other clients. We are getting into doing design, and we just started doing some construction projects. Dave Higgins just got Aditazz's construction license, and our construction name is Aditazz Assembly and Construction. The focus is our belief that we shouldn't just construct buildings. We should assemble them. Our first projects that are being built in the United States are going to be tenant improvements, predominantly to repurpose space. We have an Office of Statewide Health Planning and Development (OSHPD) project for an acute rehab remodel to an existing facility where we are going to use our post and panel wall system on the project. Because it's an existing building, we're not going to use our structural system, because the steel and the concrete slabs are existing. The exterior wall exists. We don't have the floor height to use our MEP cages, but we are going to use our wall systems. We're going to use that to learn how effective it is, whether it's good or bad. Hopefully, it will all be good. Either way, we'll improve it.

There's definitely been a lot of interest, but one of the challenges we've had is that our industry is so risk-adverse that no one wants to be first. Everyone wants to be seventh. Many of our sales have all been relationship based. It's been with someone who gets it and wants to be a part of it. That is a real challenge for our industry because I think everyone is interested in innovation, but few are confident in assuming the risk that goes along with it.

Before, we were trying to act as the service provider, where we would only engage with the owner. Now, because we're opening up our platform, there has been interest from architects and builders on leveraging our tools. It has always been our goal to enable both the end-customer and the suppliers with efficiency gains using our platform.

The reason why we've gotten more traction outside of the United States so far is the market is saturated with many very capable providers of services. There also isn't a huge demand. We're not designing or building huge projects in the United States, whereas predominantly in Asia, they have a big demand for infrastructure, and I mean buildings, not only roads and tunnels. That's definitely needed too, but we don't provide those services. Asia, especially the UAE, don't have as many providers as in this area of the world. There's just such a huge demand for services there, and there's not a lot of innovation. When they see a company like ours they're much more interested, and they're much more willing to take a risk on a newcomer.

The one thing I want to talk about is the opportunity for innovation in our industry. There are two possible responses. One is being driven by the owner, who says, "I want it and am going to pay for innovation, and this is my business model for it." Right now, the industry business model is around efficiency within silos, the silo's being the planning silo, the design silo, the construction silo, and the operations silo. The other opportunity for innovation is for industry players to come together and create a product offering that the owners haven't even conceived of. Let's say a group of planners, designers, builders, and operators go to an owner and say, "We will provide you a building, and it will cost X. That would be a game changer."

Up to now the building suppliers, meaning architects, engineers, and contractors, haven't had the need. For example, the reason why the chip industry actually happened was because of the race to the moon. The government needed a very efficient, lightweight, and reliable way to do switching to run their rocket ship to the moon. Russia shamed the United States by being the first ones in space. President Kennedy said, "We're going to go to the moon. We're going to send an astronaut there and bring them back safely." Because they had this race, putting someone on the moon and back, they needed a way. The sky was literally the limit. The space race is what enabled all this technology, and very few people are aware of it. We don't have the equivalent of a space race in the AEC industry. What the technology has done is it's taken a barrier down for innovation, but it hasn't incentivized the innovation. Until a client sees that there's less risk for doing it, very few people are willing to take the risk to do it differently.

The challenge with that is it takes a huge amount of resources and investment to do that. The owner has to say, "I want this," or a company will come around and say, "I'll provide this for you." Until one of the two things happens, the innovation is not going to occur. Aditazz is hoping to incubate that innovation more on the supply side (Zigmund Rubel, personal communication, November 25, 2015).

18.4.7   Interconnections and Reflections

The starting point for the Simple Framework is the product—a building that provides high-performance measured against the four categories of criteria for the value stakeholders seek. It follows that all of the Simple Framework elements support delivery of the building as a product. The examples we've presented throughout the book prove that teams can implement these elements. It's also clear that the technology is advancing and supporting the Simple Framework. The demand seems apparent; owners struggling to respond to higher business performance expectations from investors, employees, and customers need to deliver much better buildings faster, with higher quality and lower cost. Building users, especially younger people, are choosing employers who can give them healthy workplaces. Most compelling is the absolute necessity to drastically cut the embodied carbon content of building materials, building emissions, and the waste of materials and resources created by construction.

It might seem that greater adoption of IPD will inevitably result in teams offering to deliver high-performance buildings as products. Indeed, top managers responsible for design and construction for major owners such as Disney and Intel are promoting Lean construction and IPD and have been challenging teams to continuously improve. But the current practice of creating new project teams for each new project requires reinvesting time and resources to create the structures, resources and processes that will not live beyond the project. To consistently create the high-performing building as a project, the participants need to invest in strengthening capabilities and creating processes and systems that will span (and be amortized over) multiple projects. In an article for *DesignIntelligence* titled "Beyond IPD: The Integrated Enterprise Challenge," Peter Beck (2013) stated:

> The ultimate opportunity is to overcome the status quo and to merge the disciplines into one firm, thereby assuming responsibility for the complete design and delivery of the project and mitigating the perceived risk through the sophisticated use of these emerging technologies. Not only will the combined disciplines cooperate at levels never seen before to ensure complete and timely information along with accurate pricing and scheduling, but such an integrated enterprise will also be motivated to invest in mutually accepted design components for detailing and estimating, common mapping protocols between different BIM technologies, necessary cross-training to radically improve efficiencies at the project level, and so on. Such investments are expensive and must be amortized over many projects through an integrated enterprise model and cannot be cost effective in an IPD environment with no assurance that the parties will work together enough times to amortize the investment sufficiently. These emerging technologies offer the opportunity to improve across the disciplines, not simply within them.

Beck understands that investments must be made to integrate information and leveraging BIM across all disciplines. Without that, there is no capacity to model, simulate, and digitally prototype extensively and accurately enough to make predictions with sufficient confidence to make offers to deliver high-performing buildings as products. In addition, reliable information for the cost of production, rework, delivery time, and especially building performance is hard to extract from multiple organizations. As Andrew Arnold explained, amassing just building performance data is no small undertaking.

To build this knowledge base and capability, Beck has invested millions in creating software tools for integrating information, and organized Beck Building Systems to share knowledge with vendors and specialty contractors. He merged of his company with Urban Architecture, a well-respected Dallas architecture firm, in 1999 to create the integrated design-build enterprise he describes in *DesignIntelligence*. The increased vertical integration supports increased investment.

Beck's solution—increasing vertical integration—is a solution that will work for some firms but not for others. Vertical integration works best when demand is relatively consistent and the integrated firm can specialize in a limited number of project types. This allows the firm to hire the necessary personnel and keep them relatively busy. But where work volumes fluctuate, or where new and different expertise is needed from project to project, it is more efficient to create a virtual organization comprised of the precise elements, in the right amount, needed for the project. Choosing the right blend of virtual and vertical integration, because of the benefits and drawbacks of each, is a problem of optimization.

Supply chain management is another strategy for gaining the benefits of vertical integration within a more diverse structure. Womack, Jones, and Roos explain in the chapter "Coordinating the Supply Chain" in *The Machine That Changed the World* how Toyota overcame the problems of poor coordination with suppliers that plagued other automakers by closely collaborating with its best performers. The Big Three—General Motors, Ford, and Chrysler—continued their practice of awarding to the lowest bidder and pressuring them to reduce price through the term of their contract. Toyota invited these suppliers to collaborate with them to develop better solutions across vehicle platforms and sent Lean experts to help them improve management, engineering, and production. The suppliers, in turn, agreed to reduce their prices throughout the contract period, which they were able to do based on increased productivity. Toyota continues to collaborate closely with suppliers to this day. In turn, the suppliers can invest in the equipment, processes and training that allow them to better serve Toyota. Each party gains in this stable and collaborative supply chain.

The Toyota approach cannot be directly copied in construction because no owner is continuously building enough projects in a locality to justify its vendors developing special teams and processes for that owner. But it does suggest that focusing on the supply chain and repeat work may have advantages. Lawrence and Memorial, a hospital system in Connecticut, decided to have the same team build multiple IPD projects. The advantages in consistent processes, established relationships, and understanding of L&M's requirements greatly outweighed any perceived advantage of selecting new teams. An international financial institution is currently employing a similar strategy on renovating four office buildings on its corporate campus.

And a few industry leaders have developed programs to help educate and train their vendors in the skills needed to execute IPD projects. Herrero Builders has trained many architects, engineers, and trade contractor personnel in regular classes over the past five years. Graduates of their program created a community to meet quarterly to work on issues together and exchange learning with one single purpose: "Deliver exceptional value to all clients in a project or organization." Paulo Napolitano, Herrero's chief operating officer, Lean sensei, and director of learning, has described how this community defined as being a "Meta Organization" has enhanced Herrero's presence in the marketplace and ability to deliver greater value (Napolitano and Cervero-Romero 2012).

Renate Fruchter, a Professor in the Stanford's Civil and Environmental Engineering school, has for the past 23 years organized the AEC Global Teamwork[2] program to join students from across the globe to virtually design and build projects that respond to specific program goals. The students take on the roles of the integrated team members and must jointly address all aspects of design and construction. This program is a leading training ground for the high-performing project.

While at Universal Health Services, Bill Seed developed a program to create owner project managers (OPMs) and design and construction integrated project managers (IPMs) he described in the paper, "Integrated Project Delivery Requires a New Project Manager" (Seed, 2014). Walt Disney Imagineering has invited design and construction firms to participate in workshops where members share what they have learned about BIM and VDC, Lean Construction, and IPD. This builds capability and community. While these steps may not be as effective as Toyota's deeper integration of vendors, they help build capability between projects as well as on projects.

Zig Rubel from Aditazz reflected that the current project pressures leave little room for R&D investments. However, some firms—even though they may be competitors—are cooperating through academic research and training programs. Beck, DPR, and Mortenson are CIFE member companies along with several large European builders. As such, they support research financially and help set its agenda along with design firms, software companies such as Autodesk, and large consumers of construction such as GSA and Disney. CIFE especially, along with Georgia Tech and Penn State, is focused on leveraging BIM and VDC to improve the industry's ability to make reliable predictions about building and project team performance. DPR, Herrero, and other Bay Area contractors also support UC Berkley's Project Production Systems Laboratory (P2SL) led by Iris Tommelein and Glenn Ballard, two of the founders of the Lean Construction Institute. Recently P2SL completed a five-year study of target value design on five DPR projects underwritten by a group of design and construction firms DPR brought together (Denerolle, 2013). Support and participation in these types of programs is another strategy to effectively build capability between projects and leverage limited resources.

Undoubtedly, there are other initiatives of which we are not aware. And then there is Aditazz and a few other firms trying to think very differently. We certainly don't pretend to know if and when the tipping point from project to product delivery will come. We just know that much more and better work will need to be done between projects and before each industry firm starts their next project, and the ones to follow.

18.5 SUMMARY

The high-performing building as a product offers owners a reliable method to procure buildings that will meet their needs within their time and budgetary constraints. Unlike design and construction as a service, project outcome is guaranteed. But to create products, the IPD team will need to be confident of the outcome. Confidence, in this context, is measured by the difference between perfect knowledge and what the team has validated. As this gap lessens through the validation process, the team will reach a point where it is sufficiently confident to commit to a guaranteed outcome.

The tools and processes required to increase confidence require continual work by all team members. But current practices burden an individual project with the cost of developing the necessary systems. Moreover, because some of the skills are personal, changes to personnel between project result in losing knowledge between projects.

The teams creating products will need to develop capabilities within their staffs and their vendors in order to achieve the required confidence. Peter Beck has addressed this problem by increasing vertical integration and developing software to better integrate information. Others have created training programs or have committed repeat work to a single team. In addition, joint collaboration through academic research and training program allows firms to effectively leverage their training investments. Regardless of the strategy taken, (and some firms use multiple strategies), the goal is to increase a firm's individual capabilities and the capabilities of its strategic partners.

NOTES

1. *United States v. Spearin,* 248 US 132, 39 S.Ct. 59, 63 L.Ed 166 (1918).
2. Retrieved April 10, 2016, from http://pbl.stanford.edu/Research/ResearchEWP.html.

REFERENCES

Ballard, G., & Morris, P. H. (2010). Maximizing owner value through target value design. *AACE International Transactions, 2010,* 1–16.

Beck, H. C. III. (2013). Beyond IPD: The integrated enterprise challenge. *DesignIntelligence.* Retrieved December 12, 2013, from http://www.di.net/articles/beyond-ipd-the-integrated-enterprise-challenge/.

Best, R. E., Flager, F., & Lepech, M. D. (2015). Modeling and optimization of building mix and energy supply technology for urban districts. *Applied Energy, 159,* 161–177.

Catmull, E., & Wallace, A. (2014). *Creativity, Inc.: Overcoming the unseen forces that stand in the way of true inspiration.* New York, NY: Random House.

Cheng, R., Allison, Markku, Sturts-Dossick, C. Monson, C. (2015) *IPD: Performance, Expectations, and Future Use: A Report On Outcomes of a University of Minnesota Survey.* University of Minnesota, Retrieved October 16, 2016, from http://ipda.ca/site/assets/files/1144/20150925-ipda-ipd-survey-report.pdf.

Cheng, R., Allison, M., Sturts-Dossick, C., Monson, C., Staub-French, S., Poirier, E. (2016). *Motivation and Means: How and Why IPD and Lean Lead to Success.* University of Minnesota, Retrieved from http://arch.design.umn.edu/directory/chengr/.

Christian, D., Bredbury, J., Emdanat, S., Haase, F., Kunz, A., Rubel, Z, & Ballar, G. (2014, June 25–27). Four-phase project delivery and the pathway to perfection. In B. T., Kalsaas, L. Koskela, & T. A. Saurin (Eds.), *22nd Annual Conference of the International Group for Lean Construction* (pp. 269–280). Oslo, Norway.

Denerolle, S. (2013). Technical report: The application of target value design to the design phase of 3 hospital projects. Project Production Systems Laboratory University of California, Berkeley. Retrieved January 2013, from https://s3-us-west-2.amazonaws.com/tvdgroup/publications/Technical+Report+on+the+design+phase+of+3+TVD+projects.pdf.

Eckblad, S., Ashcraft, H., Audsley, P., Bleiman, D., Bedrick, J., Brewis, C., … Stephens, N. D. (2007). *Integrated project delivery-a working definition.* Sacramento, CA: AIA California Council.

Fischer, M. (2007). Tomorrow's workday: Spontaneous, creative, and reliable. In D. Rebolj (Ed.), *Bringing ITC knowledge to work—24th W78 Conference Maribor & 5th ITCEDU Workshop & 14th EG-ICE Workshop* (pp. 21–26). University of Maribor, Slovenia.

Fischer, M. (2013). You thought BIM was innovative—You ain't seen nothing yet: A peek over the construction technology horizon (pp. 1–26). Forum on the Construction Industry. St Regis Monarch Beach Resort, Dana Point, CA: American Bar Association.

Flager, F., & Haymaker, J. (2007). A comparison of multidisciplinary design, analysis and optimization processes in the building construction and aerospace industries. In I. Smith (Ed.), *24th International Conference on Information Technology in Construction* (pp. 625–630). Maribor, Slovenia.

Flager, F., Adya, A., Haymaker, J., & Fischer, M. (2014). A bi-level hierarchical method for shape and member sizing optimization of steel truss structures. *Computers & Structures, 131*, 1–11.

Flager, F., Soremekun, G., Adya, A., Shea, K., Haymaker, J., & Fischer, M. (2014). Fully constrained design: A general and scalable method for discrete member sizing optimization of steel truss structures. *Computers & Structures, 140*, 55–65.

Flager, F., Basbagill, J., Lepech, M., & Fischer, M. (2012). Multi-objective building envelope optimization for life-cycle cost and global warming potential. In G. Gudnason (Ed.), *9th European Conference on Product and Process Modeling*. Reykjavik, Iceland.

Hansen, M. (2009). *Collaboration: How leaders avoid the traps, build common ground, and reap big results.* Boston, MA: Harvard Business Review Press.

Lostuvali, B., Alves, T., & Modrich, R. U. (2012). Lean product development at Cathedral Hill Hospital Project. In I. D. Tommelein & C. L. Pasquire (Eds), *Proceedings of the 20th Conference of the International Group for Lean Construction Vol. 20*, pp. 1041–1050.

Marble, S. (Ed.). (2013). *Digital workflows in architecture design-assembly industry.*

McDonough, W., & Braungart, M. (2010). *Cradle to cradle: Remaking the way we make things.* New York: MacMillan.

Morgan, J. M., & Liker, J. K. (2006). *The Toyota Product Development System: Integrating people, process and technology.* New York, NY: Productivity Press.

Napolitano, P. D. T. S., & Cerveró-Romero, F. (2012, July 18–20). Meta-organization: The future for the Lean organization. In I. D. Tommelein & C. L. Pasquire (Eds.), *20th Annual Conference of the International Group for Lean Construction*. San Diego, CA.

Seed, W. R. (2014). Integrated project delivery requires a new project manager. *Proceedings of IGLC22*, 1447–1459.

Spear, S. (2010). *The high-velocity edge: How market leaders leverage operational excellence to beat the competition.* New York, NY: McGraw Hill Professional.

Womack, J. P., Jones, D. T., & Roos, D. (1990). *The machine that changed the world: The story of Lean product—Toyota's secret weapon in the global car wars.* New York, NY: Simon & Schuster.

Afterword

By J. Stuart Eckblad

FAIA, VP Major Construction, UCSF Medical Center

CREATING A "BEST FOR PROJECT" CULTURE

Over the last two decades, I have had the opportunity to develop and implement new project delivery methods. As a founder and president of the Collaborative Process Institute, and director for Project Delivery for the UCSF Medical Center Hospitals project, I've worked with other owners, designers, and builders to explore the power of integration and collaboration. The basic premise is that a truly integrated project team of multiple firms will work collaboratively, as one firm, to create a "best for project" culture. A "best for project" culture outperforms a collection of "best for firms" commitments by lowering construction cost, shortening completion times, improving quality, and reducing costly litigation. In short, a fully integrated project team can achieve extraordinary project results compared to traditional delivery methods.

Extraordinary results depend on the integrated and collaborative culture and systems the project team adopts. Success depends on creating a project-specific community. In my experience, from leading the UCSF Medical Center project, commitment to integration starts by achieving a collective passion for why the project is needed. How does it contribute to delivering patient care? Bigger than the challenge of the proposed building is the team commitment to improving patient outcomes.

The UCSF Mission Bay Hospitals case study in Chapter 8, "Integrating the Project Organization," describes many of the tools and processes we instituted to achieve project integration. The "Big Room" is an essential tool. It is the "center" and catalyst in the creation of an integrated organization. In our project, we brought 19 firms, 120 individuals full time, into a single space to complete the detail design and construction phases. We grew to over 250 individuals as we progressed into construction. A simple catalyst for integration was that team members could not phone into our center. Virtual coordination meetings are efficient, but they do not build the personal relationships required to sustain collaboration through the always-challenging construction phases. What is required is a simple and effective process to build a "community" of trade detailers, architects, engineers, builders, and suppliers committed to "best for project results."

Socializing and on-boarding new team members as the project progresses into construction is critical to optimizing the people side of integration. Special gatherings and routine on-site events are important. For our project, requiring "face-to-face" communication built a community able to cooperatively adopt to change and to challenge each other to achieve exceptional results.

The UCSF Mission Bay Hospitals project makes an important contribution to integrated project delivery (IPD). This is a public project that has been able to create an integrated organization and achieve IPD without an Integrated Form of Agreement (IFOA).

In the end, the UCSF Mission Bay Hospitals project team achieved an award winning design, a 100 year building, and LEED Gold certification. Additionally, the team achieved over $200 million dollars in savings during the design phase. We completed early. We completed without budget augmentations, delay claims, or litigation. And we continue to receive national and international inquiries regarding how we achieved these results. Just as importantly, the Mission Bay Hospitals project team successfully built an integrated organization and community committed to patient care and optimal project results.

The industry has a ways to go to perfect the integrated project organization. Today, as an owner, I am encouraged that we have seen significant interest and growth in new delivery methods and tools. Interest in IPD and IPD-ish is growing. New contract arrangements are being explored. Owners, designers, and builders are exploring business models and organization structures that support and enable integrated organizations. The continued growth in Lean continues to generate new tools and processes to make design and construction more efficient and productive.

Where do we go from here? We can achieve even greater results. Innovation and "the next big idea" will help accelerate adoption of integrated and collaborative behaviors. Significant effort is ongoing to develop new software to measure performance, new materials, larger applications of prefabricated components, and the continued growth of Lean production. There will be better and smarter tools that aid the efficiency of subcontractor fabrication and assemblies and improve and measure productivity. With additional research and experience, owners, designers, and builders can retool their business models to optimize collaboration and integration.

Moreover, to achieve optimal integrated organizations we need more research on building project-specific communities that are focused on improving team behaviors, to think of value beyond "completing under budget and ahead of schedule." We need to go way beyond traditional goal setting processes and mission statements to optimize performance of integrated organizations. I am encouraged that we are well on the way.

Afterword

By Eric R. Lamb

Management Committee, DPR Construction

The integration of the various participants of the construction industry can sometimes feel like a problem too big to solve. The fragmentation of the industry among thousands of suppliers, designers, contractors, and subcontractors is unbelievably complex and daunting. The bad behaviors and inefficiency is exactly what some suggest as a huge opportunity.

Martin, Howard, Atul, and Dean have laid out a vision, process, and technology enablers that provide a chance of making a difference to how we design and build projects in this industry. It is an international problem but is certainly exacerbated in this country with the lack of scale of any one entity or group of entities to effect change.

The antiquated distribution, business models, and practices of the construction industry present what some venture capitalists describe as a large "white space." A new business model entering this white space has no competition other than the status quo. Mark Johnson writes about business model innovation within existing industries in *Seizing the White Space,* published in 2010. IPD represents a business model innovation in the construction industry, but it is uncharted territory, and that can make it a scary place.

Integration in construction has proven to increase quality and lower cost when there is a supportive environment and trust in place. There is hope for our industry when you see some of the early results of these business model changes. Some areas that need to continue to be developed include:

- The commitment to building a team that supports the other team members as if they were all members of the same company is a critical part of the process. This development of total trust and alignment of the project goals is very important to achieving great results.

- Design simulation and automation—this will continue to develop and allow customers to optimize their design solution and predict costs as well. With builder input to these early options part of the process, it will increase the certainty of the outcome—particularly relative to cost and schedule.

- Co-location and virtual building by designers and builders become a routine part of the execution of the design prior to any construction. Integration and Lean practices with VDC (virtual design and construction) is enhancing quality and lowering total project cost. We are finding that the "process" of VDC and how it is executed is critical to its success.

- The certainty of design and 3-D models allows prefabrication and automated installation processes that were not possible only five years ago. This continued acceleration toward more "automation" is reaping huge benefits already in the early stages of a fully integrated model prior to construction activity.

- Insurance underwriters need to recognize the reduced risk of this delivery with regard to professional liability claims and defect claims.
- Power continues to move to the edge of the network. Through continued use of digital tools and technology, projects will be more decentralized with decisions made by individuals closer to the work. Large, complex projects will be managed at the edge by employees enabled with digital tools and processes. The companies that are able to adapt to this decentralized model will break away from the others.

WHERE TO NEXT?

I can visualize a vertically integrated company that provides design, fabrication, and installation of construction projects much faster and with lower cost than the current industry approach. This represents a different business model and supply chain relationship than currently exist, however. There is an argument that suggests the industry currently has three to four levels of markups, lack of visibility into the supply chain, and lack of integration of design and construction that could be solved with a single, integrated business.

We have started to see initial examples of the potential of this approach with components designed for a "design, prefabricate, and assemble" project. Our company has started to prototype and design panelized systems for structure, exterior panels, and interior walls that include framing, electrical, piping, and finish trades in an integrated solution building component. These are small steps toward integrating various trades into one installation in an automated, industrial setting. The goal is to achieve lower costs and faster cycle times while still improving on quality and safety objectives. Furthermore, there is an argument that this integrated approach could work in the field for the assembly of components and final construction of a facility. There will always be a need for specialty companies for unique work activities but their participation could be relegated to minor support roles to an integrated design-builder. The enhanced view into the supply chain, control of design and materials choices, elimination of multiple hand-offs with risk to smooth workflow, increased automation with multiple items, reduced redundancy of supervision and general conditions, and increased quality, fit, and finish of work that is designed and installed by a single entity all combine to provide a wholly different result.

The customized nature of construction and the fragmented industry represent barriers to this integrated approach. I do believe these barriers are more easily overcome with digital tools, automation, and business model changes over the next 10 years in the construction industry.

Index

Note: Page references in *italics* refer to figures and tables.